Marine Bioactive Compounds

Maria Hayes
Editor

Marine Bioactive Compounds

Sources, Characterization and Applications

Springer

Editor
Maria Hayes
Food BioSciences Department
Teagasc Food Research Centre
Ashtown, Dublin 15, Ireland
maria.hayes@teagasc.ie

ISBN 978-1-4614-1246-5 e-ISBN 978-1-4614-1247-2
DOI 10.1007/978-1-4614-1247-2
Springer New York Dordrecht Heidelberg London

Library of Congress Control Number: 2011941431

Printed on acid-free paper

Springer is part of Springer Science+Business Media (www.springer.com)

"There is hope from the sea, but none from the grave." (Irish Proverb)

Preface

The aim and scope of this book is to highlight sources, isolation, characterization, and applications of bioactive compounds from the marine environment and to discuss how marine bioactive compounds represent a major market application in the food industry and others. It discusses sustainable marine resources of macroalgal origin and gives examples of bioactive compounds isolated from these and other resources including marine by-product and fishery waste streams. In addition, it looks at the importance of correct taxonomic characterization.

Bioactive molecules and their precursors are at the very high end of the chemical products value spectrum. This book discusses how bioactive compounds have been used in functional food formulations and pharmaceutical applications, and how they alter biological activity to provide therapeutic benefits, nutritional values, and health protection.

This volume looks at the screening process for identification of bioactive molecules and describes different production methods used for bioactive compound isolation and identification. Furthermore, this book provides an insight into the market opportunities that exist for the identification and commercialization of new marine bioactive compounds. Finally, it highlights regulations in the United States, Europe, Japan, and China and what is required from the regulatory bodies in these different countries to obtain a health or novel food claim for a marine functional food product.

Maria Hayes
Teagasc, Dublin, Ireland

Contents

Contributors

Thomas B. Brück Department of Chemistry, Division of Industrial Biocatalysis, Technische Universität München (TUM), Garching bei München, Germany

Wolfram M. Brück Centre of Applied Marine Biotechnology, Letterkenny Institute of Technology, Donegal, Ireland

Bioanalytical Science, Food and Health Microbiology, Nestlé Research Center, Lausanne, Switzerland
wolfram.brueck@rdls.nestle.com

María Castro-Puyana Bioactivity and Food Analysis Department, Institute of Food Science Research (CIAL-CSIC), Madrid, Spain

A.K. Croft School of Chemistry, University of Wales Bangor, Bangor, UK

Daniel Garbe Department of Chemistry, Division of Industrial Biocatalysis, Technische Universität München (TUM), Garching bei München, Germany

W. Groenewald School of Chemistry, University of Wales Bangor, Bangor, UK

Michael D. Guiry Irish Seaweed Research Group, Ryan Institute for Environmental, Marine and Energy Research, National University of Ireland, Galway, Ireland

Maria Hayes Food BioSciences Department, Teagasc Food Research Centre, Ashtown, Dublin 15, Ireland
maria.hayes@teagasc.ie

Miguel Herrero Bioactivity and Food Analysis Department, Institute of Food Science Research (CIAL-CSIC), Nicolás Cabrera 9 Campus UAM Cantoblanco, Madrid, Spain

Rita M. Hickey Teagasc Food Research Centre, Moorepark Fermoy, Co.Cork, Ireland
rita.hickey@teagasc.ie

Elena Ibañez Bioactivity and Food Analysis Department, Institute of Food Science Research (CIAL-CSIC), Nicolás Cabrera 9 Campus UAM Cantoblanco, Madrid, Spain
elena@ifi .csic.es

Jose A. Mendiola Bioactivity and Food Analysis Department, Institute of Food Science Research (CIAL-CSIC), Nicolás Cabrera 9 Campus UAM Cantoblanco, Madrid, Spain

Steven Reisse Department of Chemistry, Division of Industrial Biocatalysis, Technische Universität München (TUM), Garching bei München, Germany

Fabio Rindi Dipartimento di Scienze della Vita e dell'Ambiente, Università Politecnica delle Marche, Ancona, Italy
f.rindi@univpm.it

Turid Rustad Department of Biotechnology, Norwegian University of Science and Technology, Trondheim, Norway
turid.rustad@biotech.ntnu.no

Anna Soler-Vila Irish Seaweed Research Group, Ryan Institute for Environmental, Marine and Energy Research, National University of Ireland, Galway, Ireland

M.S. Tierney School of Chemistry, University of Wales Bangor, Bangor, UK
Food BioSciences Department, Teagasc Food Research Centre, Ashtown, Ireland

Chapter 1
Taxonomy of Marine Macroalgae Used as Sources of Bioactive Compounds

Fabio Rindi, Anna Soler-Vila, and Michael D. Guiry

1.1 Introduction

As defined in the broadest sense, algae are oxygen-generating, photosynthetic organisms other than embryophyte land plants and lichens (Cavalier-Smith 2007). They are an artificial and highly heterogeneous aggregation of organisms belonging to many different evolutionary lineages, and therefore highly diverse from a genetic point of view. This genetic diversity is reflected in the huge diversity exhibited by algae in terms of morphological, ultrastructural, ecological, biochemical, and physiological traits.

Marine macroalgae, or seaweed, are plantlike organisms that generally live attached to rock or other hard substrata in coastal areas. They belong to three different groups, empirically distinguished since the mid-nineteenth century on the basis of thallus color: red algae (phylum Rhodophyta), brown algae (phylum Heterokontophyta (also known as the Ochrophyta), class Phaeophyceae), and green algae (phylum Chlorophyta, classes Bryopsidophyceae, Chlorophyceae, Dasycladophyceae, Prasinophyceae, and Ulvophyceae). Distinguishing these three phyla, however, involves more substantial differences than indicated by this simple designation. In addition to the pigmentation, they differ considerably in many ultrastructural and biochemical features including photosynthetic pigments, storage compounds, composition of cell walls, presence/absence of flagella, ultrastructure of mitosis, connections between adjacent cells, and the fine structure of the chloroplasts. They originated through different evolutionary processes (primary endosymbiosis

F. Rindi (✉)
Dipartimento di Scienze della Vita e dell'Ambiente, Università Politecnica delle Marche, Ancona, Italy
e-mail: f.rindi@univpm.it

A. Soler-Vila • M.D. Guiry
Irish Seaweed Research Group, Ryan Institute for Environmental, Marine and Energy Research, National University of Ireland, Galway, Ireland

M. Hayes (ed.), *Marine Bioactive Compounds: Sources, Characterization and Applications*, DOI 10.1007/978-1-4614-1247-2_1,

for green and red algae, secondary endosymbiosis for brown algae; Keeling 2010), and for this reason they are now classified in different kingdoms (green algae and red algae in the Kingdom Plantae, and brown algae in the Kingdom Chromista; Cavalier-Smith 2007).

The capacity of seaweed to produce compounds with valuable biological properties has been known for a long time and has been intensively exploited, especially for pharmaceutical purposes. There is evidence that the ancient Greeks and Romans were aware of the pharmaceutical benefits of certain species of seaweed. For example, Pedanius Dioscorides, a Greek physician, pharmacologist, and botanist who lived from about AD 40–90, mentioned the use of algae as medicines in his *Materia medica*, the precursor of all modern pharmacopeias. In China, Japan, the Philippines, and other parts of southeast Asia, which have a long history of seaweed exploitation, seaweed has long been used in folk medicine (Folmer et al. 2010).

In the last decades, increased demand for new drugs and food components based on natural products has placed seaweed at the center of large-scale screening programs and has led to the discovery of numerous new compounds. From 1970 to 1990, marine macroalgae were the group of organisms from which the highest numbers of new chemicals were discovered (Smit 2004). The diversity of the compounds produced by marine macroalgae is exceptional and covers a very wide range of biological activities, as previously described by several reviewers (Smit 2004; Bhakuni and Rawat 2005; Mayer and Gustafson 2008; Folmer et al. 2010; Tierney et al. 2010; Blunt et al. 2011; Holdt and Kraan 2011). Such vast compound diversity is once again a consequence of the great genetic diversity of these organisms and their complex evolutionary histories. Having evolved in the sea and surviving in marine coastal habitats, seaweed is involved in negative interactions with other organisms present in these environments. These interactions include competition for space, light, and nutrients with other seaweed; grazing by benthic herbivores; epibiosis by smaller-sized organisms; allelopathy by other sessile organisms, and parasitism by animal or algal parasites. Macroalgae are subjected to several forms of abiotic stresses including strong photon irradiance, particularly in the UV, and in the case of intertidal species, desiccation, variation in salinity, high or low temperatures, and sometimes freezing. Bioactive compounds have evolved as chemical weapons against other organisms and/or as responses to physical stress, and can be regarded as an adaptation to the sessile benthic lifestyle; it is not uncommon that a certain compound has multiple functions and exhibits multiple biological activities. For example, fucoidan has been described for its antioxidant (Diaz-Rubio et al. 2009), antimetastatic (Alekseyenko et al. 2007), antivenom (Angulo and Lomonte 2003) and anticoagulant activity (Cumashi et al. 2007). Inasmuch as all species of benthic algae are exposed to these interactions, it is not surprising that biological activities of some kind have been found virtually in all seaweed on which accurate biochemical screening has been performed. At present, the occurrence of chemically mediated biological activities has been documented for approximately 150 genera of benthic marine algae (Rindi 2008). The chemical diversity of bioactive compounds is a reflection of the genetic diversity of the organisms in which they are discovered, therefore the correct taxonomic characterization of the species used in screening programs is a critical requirement for appropriate characterization and exploitation of these compounds.

1.2 The Taxonomy of Benthic Marine Macroalgae

Marine algae taxonomy effectively dates from 1753, but has gone through major conceptual changes in the last 30 years. Until the 1970s, phylogenetic hypotheses in relation to seaweed were considered to have a sound basis if they were based on reproductive morphology (particularly of the development of female apparatus and its postfertilization behavior in red algae). Identification was typically based on observation (either by the unaided or microscopic) of associated suites of morphological characters considered taxonomically significant. In reality, this was not an undertaking for the faint of heart, and required many years of training and an innate ability to recognize subtle and complex patterns. In the last few decades, however, the availability of new tools (electron microscopy images and molecular data) has produced new types of data, and these have had a profound impact on the taxonomy of seaweed.

The progress of molecular systematics that has taken place in the last 20 years has revolutionized the classification of algae at all levels and has substantially reshaped taxonomic concepts in all algal groups; marine macroalgae have been among those organisms most affected by these developments. Cases of cryptic diversity have been documented for many types of seaweed (e.g., De Clerck et al. 2005; Millar and Freshwater 2005; Saunders and Lehmkuhl 2005; Andreakis et al. 2007a; Lindstrom 2008; Le Gall and Saunders 2010; Tronholm et al. 2010a, b), showing that identifications based only on morphological characters may lead to a gross underestimation of species diversity. Molecular data have also highlighted the effect of phenotypic plasticity on these organisms. The morphology of many marine macroalgae is strongly affected by environmental factors such as temperature, salinity, light irradiation, wave intensity, and interactions with other benthic organisms (Kubler and Dudgeon 1996; Blomster et al. 1998; Domis et al. 2003; Fowler-Walker et al. 2006; Díaz-Pulido et al. 2007). This is a particularly serious problem with morphologically simple species in which environmentally driven morphological variations often lead to misidentification (Domis et al. 2003; Stam et al. 2006; Verbruggen et al. 2007; Leliaert et al. 2009a). For these reasons, species identification based on short sequences of DNA (DNA barcoding; Saunders 2005), a practice originally developed for other groups of organisms, has been recently extended to seaweed and is rapidly gaining great popularity. The mitochondrial *cox*1 gene was shown to be a suitable marker for DNA barcoding in red and brown seaweed (Saunders 2005; Robba et al. 2006), although there is not yet a marker that is universally accepted as a barcode for green seaweed (Kucera 2010). It is now accepted that for several genera of marine algae molecular data are a mandatory requirement for reliable identification (*Ulva* is a typical example: Heesch et al. 2009; Hofmann et al. 2010; O'Kelly et al. 2010).

1.3 An Overview of the Taxonomy of Marine Algae Used as Sources of Bioactive Compounds

In recent years, the taxonomy of several algal genera and species known to produce valuable bioactive compounds has undergone major rearrangements. These rearrangements have important practical implications as they affect the interpretation of many studies concerning biological activities in seaweed. In particular, species circumscriptions and characterizations based on molecular data suggest that for some records of bioactivities available in the literature the species involved are likely to have been misidentified. We present here an overview of the taxonomy of the marine algae for which biological activities have been reported and/or new compounds have been described. As mentioned above, the number of algal taxa for which such reports are available is very large and a complete discussion would require much more extensive treatment. We therefore focus on taxa for which new data produced in recent years have resulted in important taxonomic rearrangements (and for which these changes have important implications with regard to the identification of the seaweed strains used in bioactivity studies). The classification, taxonomic arrangement, and species authorities used here are based mainly on AlgaeBase (Guiry and Guiry 2011), and species numbers are based on a search of this database carried out in May, 2011.

1.4 The Green Seaweeds (Phylum Chlorophyta, Class Ulvophyceae)

The green algae are one of the most ancient groups of photosynthetic eukaryotes, having appeared sometime between 900 and 500 million years ago (Becker and Marin 2009). As currently circumscribed, they represent a highly diverse assemblage of organisms subdivided into at least 11 different classes, distributed in marine, freshwater, and terrestrial habitats. All green seaweed belong to the classes Ulvophyceae, Bryopsidophyceae, and Dasycladophyceae, which include approximately 1,500 species currently referred to eight orders. In terms of thallus complexity and cellular sophistication, the Ulvophyceae have evolved a much higher diversity of morphologies and cytological types than any other class of green algae, ranging from microscopic unicells to macroscopic multicellular plants with unique morphological, cellular, and physiological characteristics (Mine et al. 2008). Biological activities with valuable properties have been reported for 15 genera of green seaweed (see Table 1.1); *Caulerpa*, *Codium,* and *Ulva* are those for which the highest number of records are available.

Table 1.1 Phylum Chlorophyta; list of genera with reported bioactivities. Superscripts indicate the bibliographic references in which the biological activities were reported

Chlorophyta	Bioactivity	Reference
Order Bryopsidales		
Avrainvillea	Anticancer	Chen et al. 1994
Bryopsis	Antimicrobial	Puglisi et al. 2007
Caulerpa	Anti-hepatotoxicity[1]; antitumor[2,3]; antinociceptive[4]; anti-inflammatory[5]; antiherpetic;[6] antiviral[7]	Abdel-Wahhab et al. 2006[1]; Barbier et al. 2001[2]; Cavas et al. 2006[3]; de Souza et al. 2009[4,5]; Ghosh et al. 2004[6]; Nicoletti et al. 1999[7]
Codium	Antioxidative and antigenotoxic[1]; antiviral[2]; anticoagulant[3]	Celikler et al. 2009[1]; Hudson et al. 1998[2]; Jurd et al. 1995[3]; Matsubara et al. 2001
Derbesia	Proliferation of thymus cells	Youngwan et al. 2006
Halimeda	Anticoagulant[1]; antimicrobial[2]; antioxidant[3];	De Lara Isassi and Alvarez Hernandez 1995[1]; Engel et al. 2006[2]; Fallarero et al. 2003[3]
Penicillus	Antimicrobial	Engel et al. 2006
Tydemania	Anticancer	Govindan et al. 1994
Order Cladophorales		
Chaetomorpha	Anti-atherosclerotic	Shi et al. 2005
Cladophora	Antiprotozoal	Spavieri et al. 2010
Cladophoropsis	Cytotoxicity	Harada et al. 1997
Order Dasycladales		
Cymopolia	Antimicrobial[1]; antimutagenic[2]	Gonzalez del Val et al. 2001[1]; Wall et al. 1989[2]
Order Ulotrichales		
Acrosiphonia	Antimicrobial	Shanmughapriya et al. 2008
Order Ulvales		
Ulva/Enteromorpha	Antioxidant[1]; antiviral[2]; anti-inflammatory[3,10]; antihyperlipidemic[5]; antitumor[6,12]; antimicrobial[7,8]; antiprotozoal[9]; antibacterial[11]; immunomodulating[12];	Ganesan et al. 2011[1]; Garg et al. 1992[2]; Jin et al. 2006[3]; Kajiwara et al. 2006[4]; Pengzhan et al. 2003[5]; Xu et al. 2004[6]; Tuney et al. 2006[7]; Tanaka et al. 1998[6]; Sukatar et al. 2006[8]; Orhan et al. 2006[9]; Okai and Higashi-Okai 1997[10]; Ismail-Ben et al. 2009[11]; Jiao et al. 2010[12]
Monostroma	Antithrombin[1]; antiherpetic[2]; anticoagulant[3];	Harada and Maeda 1998; Lee et al. 2010[2]; Zhang et al. 2008[3]; Shanmugam et al. 2001[3]

1.5 The Genus *Caulerpa* J.V. Lamouroux

The genus *Caulerpa* (class Bryopsidophyceae) currently includes 86 species (Guiry and Guiry 2011) distributed in tropical and warm-temperate waters of the world. The thallus is comprised of horizontal stolons anchored by colorless rhizoids, bearing erect photosynthetic fronds of extremely diverse morphology including threadlike,

Fig. 1.1 Example of green and brown algal genera exploited for bioactivities. *Caulerpa holmesiana* (fronds 4–5 cm tall)

bladelike, pinnate, spongy, and vesicular structures (Fig. 1.1). The structure of the thallus is siphonous (i.e., the body of the alga consists of a single giant cell).

Some species of *Caulerpa* have been consumed as food in eastern Asia for a long time, especially in salads (Trono 1999). These algae are also among the most popular ornamental plants used in marine aquaria (Stam et al. 2006). In recent years, however, some species of this genus have become infamous for their nuisance value. Due to their fast growth, some *Caulerpa* entities tend to spread in an aggressive and uncontrolled manner and for this reason have caused some of the most spectacular biological invasions known in the marine environment. An aquarium strain of *Caulerpa taxifolia* originating in Australia was accidentally released in the Mediterranean Sea, from the *Musée Océanographique* in Monaco around 1984; by 2000, it covered 131 km^2 of coastal waters bordering six Mediterranean countries (Meinesz et al. 2001; Stam et al. 2006). Early in the present century, an invasive strain of *Caulerpa racemosa* was also identified in the Mediterranean Sea (Verlaque et al. 2000). Subsequently, this alga has spread aggressively and now represents the most problematic biological invasion in the Mediterranean Sea, forming dense populations along at least 750 km of shores of 11 countries by the end of 2003 (Piazzi et al. 2005). Once again, a human-mediated introduction is suspected, even though its vector is still unknown (Verlaque et al. 2003; Stam et al. 2006). Signs of potentially invasive behavior have also been noted in other species of *Caulerpa*; although not considered invasive, populations of *Caulerpa brachypus*, *C. scalpelliformis,* and *C. verticillata* tend to grow in dense and very large monospecific stands (Stam et al. 2006).

Studies of biological activities in *Caulerpa* have led to the discovery of compounds with many different biological properties. The majority of work has been done using crude aqueous extracts and fractions of them (Smit 2004), often using solvents of increasing polarity. An extract of *Caulerpa* with cold methylene chloride and methanol was shown to reduce hepatocarcinogenicity of aflatoxins, a secondary metabolite from a fungus found in foods (particularly nuts), that are stored incorrectly (e.g., Abdel-Wahhab et al. 2006). Also, a crude extract with chloroform–methanol, from *C. taxifolia* showed antiviral activity against the feline immunodeficiency virus (FIV; Nicoletti et al. 1999). Polysaccharide extractions have also been widely used. A study of various species of *Caulerpa* (*C. racemosa*, *C. taxifolia*, *C. scalpelliformis*, *C. veravalensis,* and *C. peltata*) showed high anticoagulant activity which was associated with their sugar, sulphate content, and molecular weight of the extracts (Shanmugam et al. 2001).

Some metabolites synthesized by *Caulerpa* have been suggested to be responsible for many bioactivities. An example is caulerpenyne, a metabolite that showed cell-growth inhibitory effects (Fischel et al. 1995), antiproliferative activity in tumor cell lines (Barbier et al. 2001), and has been suggested as a possible alternative source of antitumor drugs (Cavas et al. 2006). The metabolite caulerpin has also been reported to have many in vitro biological activities, such as antitumor, growth regulator, and plant root growth stimulant; recently, in vivo bioactivities such as anticonceptive (i.e., analgesic) and anti-inflammatory have also been described (de Souza et al. 2009).

Unsurprisingly, studies on *Caulerpa* carried out in recent years have focused mostly on the invasive strains. Olsen et al. (1998) demonstrated that the invasive form of *Caulerpa taxifolia* was not conspecific with *C. mexicana*, a species that had been reported in the Mediterranean since 1941. The invasive *Caulerpa racemosa* was identified and named *Caulerpa racemosa* var. *cylindracea* by Verlaque et al. (2003), who demonstrated that the population derived from a strain originally indigenous to southwestern Australia. The large body of new molecular data produced in the last decade has clarified phylogenetic relationships in *Caulerpa*, highlighting the problems of morphological identification caused by morphological plasticity and the presence of multiple intermediate forms that are not environmentally stable (e.g., Famà et al. 2002; Domis et al. 2003; Meusnier et al. 2004; Stam et al. 2006). Because of these problems and the fact that invasive strains of *Caulerpa* can be identified unambiguously only by molecular tools, and also that the potential capacity for invasive behavior in other *Caulerpa* species is unclear, Stam et al. (2006) concluded that this genus should be banned completely from commerce, particularly in the aquarium trade.

Concerning studies on biological activities, it is possible that in some cases the identification of the *Caulerpa* species used are not correct. In particular, it is unfortunate that very few details are given when the species involved is *Caulerpa racemosa*; this species includes a great number of forms and varieties, some of which might be recognized in the future as separate species; it would therefore be very important to specify the infraspecific taxa and provide details of their morphologies. Where invasive strains are involved, harvesting may represent a convenient way to limit the impact of these algae on native benthic communities. However, harvesting of *Caulerpa* is a delicate operation that should be performed with great care and only by well-trained personnel. Thalli of *Caulerpa* have a great capacity to reproduce by vegetative fragmentation (Ceccherelli and Cinelli 1999; Wright 2005). Fragments a few centimeters in length generally easily regenerate a complete plant. Therefore, removal of *Caulerpa* performed without adequate experience may result in further propagation.

1.6 The Genus *Codium* Stackhouse

Codium (class Bryopsidophyceae) is a genus of intertidal and subtidal green seaweed currently of 130 species (Guiry and Guiry 2011), widely distributed in all seas of the world, with the highest species diversity in subtropical regions

Fig. 1.2 Example of green and brown algal genera exploited for bioactivities. *Codium platyclados* (specimen about 20 cm tall)

(Verbruggen et al. 2007). The thallus is composed of a single, giant, branched tubular cell containing multiple nuclei and its habit varies greatly in different species (crustose, globose, or erect, dichotomously branched, cylindrical, or flattened tubes; Fig. 1.2).

Bioactivity studies in this genus are numerous, but they have usually focused on a few taxa (see Table 1.1). A crude methanol extract of *C. fragile* showed antiviral activity by inactivation of virus particles (Hudson et al. 1998). One of the first reports of anticoagulant activity associated with marine green algae concerned a *Codium*, *C. fragile* ssp. *tomentosoides* (Deacon-Smith et al. 1985). Since then, many reports have appeared describing the anticoagulant activity in *Codium* taxa (e.g., Rogers et al. 1990; Matsubara et al. 2000; 2001). Furthermore, high molecular weight proteoglycan (a sulphated polysaccharide isolated from *C. fragile* ssp. *atlanticum*) has been described and its anticoagulant activity tested (Rogers et al. 1990; Jurd et al. 1995). Recent work showed the antigenotoxicity and antioxidative capacity of a crude ethanolic extract of *C. tomentosum* (Celikler et al. 2009) and the antibacterial activity of *C. decorticatum* (Sunilson et al. 2009).

Recent molecular data have provided a better understanding of the difficulties in defining species within this genus. Identifying *Codium* specimens using only morphological characters can be challenging; even though a few distinctive species clearly stand out from the rest, most collections are very difficult to identify (Verbruggen et al. 2007). Whereas the species of *Codium* along the coasts of North America, Europe, South Africa, and southern Australia are mostly well characterized, specimens collected elsewhere are often difficult or nearly impossible to identify, and an accurate identification would be reliant on molecular data. It has been demonstrated that the morphology typical of certain species, such as *Codium fragile* and *C. geppiorum*, has evolved separately in different evolutionary lineages (Verbruggen et al. 2007). In the case of *Codium fragile* this has an important implication: molecular data are necessary to identify unambiguously invasive strains attributed to this species. These were formerly identified as *Codium fragile* subsp. *tomentosoides*; Maggs and Kelly (2007), however, reassessed the taxonomy and nomenclature of *C. fragile* and revised the concept of the subspecies defined previously on morphological grounds alone. This alga is one of the most widely introduced seaweeds on a global scale and has been the subject of several bioactivities studies (Trowbridge 1998).

1.7 The Genus *Ulva* Linnaeus

The genus *Ulva* (class Ulvophyceae) currently includes 100 species (Guiry and Guiry 2011) commonly found in the intertidal and shallow subtidal zones of rocky shores all over the world. Thalli of *Ulva* consist either of two-layered blades of various shapes and dimensions ("Ulva" morphotypes; Fig. 1.3) or of single-layered tubular plants, either branched or unbranched ("Enteromorpha" morphotypes). The life history theoretically consists of an isomorphic alternation of a diploid sporophyte and a haploid gametophyte, reproducing, respectively, by biflagellate zoospores and gametes; however, most populations appear to reproduce parthenogenically. Specimens of *Ulva* normally grow attached to hard substrata in coastal areas; in some species, however, the thallus can survive and continue to grow after detachment from the substratum. When rapid growth is triggered by high nutrient loads, unattached *Ulva* fronds accumulate in large amounts that may become a considerable practical nuisance, a phenomenon often called "green tides." Particularly well known is the case of an enormous green tide produced by a species of *Ulva* that in July 2008 threatened the Olympic sailing competition at Qingdao, China (Leliaert et al. 2009b).

Overall, *Ulva* is one of the most common and widely distributed genera of marine macroalgae on a global scale. In recent years it has also been one of the most investigated, and at the same time it has been the genus for which the most substantial taxonomic rearrangements have taken place. Until 2003, leafy forms and tubular forms were generally attributed to two separate genera, *Ulva* and *Enteromorpha*, respectively. However, intermediate forms between *Ulva* and *Enteromorpha* were observed in some circumstances and there was evidence that under certain environmental conditions some species could switch from one morphotype to the other (Tan et al. 1999). Using a molecular dataset based on two markers (ITS nrDNA and *rbc*L), Hayden et al. (2003) concluded that the separation of *Ulva* and *Enteromorpha* was not justified and proposed that the latter should be reintegrated into the former, which has nomenclatural priority. Other important taxonomic changes at genus level were the erection of the genus *Umbraulva* by Bae and Lee (2001) and the

Fig. 1.3 Example of green and brown algal genera exploited for bioactivities. *Ulva pertusa* (specimen about 20 cm wide) © Chiba University

reinstatement of the genus *Gemina* proposed by Heesch et al. (2009) for five New Zealand species.

At species level, the molecular data produced in the last 10 years have led to a much better characterization of species concepts and circumscriptions. A great deal of morphological plasticity has been demonstrated in several species, including some common ones. For example, Blomster et al. (1998) showed that the morphology of two widespread intertidal species, *Ulva compressa* and *U. intestinalis* (as *Enteromorpha compressa* and *E. intestinalis*), is considerably affected by salinity and in conditions of low salinity *U. intestinalis* often shows the branched habit typical of *U. compressa*. Recent studies based on molecular data carried out in different parts of the world invariably have shown a higher species diversity than suggested by morphology, revealing cryptic species and introduced taxa not previously detected (e.g., Loughnane et al. 2008; Heesch et al. 2009; Hofmann et al. 2010; Kraft et al. 2010; O'Kelly et al. 2010). Several new species were described (Hiraoka et al. 2004; Kraft et al. 2010) and a sharp biogeographic differentiation between taxa of the Pacific Islands and other parts of the world was demonstrated (Heesch et al. 2009; O'Kelly et al. 2010). It is now clear that molecular data are mandatory for identification of most *Ulva* species and that this species complex requires considerable further work, focused in particular on the characterization of type specimens. Recent improvements in DNA extraction procedures have made it possible to obtain partial sequences from type specimens, and in some cases this has led to unexpected results. For example, it has been shown that the molecular identity of the type specimen of *Ulva lactuca* (the generitype, as well as the most widely recorded) does not match that of specimens which have since been assigned to this name, and corresponds to the entity named *Ulva fasciata* (O'Kelly et al. 2010).

A consequence of these difficulties is that all the information concerning bioactivities in species of *Ulva* needs to be reassessed. All studies known to us reporting bioactivities in *Ulva* have been based on morphological identifications. In most cases these identifications were not based on a careful scrutiny by a professional algal taxonomist and it is probable that the authors simply attached to their specimens the names of the most common species that superficially resembled their material. The number of bioactivities described for *Ulva* is extensive and only some are included in Table 1.1. We know now that identifications based only on morphology are generally not reliable, and the species' names used in most of these studies are very likely to be incorrect, especially when supposedly cosmopolitan or widely distributed species are reported (such as *Ulva lactuca*, *U. intestinalis*, *U. prolifera*).

1.8 The Brown Seaweeds (Phylum Heterokontophyta, Class Phaeophyceae)

The brown algae form the class Phaeophyceae, which is believed to have arisen between 200 and 150 my ago (Silberfeld et al. 2010). This class is represented by 1,760 species (Guiry and Guiry 2011), currently arranged in 18 orders. The brown seaweed shows a huge variation in habit and size, ranging from thin, uniseriate filaments to complex pseudoparenchymatous algae up to 60 m in length. They are most diverse and abundant in cold seas and include the largest of all algae. They are also the group of macroalgae for which the largest number of studies concerning biological activities is available. There is no doubt that the large size of much brown seaweed has made them very suitable subjects to test biological activities and allows for extraction, in large amounts, of the associated bioactive compounds. Biological activities with valuable properties have been reported for 53 genera of brown algae, mostly from the orders Dictyotales, Fucales, and Laminariales (Table 1.2).

1.9 The Order Dictyotales

The Dictyotales (class Phaeophyceae) represent one of the earliest-diverging lineages in the Phaeophyceae (Silberfeld et al. 2010) and includes a single family, the Dictyotaceae, with 18 genera containing 220 species (Guiry and Guiry 2011) distributed mostly in tropical and temperate waters. The thallus is laminar, formed by a variable number of cell layers, with habit varying from an undivided, fan-shaped blade with smooth margin (e.g., *Padina*) to blades with regular dichotomous branching (e.g., *Dictyopteris*, *Dictyota*; Fig. 1.4).

The number of bioactivities described for the Dictyotales is extensive and spans ten different genera. Antimicrobial activity has been reported for hexane extracts from *Dictyopteris membranacea* (Ozdemir et al. 2006) and methanol extracts of *Padina pavonica* (Dulger et al. 2009). Fractions of ethanol extracts of *Spatoglossum asperum* have showed antifungal activity against the destructive plant pathogen, *Macrophomina phaseolina*, a soil-borne fungus (Ara et al. 2005). Crude extracts of *Canistrocarpus cervicornis* and *Dictyota pfaffi* have been reported to inhibit the venom of *Lomonia obliqua*, a moth caterpillar dangerous to humans found in South America (Domingos et al. 2009). Many secondary metabolites have been described from the genus *Dictyota*, which is the most species-rich (76 species; Guiry and Guiry 2011). *Dictyota* species produce the diterpene dolabellane, a secondary metabolite with defensive function having an inhibitory effect against herbivores (Barbosa et al. 2004). Various diterpenoids with antifungal activity have also been isolated from *Dilophus ligulatus*, including acetyldictyolal, epoxyoxodolabelladiene, neodictyolactono, the acetals 6B, and isoacetoxycrenulatin (Bouaïcha et al. 1993). Sulfated polysaccharides are also common in this order. Polysaccharide extracts from *Dictyota cervicornis* showed anticoagulant and antioxidant activities (Costa et al. 2010); sulphated polysaccharides from *Dictyopteris delicatula* showed antioxidant activity and antiproliferative activity in cancer cell lines in vitro a positive correlation with the sulphur content in the molecules. Anti-inflammatory activity of fucan (a common sulphated polysaccharide in Phaeophyta) obtained from *Lobophora* has been reported by Siqueira et al. (2011). Furthermore, polysaccharides from *Padina arborescenos* showed anticancer potential, and fucans from *Stoechospermum marginatum* exhibited antiviral activity mainly as a inhibitor of herpes simplex virus type 1 and type 2 (Adhikari et al. 2006).

Table 1.2 Phylum Ochropyta; list of genera with reported bioactivities. Superscripts indicate the bibliographic references in which the biological activities were reported

Ochrophyta	Bioactivity	Reference
Order Dictyotales		
Canistrocarpus	Antivenom	Domingos et al. 2009
Dictyopteris	Antimicrobial	Ozdemir et al. 2006
Dictyota/Dilophus	Antivenom[1]; anticoagulant[2]; antiviral[3]; antibacterial[4]; antifungal[5,4]; antioxidant[6]	Domingos et al. 2009[1]; Albuquerque et al. 2004[2]; Barbosa et al. 2004[3]; Moreau et al. 1984[4]; Bouaïcha et al. 1993[5]; Costa et al. 2010[6];
Lobophora	Anti-inflamatory	Siqueira et al. 2011
Padina	Antimicrobial[1]; antioxidant[2]; cytotoxicity[3]; anticoagulant[4]; anti-cancer[5]	Dulger et al. 2009[1]; Rocha de Souza et al. 2007[2]; Ktari and Guyot 1999[3]; Silva et al. 2005[4]; Wang et al. 2008[5];
Spatoglossum	Antifungal[1]; antithrombotic[2]; hypoglycemic[3]	Ara et al. 2005[1]; Barroso et al. 2008[2]; Teixeira et al. 2007[3];
Stoechospermum	Antiviral	Adhikari et al. 2006
Stypopodium	Antivenom[1]; antiproliferative[2]; antimitotic[2]; cytotoxic[3]; antimicrobial[4]	Domingos et al. 2009[1]; Rocha de Souza et al. 2007[3]; Rovirosa and San-Martin 1997[4];
Zonaria	Antimicrobial	Vlachos et al. 1997
Order Ectocarpales		
Adenocystis	Antiviral	Ponce et al. 2003
Cladosiphon	Induces apoptosis[1]; anticancer[2]; gastric mucosal protection[3]	Haneji et al. 2005[1]; Kawamoto et al. 2006[2]; Shibata et al. 2000[3];
Colpomenia	Antimicrobial[1]; induction of apoptosis[2]	Dulger et al. 2009[1]; Huang et al. 2005a[2];
Dictyosiphon	Antiviral	Tsuda et al. 2007
Leathesia	Antiviral	Feldman et al. 1999
Nemacystus	Anticoagulant	Kitamura et al. 1992
Punctaria	Antitumor	Xu et al. 2004
Order Fucales		
Ascophyllum	Anti-inflammatory[1]; antithrombotic[2]; anticoagulant[3]	Blondin et al. 1996[1]; Boisson-Vidal et al. 2000[2]; Cumashi et al. 2007[3];
Bifurcaria	Antifouling activity[1]; anti-proliferative effect[2]; antimitotic[3]	Meréchal et al. 2004[1]; Moreau et al. 2006[2]; Valls et al. 1993[3];
Caulocystis	Anti-inflammatory	Buckle et al. 1980
Cystoseira	Antimicrobial[1,6]; antibacterial[2]; antifungal[3]; antioxidant[4]; antiviral[5]	Dulger et al. 2009[1]; Abourriche et al. 1999[1]; Badea et al. 2009[2]; Bennamara et al. 1999[2,3]; Foti et al. 1994[4]; Mandal et al. 2007[5]; Ozdemir et al. 2006[6];
Durvillaea	Anticoagulant	Matsuhiro et al. 1996
Fucus	antioxidant[1]; antitumor; antimetastatic[2]; antivenom;[3] anticoagulant[4]; antithrombin[5]	Diaz-Rubio et al. 2009[1]; Rocha de Souza et al. 2007[1]; Alekseyenko et al. 2007[2]; Angulo and Lomonte 2003[3]; Cumashi et al. 2007[4,5];
Himanthalia	Analgesic activity	Anca et al. 1990

(continued)

Table 1.2 (continued)

Ochrophyta	Bioactivity	Reference
Hizikia	Anticoagulant[2]; antioxidant[2]; immunomodulating[3]	Dobashi et al. 1989[1]; Iwahori et al. 1999[2]; Okai and Higashi-Okai 1997[3];
Myagropsis	Hepatoprotective	Wong et al. 2004
Notheia	Anti-parasitic	Capon et al. 1998
Pelvetia	Anticoagulant[1]; antioxidant[2]; anti-diabetic[3]	Colliec et al. 1991[1]; Lee et al. 2003[2], 2004[3];
Sargassum	Anticancer[1]; antiviral[2]; antioxidant[3,5,10]; antibacterial[4]; anti-inflammatory[6]; antiangiogenic[7]; antitumoral[8,9]; antiviral[10]; antiherpetic[11]	Gamal-Eldeen et al. 2009[1]; Ahn et al. 2002[2]; Anggadiredja et al. 1997[3]; Arunkumar et al. 2005[4]; Bazes et al. 2009[4]; Costa et al. 2010[5]; Dar et al. 2007[6]; Dias et al. 2005[7,8]; Itoh et al. 1993[9]; Iwashima et al. 2005[10]; Majczak et al. 2003[11];
Turbinaria	Antioxidant[1]; anti-inflammatory[1]; cytotoxic[2]	Ananthi et al. 2010[1]; Sheu et al. 1997[2];
Order Ishigeales		
Ishige	Antiviral	Ahn et al. 2002
Order Laminariales		
Alaria	Antiviral	Pardee et al. 2004
Ecklonia	Antioxidant[1,6]; antiviral[2]; antibacterial[3,6]; anticoagulant[4]; anti-tumor[5]	Ahn et al. 2007[1], 2002[2]; Choi et al. 2010[3]; Nishino and Nagumo 1991[4]; Park et al. 1998[5]; Kuda et al. 2007[6];
Egregia	Antiviral	Pardee et al. 2004; Todd et al. 1994;
Eisenia	Antioxidant[1]; antiviral[2]; anti-allergic[3]	Nakamura et al. 1996[1]; Kamei and Aoki 2007[2]; Sugiura et al. 2006[3];
Kjellmaniella	Antimicrobial	Kajiwara et al. 2006
Laminaria	Anticancer[1]; anti-oxidative; anti-inflammatory[2]; anticoagulant[3]; antithrombin[4]; neuroprotective[5]; antioxidant[6]; antiproliferative[6]	Bespalov et al. 2005[1]; Choi and Yea 2009[2]; Cumashi et al. 2007[3,4]; Tian et al. 1997[5]; Yuan and Walsh 2006[6]; Wang et al. 2008;
Landsburgia	Antifungal; cytotoxic	Perry et al. 1991
Lessonia	Anticoagulant activity	Chandia and Matsuhiro 2008
Macrocystis	Anticolesterol[1]; antioxidant[2]; antiproliferative[2]	Lee et al. 1998[1]; Yuan and Walsh 2006[2];
Nereocystis	Antioxidant[1]; antiproliferative[1]	Yuan and Walsh 2006[1]
Undaria	Anti-aging[1]; antiviral[2,3]; antioxidant[4]; anti-inflammatory[5]; antitumor[6]	Choi et al. 1992[1]; Hayashi et al. 2008[2]; Hemmingson et al. 2006[3]; Hu et al. 2010[4]; Khan et al. 2007[5]; Maruyama et al. 2006[6];
Order Ralfsiales		
Ralfsia	Antiviral	Pardee et al. 2004
Order Scytosiphonales		
Analipus	Antiviral	Kim et al. 1997
Endarachne	Antiviral	Wong et al. 2008
Hydroclathrus	Anti-cancer[1]; antiviral[2]	Wang et al. 2008[1], 2007[2];

(continued)

Table 1.2 (continued)

Ochrophyta	Bioactivity	Reference
Jolyna	Anti-bacterial	Atta-Ur-Rahman et al. 1997
Petalonia	Antiviral	Tsuda et al. 2007
Scytosiphon	Antimicrobial[1]; antitumor[2]	Dulger et al. 2009[1]; Xu et al. 2004[2];
Order Sphacelariales		
Cladostephus	Antimicrobial	Dulger et al. 2009
Halopteris	Antimicrobial	Dulger et al. 2009
Stypocaulon	Antioxidant	Lopez et al. 2011
Order Sporochnales		
Sporochnus	Antimicrobial	Gunasekera et al. 1995

Fig. 1.4 Example of green and brown algal genera exploited for bioactivities. *Dictyota dichotoma* (specimen about 20 cm wide)

The classification of the Dictyotales has been re-examined in recent years using molecular data. De Clerck et al. (2006) established the new genera *Canistrocarpus* and *Rugulopteryx* and merged *Glossophora*, *Glossophorella*, and *Pachydictyon* into *Dictyota*; subsequently, Bittner et al. (2008) showed that some common genera defined on a morphological basis, such as *Dictyopteris* and *Zonaria*, do not represent natural taxa and need to be split into several separate genera. Kraft (2009) established the monotypic genus *Herringtonia* for a single endemic species from Lord Howe Island in the western Pacific.

With 76 species (Guiry and Guiry 2011), *Dictyota* is the most species-rich genus in the Dictyotales. Although recognition of this genus is relatively straightforward, identification of species is often complicated, due to the wide range of phenotypic variation exhibited by many species (a good example is represented by the type species *D. dichotoma*; Tronholm et al. 2010b). In addition, morphological differences between some species are subtle and easily go unnoticed, particularly if specimens are not examined with sufficient care. For example, using a combination of morphological and molecular data Tronholm et al. (2010a) showed that in the Canary Islands *Dictyota cymatophila* was confused with *D. dichotoma* for a long time. The two species are frequently found mixed in the field, where they are usually impossible to distinguish; they can be recognized, however, based on differences in the phenology and in microscopic morphological characters. Consequently, species identifications of *Dictyota* for which biological activities are reported should be treated with caution; a reassessment would be desirable, especially for species reported to have a wide geographical distribution, particularly the generitype *D. dichotoma*.

1.10 The Order Fucales

The order Fucales (class Phaeophyceae) currently includes 520 species (Guiry and Guiry 2011) of brown seaweed found in the intertidal and shallow subtidal of rocky shores all around the world, particularly in cold-temperate regions. This is a group of special importance from both an ecological and economic point of view. Some fucalean seaweed (particularly species of *Fucus*, *Ascophyllum*, *Cystoseira,* and *Hormosira*) form dense populations in the intertidal zone of many temperate rocky shores, producing almost monospecific belts. Due to their relatively large size (20 cm to 2 m in length) they are habitat builders, the presence of which greatly modifies the colonized environment and promotes biological diversity (Eriksson et al. 2006; Mangialajo et al. 2008); the belts of *Fucus* in the northern Atlantic, *Cystoseira* in the Mediterranean and Black Seas and *Hormosira* in Australia and New Zealand are well-known examples. Several species of fucalean algae and their extracts are used for many applications (fertilizers, food products, drugs, cosmetics), and for this reason they are harvested commercially in several countries (Briand 1991).

The Fucales is the algal order for which the largest number of published reports on bioactivities and natural products is available. Studies concern species belonging to 13 different genera (see Table 1.2); *Cystoseira*, *Fucus,* and *Sargassum* are those to which most of the available information refers.

1.11 The Genus *Cystoseira* C. Agardh

About 44 *Cystoseira* species are currently recognized (Guiry and Guiry 2011), which are distributed mainly in temperate seas, with a few tropical representatives. Although the geographical distribution of this genus is widespread, its center of diversity is the Mediterranean and Black Seas; about 30 species occur in these seas and several are endemic to them. Thalli of *Cystoseira* are bushy plants up to 50 cm tall, consisting of one or more main axes on which numerous branches (with habit and arrangement varying in different species) are borne (Fig. 1.5). This genus has traditionally been a valuable resource for discovery of bioactivities and novel

Fig. 1.5 Example of green and brown algal genera exploited for bioactivities. *Cystoseira foeniculacea* (plant 30–35 cm tall)

natural products. The main bioactivities described for this genus are antibacterial (Badea et al. 2009; Dulger et al. 2009), antiviral (Mandal et al. 2007), and antifungal (Abourriche et al. 1999; Bennamara et al. 1999). Secondary metabolites of these algae have been used for chemotaxonomic purposes and have proven useful to distinguish genera within the family Cystoseiraceae (equals Sargassaceae in recent classification schemes; Amico 1995; Valls and Piovetti 1995).

From the taxonomic point of view, *Cystoseira* is one of the macroalgal genera that most requires a modern reassessment. Molecular data have become available only very recently for *Cystoseira* and related genera, and have revealed a very complex evolutionary scenario. Draisma et al. (2010) showed that *Cystoseira*, as circumscribed morphologically, is polyphyletic, and this is also the case for other closely related genera such as *Bifurcaria* and *Halidrys* (hypothesis that had already been suggested by Harvey and Goff 2006). For this reason these authors split *Cystoseira* in four different genera, reinstating *Sirophysalis* for the Indo-Pacific species *Cystoseira trinodis*, *Polycladia* for the western Indian Ocean species *Cystoseira indica* and *Cystoseira myrica*, and *Stephanocystis* for the North Pacific *Cystoseira* and *Halidrys* species. Draisma et al. (2010) concluded that European *Cystoseira* will also need further study to discover diagnostic characters for different lineages. Many species of *Cystoseira* are very variable in morphology; for Mediterranean species, in particular, it is not uncommon to find specimens with intermediate characteristics that are impossible to identify unambiguously on the basis of morphology. The matter is further complicated by the fact that hybridization between species is believed to occur, as is the case in several other fucalean genera, such as *Fucus*, and the occurrence of chimeras is also possible. For these reasons, additional molecular data based on larger taxon sampling are required, and the development of identifications based on DNA barcoding is particularly desirable for this genus.

1.12 The Genus *Fucus* Linnaeus

Fucus (class Phaeophyceae) is one of the oldest described genera of macroalgae and currently includes 14 species (Guiry and Guiry 2011) of intertidal seaweed (commonly called wracks) distributed in cold-temperate waters of the northern hemisphere. The thallus of these algae consists of a terete stipe expanding into a leathery flattened frond with more or less dichotomous branching (Fig. 1.6). Reports of biological activities in *Fucus* are frequent and refer mainly to *F. serratus* and *F. vesiculosus* (e.g., antioxidant; Diaz-Rubio et al. 2009, anti-inflammatory and anticoagulant properties of fucoidans; Cumashi et al. 2007, Table 1.2).

Overall, the taxonomy of *Fucus* is fairly settled, and recent studies have focused more on reproductive and population biology than on taxonomic matters. However, the morphological polymorphism of some members of this genus may sometimes lead to underestimation of species diversity. As a result, new species are occasionally described; this is the case of *Fucus radicans* (Bergström et al. 2005), which in the Baltic Sea was long confused with *F. vesiculosus*, and of *F. guiryi*, found to be

Fig. 1.6 Example of green and brown algal genera exploited for bioactivities. *Fucus ceranoides* (specimen about 25 cm tall)

Fig. 1.7 Example of green and brown algal genera exploited for bioactivities. *Sargassum elegans* (plants about 50 cm tall)

widely distributed from the Canary Islands and northern Morocco to Ireland and Scotland in the high intertidal zone (Zardi et al. 2011).

1.13 The Genus *Sargassum* C. Agardh

Sargassum, with more than 338 species (Guiry and Guiry 2011), is the species-richest genus of the Fucales, as well as being one of the largest non-diatom. marine genera. The genus is distributed worldwide, but it is especially well represented in tropical and subtropical regions where it may develop dense submarine forests of considerable ecological significance (e.g., Mattio and Payri 2011). Specimens of *Sargassum* are linear or bushy plants ranging from a few centimeters to several meters in length, formed by a holdfast and one to several main axes ramified into branches of several orders that end in foliar appendages with a leafy habit (Fig. 1.7). Besides its floristic and ecological importance, *Sargassum* is also used for several applications and is the macroalgal genus for which the largest number of published studies on bioactivities is available. The range of biological activities reported for this genus is very wide. Antioxidant activity has been reported for extracts from fresh samples of *Sargassum polycystum*; a crude methanol extract showed more activity than diethyl ether or hexane extracts (Anggadiredja et al. 1997). A dichloromethane extract from *Sargassum muticum* showed antifouling properties; identification by NMR and GC/MS of the active compound showed the presence of palmitic acid, a fatty acid with antibacterial activity (Bazes et al. 2009). Anticoagulant and antioxidant activities have been described in *S. filipendula* from Brazil (Costa et al. 2010). Other bioactivities, such as anti-HIV activity, were described for *S. confusum, S. hemiphyllum, and S. ringgoldianum* from the coast of Korea (Ahn et al. 2002). A polysaccharide extracted from *Sargassum stenophyllum* was described for its antiangiogenic and antitumor properties (Dias et al. 2005) and a range of water-soluble polysaccharide extracts from *Sargassum latifolium* showed cancer chemopreventive properties (Gamal-Eldeen et al. 2009).

There are not many studies describing the seasonal variation of biological activities. An example is the study of Dar et al. (2007), in which the anti-inflammatory activity of seasonal extracts of *Sargassum wightii* is compared. This study showed significant differences in the anti-inflammatory activity between winter and summer extracts, which were related to seasonal variation due to nutrient availability thereby affecting synthesis of chemical constituents needed for the growth of the alga (Dar et al. 2007).

Sargassum has been traditionally regarded as a difficult genus in need of comprehensive taxonomic revision. Taxonomic difficulties are found at two levels: ambiguities in species distinction, and level of classification, that is, uncertainties in the attribution of a particular species to one of the genus's numerous subdivisions (Mattio and Payri 2011). These difficulties are due to a combination of several problems, in particular high polymorphism and phenotypic plasticity in many species, possible hybridization that produces individuals with intermediate characteristics, excessive importance being attached to highly variable characters, inadequate original descriptions for many species, and type material that does not adequately reflect the range of variation. Molecular data produced in recent studies have significantly contributed to the understanding of the diversity and phylogenetic relationships in this genus. Draisma et al. (2010) showed that *Sargassum* was polyphyletic and reinstated the genus *Sargassopsis* Trevisan (non Nizamuddin et al.) for *S. decurrens*. Examining the diversity of *Sargassum* in the south Pacific, Mattio et al. (2008, 2009); Mattio and Payri (2009) reassessed the status of numerous species, resolved several taxonomic incongruities, and provided an advanced revision of several sections. It is desirable that equally detailed studies become available for other geographical regions and that an accurate characterization of the material used is made in the studies concerning bioactivities.

1.14 The Order Laminariales

The order Laminariales (class Phaeophyceae) includes about 30 genera and 123 species (Guiry and Guiry 2011) of large-sized brown seaweed commonly called kelp. The largest species of algae belong to this group; the giant Pacific kelp *Macrocystis* and *Nereocystis* may reach lengths of 60 m. Kelp thalli consist of a

Fig. 1.8 Example of green and brown algal genera exploited for bioactivities. *Laminaria hyperborea* (a group of plants 1–1.5-m tall)

holdfast attached to the rocky bottom by haptera or stolons, a stipe, and a blade of various shape (Fig. 1.8); structurally these organisms are the most complex of marine algae, having developed specialized cells (called trumpet hyphae) for the transport of nutrients, which are stored mainly as laminaran and mannitol (Lane et al. 2006). The Laminariales occur in cold-temperate and polar waters of both hemispheres but have their center of diversity in the north Pacific (>40 species). From an economic point of view, they are among the most important marine algae. Species of kelp are used as food in Asia, where they have been farmed for this purpose for a long time (Lüning and Pang 2003). Alginates and other extracts obtained from these algae are widely used in food, cosmetic, and medical products (Guiry and Blunden 1991). Reports of bioactivities in the Laminariales are numerous and concern species belonging to 11 genera, most of them referring to species of *Ecklonia*, *Eisenia*, *Laminaria,* and *Undaria.* For *Ecklonia*, in particular the records are extensive. Ahn et al. (2007) purified from *Ecklonia cava* three phlorotannins (phloroglucinol, eckol, and dieckol) and evaluated their antioxidant potential in cosmetics, food, and drug products. Shibata et al. (2007) isolated from *Ecklonia cava, Ecklonia kurome,* and *Eisenia bicyclis* various phlorotannins with potential anti-inflammatory activity. *Eisenia arborea*, an edible brown alga used in folk medicine and in gynecopathy in Japan for its antiallergic properties (Sugiura et al. 2006), has been the focus of many studies in Japan (Nakamura et al. 1997; Kamei and Aoki 2007). The properties of the fucoidan of *Laminaria* have been studied in detail; Cumashi et al. (2007) extracted fucoidan from *Laminaria saccharina* and *L. digitata* and described anti-inflammatory, antiangiogenic, anticoagulant, and antiadhesive activities for them. Bioactivities are often associated with the polyphenol content of these algae (e.g., *Laminaria setchellii* in Yuan and Walsh 2006). The synthesis of sulphated polysaccharides in this order has also been examined in detail; for example, some sulfated polysaccharide derivatives (oversulfated, acetylated, and benzoylated fucoidan) were isolated by Wang et al. (2009) and their antioxidant activity described.

The Laminariales represent a good example of the fact that even algal groups that are intensively studied and exploited are often poorly understood from a taxonomic point of view. Lane et al. (2006), based on a molecular dataset of five different markers, was the first to assess phylogenetic relationships in the order with strong bootstrap support. The authors substantially rearranged the classification of the order, erecting the new family Costariaceae for four Pacific genera (*Costaria, Agarum, Dictyoneurum,* and *Thalassiophyllum*) and emending the circumscription of *Laminaria*. The genus *Saccharina* Stackhouse was reinstated to include 18 species formerly included in *Laminaria*. Two of these were *Laminaria japonica* and *L. saccharina*, species that are among the most studied for biological activities; they were renamed *Saccharina japonica* and *S. latissima*, respectively.

1.15 The Red Seaweeds (Phylum Rhodophyta)

The red algae, or Rhodophyta, are one of the most ancient groups of eukaryotic organisms (fossils of *Bangiomorpha pubescens*, which is believed to be the oldest red alga, are approximately 1.2 billion years old; Blouin et al. 2011). Apart from a small number of freshwater species, the red algae are predominantly marine. With more than 6,135 species (Guiry and Guiry 2011), they account for the vast majority of seaweed species currently known and represent the dominant group in terms of biodiversity in all seaweed floras of the world. Of all the macroalgae, the Rhodophyta are the phylum for which the most substantial classification rearrangements have taken place in recent years. They have been subdivided in two subphyla (Cyanidiophytina and Rhodophytina), seven classes (Cyanidiophyceae, Bangiophyceae, Compsopogonophyceae, Florideophyceae, Porphyridiophyceae, Rhodellophyceae, and Stylonematophyceae) and 33 orders (Saunders and Hommersand 2004; Yoon et al. 2006; see Guiry and Guiry 2011, for more detail).

Given their long and complex evolutionary history the red algae are characterized by great genetic and morphological diversity, which is also reflected in the vast diversity of biological activities and secondary metabolites exhibited by these algae (see Table 1.3). Valuable biological activities have been reported for 82 genera, belonging mainly to the orders Ceramiales, Gigartinales, and Halymeniales (Rindi 2008). Here we discuss in detail the taxa for which recent studies have provided new and significant insights.

1.16 The Genus *Porphyra* C. Agardh

Porphyra (class Bangiophcyeae) is a cosmopolitan genus of predominantly intertidal red algae generally with a heteromorphic haplodiplontic life cycle. It is estimated to have somewhere between 115 (Guiry and Guiry 2011) and 150 species (Brodie et al. 2008) and has its center of diversity in the northern Pacific. The gametophyte is a foliose blade, one or two cells in thickness, ranging in morphology from orbicular to linear, with margins smooth, dentate, or ruffled depending on the species (Fig. 1.9). The sporophyte (*Conchocelis* phase) is a microscopic alga

Table 1.3 Phylum Rhodophyta; list of genera with reported bioactivities. Superscripts indicate the bibliographic references in which the biological activities were reported

Rhodophyta	Bioactivity	Reference
Order Bangiales		
Porphyra	Anti-inflammatory[1]; anticancer[2]; induction of apoptosis[3]; macrophage stimulation[4]; antioxidant[5]	Kazlowska et al. 2010[1]; Ichihara et al. 1999[2]; Kwon et al. 2007[3]; Yoshizawa et al. 1995[4]; Zhao et al. 2006[5]; Allmendinger et al. 2010
Order Bonnemaisoniales		
Asparagopsis	Antiviral[1]; antibacterial[2]; antimicrobial[3]; anti-HIV[4];	Bouhlal et al. 2010[1]; Bansemir et al. 2006[2]; González Del Val et al. 2001[3]; Haslin et al. 2001[4]
Bonnemaisonia	Angiotensin-I-converting enzyme (ACE) inhibitory activities	Cha et al. 2006
Delisea	Anti-phytopathogenic	Manefield et al. 2001
Order Ceramiales		
Acanthophora	Antiviral	Duarte et al. 2004
Aglaothamnion	Antioxidant	Zubia et al. 2009
Amansia	Lymphocyte transformation[1]; antinociceptive[2]	Lima et al. 2008[1]; Neves et al. 2007[2];
Boergeseniella	Antiviral	Rhimou et al. 2010
Bostrychia	Antiprotozoal[1]; antifungal[2]; antiviral[3]	de Felicio et al. 2010[1,2]; Duarte et al. 2001[3]
Bryothamnion	Antioxidant[1]; antinociceptive[2]	Fallarero et al. 2003, 2006[1]; Viana et al. 2002[2]
Ceramium	Antiviral[1]; antiprotozoal[2]; antibacterial[3]	Rhimou et al. 2010[1]; Allmendinger et al. 2010[2]; Bansemir et al. 2006[3]
Centroceras	Antibacterial; anticancer	Villarreal-Gomez et al. 2010
Chondria	Anthelmintic	Davyt et al. 2001
Cottoniella	Mammalian insulin release modulator	Moghaddam et al. 1991
Cryptopleura	Antiviral	Carlucci et al. 1997
Delesseria	Inhibit the release of inflammatory cytokines[2]; anticoagulant[3]	Gruenewald and Alban 2007[1]; Partschefeld and Alban 2007[2]; Potin et al. 1992[3]
Drachiella	Antibacterial	Bansemir et al. 2006

(continued)

Table 1.3 (continued)

Rhodophyta	Bioactivity	Reference
Digenea	Central nervous system stimulant[1]; anti-HIV[2]	Martinez-Lozano et al. 2000[1]; Sekine et al. 1995[2]
Griffithsia	Antimicrobial[1]; anti-HIV[2]; antiviral[3];	Emau et al. 2007[1] ;Mori et al. 2005[2]; Ziolkowska and Wlodawer 2006
Halopitys	Antiviral[1]; antibacterial[2]	Rhimou et al. 2010[1]; Bansemir et al. 2006[2]
Haraldiophyllum		Güven et al. 2010
Heterosiphonia	Antioxidant	Zubia et al. 2009
Laurencia/Chondrophycus/ Osmundea/Palisada/ Yuzurua	Anti-hepatotoxicity[1]; antioxidant[2]; antiviral[3]; antibacterial[4]; antifungal[5]; anti-tumoral[6]	Abdel-Wahhab et al. 2006[1]; Anggadiredja et al. 1997[2]; Lliopoulou et al. 2002[3]; Koenig and Wright 1997[4]; Morales et al. 2006[4,5]; Norte et al. 1996[6]
Martensia	Anti-lipid peroxidation	Takahashi et al. 1998
Melanothamnus	Antileishmanial	Sabina et al. 2005
Neorhodomela	Antioxidant; anti-inflammatory	Lim et al. 2006
Neurymenia	Antibacterial	Stout et al. 2009
Odonthalia	Antimicrobial[1]; antifungal[2]	Oh et al. 2008[1]; Tariq 1991[2]
Osmundaria/Vidalia	Antimicrobial	Barreto and Meyer 2006
Platysiphonia	Mammalian insulin release modulator	Moghaddam et al. 1991
Polysiphonia	Antioxidant[1]; anti-cancer[2]; anti-herpes[3]; cytotoxic[4]	Zubia et al. 2009[1]; Gwak 2010[2]; Serkedjieva 2000[3]; Shoeib et al. 2003[4]
Pterosiphonia	Antiviral	Rhimou et al. 2010
Rhodomela	Antioxidant[1]; antibacterial[2]	Huang and Wang 2004[1]; Xu et al. 2003[2]
Spyridia	Antibiotic[1]; antimicrobial[2]	Robles Centeno and Ballantine 1999[1]; Zamora and Ballantine 2000[2]
Symphyocladia	Antioxidant[1]; anticancer[2]; antiviral[3]	Huang and Wang 2004[1]; Lee et al. 2007a[2]; Park et al. 2005[3]
Order Corallinales		
Amphiroa	Anti-tumor	Nakamura et al. 1997
Corallina	Antiprotozoal[1]; induction of apoptosis[2]	Allmendinger et al. 2010[1]; Kwon et al. 2007[2]
Jania	antitumor[1]; antimmicrobial[2]	Kamei and Sagara 2002[1]; Karabay-Yavasoglu et al. 2007[2]

Lithophyllum	Angiotensin-I-converting enzyme (ACE) inhibitory activities	Cha et al. 2006
Lithothamnion		Navarro et al. 2011
Marginisporum	Antitumor	Hiroishi et al. 2001
Order Gelidiales		
Gelidiella	Contraceptive	Premakumara et al. 1996
Gelidium	Antiviral[1]; growth-inhibitory effects of cells[2]; anticoagulant[3]	Rhimou et al. 2010[1]; Chen et al. 2004[2]; Qi et al. 2008[3]
Pterocladia/Pterocladiella	Antiviral[1]; antiherpetic[2]	Damonte et al. 1994[1]; Pujol et al. 1996[2]
Order Gigartinales		
Agardhiella	Anti-HIV[1]; antiviral[2]	Tziveleka et al. 2003[1]; Witvrouw et al. 1994[2]
Ahnfeltiopsis	Angiotensin-I-converting enzyme (ACE) inhibitory activities	Cha et al. 2006
(Callophycus)	Antineoplastic	Kubanek et al. 2005
Callophyllis	Anti-HIV	Nakamura et al. 1994
Chondrus	Immunomodulatory[1]; antitumor[1]	Nazarova et al. 1998[1]; Zhou et al. 2006[1]
Dilsea	Antifungal	Tariq 1991
Eucheuma	Anti-tumor	Fukuda et al. 2006;
Furcellaria	Antioxidant	Zubia et al. 2009
Gigartina	Antiviral[1]; anticoagulant[2]	Barabanova et al. 2006[1]; Carlucci et al. 1999[1]; 1997[2]
Gloiopeltis	Therapeutic potential in hepatoma cancer[1]; anti-tumor[2]; antiviral[3];	Bae and Choi 2007[1]; Ren et al. 1995[2]; Tsuda et al. 2007[3]
Gymnogongrus	Antiviral	Talarico et al. 2005
Hyalosiphonia	Antiviral	Tsuda et al. 2007
Hypnea	Antiviral[1]; PPE elastase inhibition[2]	Rhimou et al. 2010[1]; Bultel-Ponce et al. 2002[2]
Kappaphycus	Anti-tumor	Yuan and Song 2005
Meristiella	Antiviral	de SF-Tischer et al. 2006
Meristotheca	Cell stimulating activity	Liu et al. 1997
Portieria	Anti-tumor[1]; antimicrobial[2];	Fuller et al. 1994[1]; Puglisi et al. 2007[2];

(continued)

Table 1.3 (continued)

Rhodophyta	Bioactivity	Reference
Solieria	Antibacterial[1]; antifungal[2];	Holanda et al. 2005[1]; Khanzada et al. 2007[2]
Sphaerococcus	Antiviral	Rhimou et al. 2010
Stenogramma	Antiherpetic	Caceres et al. 2000
Tichocarpus	Antiviral	Reunov et al. 2004
Order Gracilariales		
Gracilaria	Antibacterial[1]; angiotensin-I-converting enzyme (ACE) inhibitory activities[2]; antitumor[3]; anti-inflammatory[4]; antileishmanial[5]; antimicrobial[6]	Bansemir et al. 2006[1]; Cha et al. 2006[2]; Fernandez et al. 1989[3]; Lee et al., 2007[4]; Sabina et al. 2005[5]; Tuney et al. 2006[6]; Shanmughapriya et al. 2008[6]
Order Halymeniales		
Carpopeltis	Atopic allergic reaction	Na et al. 2005
Cryptonemia	Antiviral	Talarico et al. 2005
Grateloupia	Angiotensin-I-converting enzyme (ACE) inhibitory activities[1]; antiviral[2,5]; anticoagulant[3]; antioxidant[4]; anti-HIV[6]	Cha et al. 2006; Chattopadhyay et al. 2007[2]; Given et al. 1991;[3] Heo et al. 2006[4]; Tsuda et al. 2007[5]; Wang et al. 2007[6]
Halymenia		Baricuatro 1997
Pachymeniopsis	anticoagulant	Ekanayake et al. 2007
Order Nemaliales		
Galaxaura	Induction of apoptosis[1]; anti-inflammatory[2]	Huang et al., 2005[1]; Rozas and Freitas 2007[2]
Liagora	Topical antiedematous	Mendonca and Freitas 2000
Nothogenia	Antiviral[1]; antiherpetic[2]; anticoagulant[2]; anti-HIV[3]	Damonte et al. 1994[1]; Kolender et al. 1995[1]; 1997[2]; Tziveleka et al. 2003[3]
Scinaia	Antileishmanial	Sabina et al. 2005
Order Nemastomatales		
Schizymenia	Antiviral[1]; anti-HIV-1[2]; anticoagulant[3]	Bourgougnon et al. 1993[1]; Bourgougnon et al. 1996[2]; Ekanayake et al. 2007[3]

Order Palmariales		
Palmaria	Antioxidant[1]; antiproliferative[1]	Yuan et al. 2005[1]
Order Peyssonneliales		
Peyssonnelia	Antiviral[1]; DNA Methyl Transferase inhibitory Activity[2]	Loya et al. 1995[1]; McPhail et al. 2004[2]
Order Plocamiales		
Plocamium	Antiviral[1]; cytotoxic[2]	Bouhlal et al. 2010[1]; de Ines et al. 2004[2]
Order Rhodymeniales		
Botryocladia	Antithrombotic[1]; antifilarial[2]	Farias et al. 2001[1]; Lakshmi et al. 2004[2];
Champia	Antitumor[2]	Sampaio Assreuy et al. 2008[1]; Lins et al. 2009[2]
Lomentaria	Angiotensin-I-converting enzyme (ACE) inhibitory activities[1]; anticoagulant[2]	Cha et al. 2006[1]; Pushpamali et al. 2008[2]

9

Fig. 1.9 Example of red algal genera exploited for bioactivities. *Porphyra umbilicalis* (specimen about 10 cm wide)

consisting of short, uniseriate, irregularly branched filaments that penetrate calcareous substrata such as oyster shells and barnacles. Due to its usage as food, *Porphyra* is one of the most important commercial seaweed, producing a crop valued at $1.3 billion U.S. per year (Blouin et al. 2011). Species of *Porphyra* have frequently been used for studies of bioactivities; most of them refer to *P. yezoensis*, the species most used for cultivation and food consumption. The polysaccharide fraction of *P. yezoensis* has been shown to stimulate murine phagocytic functions in vivo and in vitro (Yoshizawa et al. 1993) and the sulphate groups to stimulate macrophage activity (Yoshizawa et al. 1993).

From a taxonomic point of view, *Porphyra* is among the red algal genera that have been studied best in recent years, and probably the one for which most progress has been made; new species have been described, intraspecific variation has been characterized, and assessments of species identities have been proposed for regional floras (Brodie et al. 2007, 2008; Lindstrom 2008; Kikuchi et al. 2010). Despite this, a considerable amount of work remains to be done. The status of the genus will require a reassessment, because *Porphyra* is not monophyletic (Nelson et al. 2006; Brodie et al. 2008); it is also believed that several more species will be added to the flora of certain regions (e.g., North Atlantic and Mediterranean). At the species level, a great deal of cryptic diversity has been demonstrated in certain species (Brodie et al. 2007; Lindstrom 2008), which makes these algae prone to misidentification when their examination is based only on morphological observations. It is important that future studies on bioactivity in *Porphyra* include species characterizations based on molecular data.

1.17 The Order Bonnemaisoniales

The order Bonnemaisoniales (class Florideophyceae) includes 34 species of red algae (Guiry and Guiry 2011) with a diplohaplontic heteromorphic lifecycle, and that are widely distributed in temperate and tropical regions. The gametophyte is a tufted feathery plant with a pseudoparenchymatous structure, densely branched,

Fig. 1.10 Example of red algal genera exploited for bioactivities. *Bonnemaisonia hamifera* (specimen about 20 cm tall)

and up to 40 cm long (Fig. 1.10); the sporophyte is a thin filamentous alga, either forming distinctive pom-poms or developing a thin crust on the surface of the substratum. The genera *Asparagopsis* and *Bonnemaisonia* are of particular interest and significance. They include some highly invasive seaweed species from the Pacific and Indian Oceans that have been introduced in many regions and in some cases may have caused major damage to native communities (*Asparagopsis armata*, some strains of *A. taxiformis*, and *Bonnemaisonia hamifera*). These algae are known to produce several secondary metabolites with valuable biological activities, in particular volatile halogenated compounds with strong antibacterial properties (e.g., Genovese et al. 2009). Some recent ecological studies have provided evidence that these compounds are the main mechanism that these algae use to maintain bacteria-free surfaces (Paul et al. 2006; Nylund et al. 2008). Several other bioactivities have also been described from crude extracts derived from *Asparagopsis armata* (e.g., antiviral and antibacterial: Bansemir et al. 2006; Rhimou et al. 2010) and *Asparagopsis taxiformis* (antifungal and antibacterial: Gonzalez del Val et al. 2001). Because of these properties, *Asparagopsis armata* has been farmed commercially in northern Europe and used for the preparation of cosmetic products; it is possible that the farming activities have contributed to the persistence of *Asparagopsis* populations on a local scale (Kraan and Barrington 2005).

Overall, the taxonomy of the Bonnemaisoniales has not required major recent reassessments. However, molecular studies have characterized several species, revealing a higher genetic diversity than previously understood (Andreakis et al. 2004, 2007a, 2007b; Salvador et al. 2008; Sherwood 2008). *Asparagopsis taxiformis*, a tropical seaweed found widely in nature, is a species that has been investigated in great detail. Molecular studies have revealed a well-defined biogeographical structure, with four different genetic lineages that are indistinguishable on morphological grounds (Ní Chualáin et al. 2004; Andreakis et al. 2007a). This has resulted in cryptic introductions going unnoticed in some regions (Sherwood 2008). One of these lineages has assumed an invasive behavior and has become widespread in the western Mediterranean in recent years; it is believed that the development of these traits is linked to polyploidy (Andreakis et al. 2007b). In consideration of these findings, it is highly desirable that future studies on bioactivities in the Bonnemaisoniales provide a detailed molecular characterization of the material used.

1.18 The Genus *Grateloupia* C. Agardh

Currently including 86 species (Guiry and Guiry 2011) distributed in warm temperate and tropical waters throughout the world, *Grateloupia* is one of the most species-rich genera of the order Halymeniales (class Florideophyceae). Species of *Grateloupia* consist of pseudoparenchymatous plants up to 40 cm tall, compressed to foliose, linear to lanceolate, usually branched proliferously to one or more orders, in one or more planes, with texture varying from lubricious to leathery.

Grateloupia has long been known as a taxonomically difficult genus. To assess species boundaries based on morphological discontinuities has been considered problematic because of substantial intraspecific or even within-individual variation in gross morphology (De Clerck et al. 2005; Lopez-Figueroa et al. 2007). In this regard, the type species *Grateloupia filicina* represents a striking case. Until the early years of the present century this species was considered a textbook example of a marine red alga with a more or less cosmopolitan distribution (De Clerck et al. 2005); and, throughout the world, specimens of *Grateloupia* with finely pinnate branching were referred to as *G. filicina*. Molecular data produced in the last 10 years, however, have revealed a much more complex scenario. Studies encompassing samples obtained from the entire range of distribution have revealed a large number of cryptic species and demonstrated that the "real" *Grateloupia filicina* occurs only in the Mediterranean Sea (De Clerck et al. 2005). This led to the description of several new species and the resurrection of others (Kawaguchi et al. 2001; Faye et al. 2004; De Clerck et al. 2005; Wilkes et al. 2005). Another factor that has confused the taxonomy of this genus is the fact that some species have been introduced in regions where they were initially misidentified. Such is the case for the Japanese species *Grateloupia turuturu*, introduced in the North Atlantic and confused for a long time with *G. doryphora*; Gavio and Fredericq (2002) clarified its taxonomic identity. Confusion also occurred between *Grateloupia turuturu* and *G. lanceola*, a little-known native European species; and Lopez-Figueroa et al. (2007) clarified their distinctness. The taxonomy of this genus is far from settled, as additional species continue to be described or reinstated (Wilkes et al. 2005; Lin et al. 2008).

Records of bioactivities in species of *Grateloupia* are frequent and based mostly on samples from eastern Asia. Reports referred to *Grateloupia filicina* are certainly wrong and require a reassessment based on molecular data. More generally, we suggest that the taxonomic identity of the specimens used should be reassessed for all studies in which bioactivities have been reported in this genus (Fig. 1.11).

Fig. 1.11 Example of red algal genera exploited for bioactivities. *Grateloupia subpectinata* (specimens about 10 cm tall)

1.19 The Genus *Gracilaria* Greville

The genus *Gracilaria* is the largest in the order Gracilariales (class Florideophyceae) and includes 167 species widely distributed in temperate and tropical seas (Guiry and Guiry 2011). Plants of *Gracilaria* can reach 60 cm in length and consist of pseudoparenchymatous thalli ranging from erect to prostrate and from terete to broadly flattened (Fig. 1.12); some species form seemingly articulated fronds composed of cylindrical or irregularly shaped units. Species of *Gracilaria* have been popular subjects for applied research, due mainly to their high agar content; reports of biological activities are numerous and concern several species. The most common bioactivities known in *Gracilaria* are antibacterial and antiviral, and have been best described in *G. cornea* (Bansemir et al. 2006), *G. corticata* (Shanmughapriya et al. 2008), *G. gracilis* (Tuney et al. 2006), and *G. changii* (Sasidharan et al. 2008).

The taxonomic relationships in *Gracilaria* and closely related genera have been extensively studied using DNA sequencing (e.g., Gurgel and Fredericq 2004; Gargiulo et al. 2006; Lin 2006). It is believed that the actual number of species has been underestimated (particularly in the western Atlantic) because of convergence in habit and vegetative and reproductive anatomy. Gurgel and Fredericq (2004) identified nine distinct evolutionary lineages in *Gracilaria* and reinstated the genus *Hydropuntia*, transferring to it a number of species previously belonging to *Gracilaria*. Although some species have been studied and circumscribed in detail (e.g., *Gracilaria tikvahiae*; Gurgel et al. 2004) some major taxonomic matters are still pending, in particular the characterization of *Gracilaria verrucosa*. This is one of the species most frequently recorded; Steentoft et al. (1995), however, showed that in the western Atlantic it consisted of a complex of two different species, *Gracilaria gracilis* and *Gracilariopsis longissima* and the name *Gracilaria verrucosa* has had to be rejected (Irvine and Steentoft 1995). Despite this, the name *Gracilaria verrucosa* continues to be used widely in the literature and is still found in many papers, including some on bioactivities. These records should be reassessed using molecular data.

Fig. 1.12 Example of red algal genera exploited for bioactivities. *Gracilaria gracilis* (fronds 20–25 cm tall)

Fig. 1.13 Example of red algal genera exploited for bioactivities. *Plocamium cartilagineum* (fronds 10–15 cm tall)

1.20 The Genus *Plocamium* J.V. Lamouroux

Plocamium is the largest genus of the order Plocamiales (class Florideophyceae), accounting for 40 species (Guiry and Guiry 2011) distributed mainly in cold-temperate and polar seas (a few are, however, characteristic of tropical waters). These algae consist of erect bushy plants up to 50 cm tall, with linear and flattened fronds, arising from a terete lower axis and anchored by a rhizomatous holdfast. The branching pattern is characteristic, with the main axes extending by sympodial growth and the margins of the fronds bearing alternating series of 2–6 branchlets (Fig. 1.13).

Species of *Plocamium* have been a popular subject for biochemical studies, which have led to the discovery of several natural products with valuable bioactivities. Antiviral properties (Rhimou et al. 2010) and cytotoxic activities on tumor cell lines (de Ines et al. 2004) have been studied best. In some studies, biochemical and physiological properties have been considered informative for taxonomic purposes (Yano et al. 2004; 2006). Molecular data produced in recent years have unraveled a great deal of hidden diversity, especially in the area of distribution of the generitype, *Plocamium cartilagineum*. This latter entity is generally regarded as more or less cosmopolitan and for a long time it was considered the only species of *Plocamium* present in Atlantic Europe. Saunders and Lehmkuhl (2005) showed that in this latter area *Plocamium* is instead represented by a complex of four different species; in addition to *P. cartilagineum*, they found the three new species *Plocamium maggsiae*, *P. nanum,* and *P. subtile*, and clarified the morphological differences among these entities. Yano et al. (2004, 2006) have also revealed cryptic diversity in Japanese populations of *Plocamium recurvatum* and *P. teilfariae*. In consideration of these reports, the taxonomic identity of the strains of *Plocamium* used in studies of bioactivities requires reassessment. In particular, it is likely that many records of bioactivities in *Plocamium cartilagineum* are incorrect; the identity of the material used in these studies should be re-examined using molecular data.

1.21 The Order Ceramiales

The order Ceramiales (class Florideophyceae), with about 2,400 species (Guiry and Guiry 2011) is the largest order of red algae, as well as one of the most evolutionarily advanced (Verbruggen et al. 2010). In recent years, the taxonomy of this order

Fig. 1.14 Example of red algal genera exploited for bioactivities. *Osmundea hybrida* (fronds 10–15 cm tall)

has been covered in a very uneven way; whereas some families and genera have been studied in great detail, other species-rich taxa are still waiting for a comprehensive molecular reassessment (e.g., the genus *Polysiphonia*). Accordingly, with its genetic and morphological diversity, studies on this order have shown a high diversity in biological activities and secondary metabolites. Reports are available for species of 30 different genera (Rindi 2008), but it is likely that this number will increase in the future; the small size and filamentous habit have so far been a major limitation to test bioactivities in many members of this order.

Among genera referred to the Rhodomelaceae, *Laurencia* has been the most popular for studies of natural products. This genus currently includes some 130 species (Guiry and Guiry 2011) occurring on temperate and tropical shores all around the world, consisting of a pseudoparenchymatous thallus up to 40 cm tall, sparingly to highly branched. In the last 20 years, however, its circumscription has been substantially emended. *Osmundea* (Fig. 1.14) and *Chondrophycus*, two genera previously included in generic synonymy with *Laurencia*, were reinstated by Nam et al. (1994) and Garbary and Harper (1998). More recently, Nam (2007) elevated to generic status the name *Palisada*, a former subsection of *Laurencia*. Martin-Lescanne et al. (2010) validated these conclusions using molecular data and supported the splitting of the *Laurencia* complex into five separate genera (*Laurencia*, *Osmundea*, *Chondrophycus*, *Palisada*, and *Yuzurua*). With regard to studies concerning bioactivities in *Laurencia*, a careful reassessment of the species identification should be made. Many studies describing bioactivities in this genus refer their material to species previously believed to have a very wide distribution, such as *Laurencia obtusa* and *Osmundea pinnatifida* (as *Laurencia pinnatifida*). Recent work suggests that these species are probably restricted to Europe; outside Europe, material referred to these species most probably represents misidentifications and should be reassessed using molecular data.

1.22 Conclusions and Recommendations

Correct identification and classification are critical requirements for exploitation of any organism used in applied science. Taxonomy is the discipline of describing, naming, and classifying living organisms, and is essential for understanding and cataloguing biodiversity (see Mattio and Payri 2011). Knowing organisms in detail is necessary to select species with economic potential; applied studies should always be based on a sound taxonomic characterization of the specimens used. In relation to studies on bioactivities, a misidentification will result in an incorrect selection of the algae containing the targeted molecule of interest, with the likely consequence of a considerable waste of time and scarce financial resources. Incorrect identifications will also lead to misinterpretations of published studies, spreading confusion and misleading future studies. For this reason, in applied investigations: (1) samples of adequate quality should be subjected to the scrutiny of a professional taxonomist; (2) voucher specimens should be deposited in reliable repositories (such as herbaria in national museums); and (3) DNA sequences should be established and deposited so that they are publicly available. Unfortunately, in the case of studies concerning bioactivities in seaweed this seems to have been carried out rarely. We suspect that in many cases the authors attached to their material the name of common species superficially resembling their specimens; this is almost certainly the case of species with supposedly widespread distribution (particularly *Gracilaria verrucosa*, *Grateloupia filicina*, *Laurencia obtusa*, *Osmundea pinnatifida*, and *Plocamium cartilagineum*). Using the unfortunate maxim that "a species is what a good taxonomist says it is" is – and always was in this context – inadequate both because good taxonomists are few and getting fewer, and many "good taxonomists" are simply not able to identify cryptic species without DNA evidence.

In consideration of the recent developments that have taken place in algal taxonomy, the necessity of a careful taxonomic characterization is even more critical today. The genetic diversity revealed by molecular systematic studies was impossible to imagine more than 20 years ago. The new data have substantially reshaped our understanding of species concepts and circumscriptions and it can be expected that future developments in algal genomics will lead to even more substantial changes (the first complete genome of a marine macroalga has recently been sequenced (*Ectocarpus*; Cock et al. 2010) and others (*Chondrus crispus*, *Porphyra umbilicalis*) are now close to completion. We expect that major taxonomic rearrangements will continue to be necessary in the near future, especially for the brown and green algae. The cases of the taxa discussed above show that a molecular characterization has become almost mandatory for a reliable identification of many species and it is very important that applied studies keep in mind these developments. Applied work should adopt procedures that ensure an identification of the samples as correctly as possible; even more important, the identification should be subsequently verifiable.

Verification is one of the basic principles of taxonomy; after completion of the work, it should be possible to go back to the voucher material and check it, confirming or modifying its identification (if necessary). For this reason the deposition of voucher specimens in official institutions (such as herbaria or museums) with some expectation of permanency is a practice normally adopted in most taxonomic studies, and mandatorily required by the International Code of Botanical Nomenclature for the description of new species. The authors of studies on bioactivities in seaweed should arrange to have samples conserved permanently in a form that allows verification of the species identity, either in a private or (more desirably) in a permanent public collection, and reviewers and editors of journals should be adamant about refusing publication without evidence of such deposition. In consideration of the fundamental importance that molecular data and DNA barcoding have assumed for algal identification, it is highly desirable that a part of the material be conserved in a form allowing DNA extraction for a long time after processing of the sample. This can be easily achieved by simply drying part of the material in silica gel and conserving it in sealed containers. When certain algal strains prove to be of special value for applied matters, it would also be desirable to conserve the alga alive, as cultured material or in a cryopreserved form. Public culture and cryopreservation collections can serve this purpose (Brodie et al. 2009).

Marine algae producing valuable secondary metabolites are an important natural resource; an effective and sustainable exploitation requires their conservation in the long term. Some species that are utilized for food or other applied purposes, such as species of *Porphyra*, *Laminaria,* and *Saccharina*, which are farmed on a large scale and the extensive biotechnological knowledge that has been developed for their exploitation ensures that their conservation is not a problem. For most species, however, extraction of bioactive compounds is performed on material obtained from natural populations harvested in the field. Numerous processes that operate at different spatial and temporal scales determine the distribution and persistence of an algal species in nature. Some processes operate on global spatial scales and affect the species in its entirety (e.g., processes linked to climate change, such as increased temperatures, CO_2 content, and acidification of seawater). Others operate on local or regional scales and normally affect only individual populations (e.g., pollution, nutrient enrichments caused by local sources, storms); removal of seaweed for exploitation of natural products or other applied purposes falls within this category. When conservation matters are involved, all these processes should be kept in mind and their possible interaction should be examined. An accurate knowledge of the biology of the species is obviously a critical requirement to understand the effect of these processes and draw plans for its conservation. Life history, reproductive and vegetative phenology, rates of growth, fertility, fertilization and recruitment, effects of abiotic factors on early life history stages, tolerances to high/low temperatures, salinities, light irradiation, and UV radiation, are details essential to predict the responses of seaweeds to disturbances; it is highly desirable, if not critical, that they be available for populations of macroalgae harvested for bioactivities (and other applications). In general, algae are not regarded as endangered organisms in need of protection.

Seaweed has never been a major focus in conservation programs, as reflected by the fact that relatively few species are currently included in the IUCN red list (75 species, mainly from the Galapagos Islands; http://www.iucnredlist.org/). To date, only one seaweed species is reputed to have become extinct due to human activities, the Australian red alga *Vanvoorstia bennettiana* described from Port Jackson [Sydney] in 1855 and not seen since (Millar 2003). The wide-spread diffusion of seaweed on rocky shores gives the general impression of commonness in the oceans and thus protection in numbers (Brodie et al. 2009). The fact that seaweed disperses by spores, gametes, and often by vegetative fragments released in the seawater, gives them a high dispersal capacity, which may limit the risk of extinction of local populations. However, we know now that many algae have a more or less restricted geographical distribution (Brodie et al. 2009). This has become particularly evident as molecular studies have revealed cryptic diversity at species and infraspecific levels, with many taxa having a more restricted distribution than previously believed; so, once again, the importance of a correct taxonomic characterization based on molecular data cannot be overstated.

A restricted distribution implies that these organisms may be vulnerable to stochastic environmental and anthropogenic events, and even more to long-term degradation of their habitats due to human influence. Some species of *Cystoseira*, which are a valuable source of interesting natural products as described above, are a textbook example of seaweed threatened by habitat degradation operating at a local or regional scale. Most species of *Cystoseira* occur in the intertidal and upper subtidal zone, and in coastal rockpools; some Mediterranean and Black Sea species known to exhibit valuable bioactivities (*Cystoseira barbata*, *C. brachycarpa* var. *balearica*, *C. crinita*, *C. mediterranea*) form dense canopies in these habitats. These algae are known to be very sensitive to changes in environmental conditions (especially increases in sediment loads; Irving et al. 2009) and there is now strong evidence that anthropogenic disturbance causes loss of *Cystoseira* populations (Benedetti-Cecchi et al. 2001; Mangialajo et al. 2008; Perkol-Finkel and Airoldi 2010). In some areas, this phenomenon is exacerbated by extreme storm events and can reach a point at which recovery becomes almost impossible (Perkol-Finkel and Airoldi 2010). Therefore, harvesting of *Cystoseira* for natural products should be performed only at sites with good environmental conditions, where healthy populations occur. The harvesting should be performed in a sustainable way and ensure a viable turnover in the populations. Once again, a detailed knowledge of growth, fertility, recruitment rates, and early life history stages are essential in this regard. Several species of fucalean algae have been reported to have low recruitment rates, limited dispersal, and high vulnerability to catastrophic events (Ballesteros et al. 2009; Olsen et al. 2010); so, extreme care should be used when harvesting natural populations. Similar recommendations apply to many other species of seaweed used for applied purposes.

Acknowledgments The authors are very grateful to the Marine Institute (Ireland) for financial support received under the Beaufort Award, within the framework of the Marine Biodiscovery Program and the Marine Functional Foods Research Initiative (NutraMara project).

References

Abdel-Wahhab, M.A., H.H. Ahmed, and M.M. Hagazi. 2006. Prevention of aflatoxin B-1-initiated hepatotoxicity in rat by marine algae extracts. *Journal of Applied Toxicology* 26(3): 229–238.

Abourriche, A., M. Charrouf, M. Berrada, et al. 1999. Antimicrobial activities and cytotoxicity of the brown alga *Cystoseira tamariscifolia*. *Fitoterapia* 70(6): 611–614.

Adhikari, U., C.G. Mateii, K. Chattopadhyay, et al. 2006. Structure and antiviral activity of sulfated fucans from *Stoechospermum marginatum*. *Phytochemistry* 67(22): 2474–2482.

Ahn, G.N., K.N. Kim, S.H. Cha, et al. 2007. Antioxidant activities of phlorotannins purified from Ecklonia cava on free radical scavenging using ESR and H2O2-mediated DNA damage. *European Food Research and Technology* 226: 71–79.

Ahn, M.J., K.D. Yoon, C.Y. Kim, et al. 2002. Inhibition of HIV-1 reverse transcriptase and HIV-1 integrase and antiviral activity of Korean seaweed extracts. *Journal of Applied Phycology* 14(5): 325–329.

Alekseyenko, T.V., S.Y. Zhanayeva, A.A. Venediktova, et al. 2007. Antitumor and antimetastatic activity of fucoidan, a sulfated polysaccharide isolated from the Okhotsk Sea *Fucus evanescens* brown alga. *Bulletin of Experimental Biology and Medicine* 143: 730–732.

Albuquerque, I.R.L., K.C.S. Queiroz, L.G. Alves, et al. 2004. Heterofucans from *Dictyota menstrualis* have anticoagulant activity. Conference Information: 18th Annual Meeting of the Federacao-de-Sociedades-de-Biologia-Experimental, Date: Aug 27–30, 2003 Curitiba. *Brazilian Journal of Medical and Biological Research* 37(2): 167–171.

Allmendinger, A., J.M. Spavieri, et al. 2010. Antiprotozoal, antimycobacterial and cytotoxic potential of twenty-three British and Irish red algae. *Phytotherapy Research* 24(7): 1099–1103.

Amico, V. 1995. Marine brown of algae of the family Cystoseiraceae: chemistry and chemotaxonomy. *Phytochemistry* 39: 1257–1279.

Ananthi, S., H.R.B. Raghavendran, et al. 2010. In vitro antioxidant and in vivo anti-inflammatory potential of crude polysaccharide from *Turbinaria ornata* (Marine Brown Alga). *Food and Chemical Toxicology* 48(1): 187–192.

Anca, J.M., M. Lamela, and J.M. Cadavid. 1990. Effects of *Himathalia elongata* on the central nervous system of mice. *Journal of Ethnopharmacology* 29(2): 225–231.

Andreakis, N., G. Procaccini, and W.H.C.F. Kooistra. 2004. *Asparagopsis taxiformis* and *Asparagopsis armata* (Bonnemaisoniales, Rhodophyta): genetic and morphological identification of Mediterranean populations. *European Journal of Phycology* 39: 273–283.

Andreakis, N., G. Procaccini, C.A. Maggs, and W.H.C.F. Kooistra. 2007a. Phylogeography of the invasive seaweed *Asparagopsis* (Bonnemaisoniales, Rhodophyta) reveals cryptic diversity. *Molecular Ecology* 16: 2285–2299.

Andreakis, N., W.H.C.F. Kooistra, and G. Procaccini. 2007b. Microsatellite markers in an invasive strain of *Asparagopsis taxiformis* (Bonnemaisoniales, Rhodophyta): insights in ploidy level and sexual reproduction. *Gene* 406: 144–151.

Anggadiredja, J., R. Andyani, et al. 1997. Antioxidant activity of *Sargassum polycystum* (Phaeophyta) and *Laurencia obtusa* (Rhodophyta) from Seribu Islands. *Journal of Applied Phycology* 9(5): 447–479.

Angulo, Y., and B. Lomonte. 2003. Inhibitory effect of fucoidan on the activities of crotaline snake venom myotoxic phospholipases A(2). *Biochemical Pharmacology* 66(10): 1993–2000.

Ara, J., V. Sultana, R. Qasim, et al. 2005. Biological activity of *Spatoglossum asperum*: a brown alga. *Phytotherapy Research* 19(7): 618–623.

Arunkumar, K., N. Selvapalam, and R. Rengasamy. 2005. The antibacterial compound sulphoglycerolipid 1–0 palmitoyl-3-0 (6′-sulpho-alpha-quinovopyranosyl)-glycerol from *Sargassum wightii* Greville (Phaeophyceae). *Botanica Marina* 48(5–6): 441–445.

Atta-Ur-Rahman, M.I.C., A. Majeed, M. Shabbir, et al. 1997. A succinylanthranilic acid ester and other bioactive constituents of *Jolyna laminarioides*. *Phytochemistry (Oxford)* 46(7): 1215–1218.

Badea, V., D.P. Balaban, et al. 2009. The antibacterial activity evaluation of *Cystoseira barbata* biomass and some alginates upon bacteria from oropharyngeal cavity. *Romanian Biotechnological Letters* 14(6): 4851–4857.

Bae, S.J., and Y.H. Choi. 2007. Methanol extract of the seaweed *Gloiopeltis furcata* induces G2/M arrest and inhibits cyclooxygenase-2 activity in human hepatocarcinoma HepG2 cells. *Phytotherapy Research* 21(1): 52–57.

Bae, E.H., and I.K. Lee. 2001. *Umbraulva*, a new genus based on *Ulva japonica* (Holmes) Papenfuss (Ulvaceae, Chlorophyta). *Algae* 16: 217–231.

Ballesteros, E., J. Garrabou, B. Hereu, M. Zabala, E. Cebrian, and E. Sala. 2009. Deep-water stands of *Cystoseira zosteroides* C. Agardh (Fucales, Ochrophyta) in the Northwestern Mediterranean: Insights into assemblage structure and population dynamics. *Estuarine, Coastal and Shelf Science* 82: 477–484.

Bansemir, A., M. Blume, S. Schroder, et al. 2006. Screening of cultivated seaweeds for antibacterial activity against fish pathogenic bacteria. *Aquaculture* 252(1): 79–84.

Barabanova, A.O., I.M. Yermak, A.V. Reunov, et al. 2006. Carrageenans – sulphated polysaccharides of red algae as inhibitors of tobacco mosaic virus. *Rastitel'nye Resursy* 42(4): 80–86.

Barbier, P., S. Guise, P. Huitorel, et al. 2001. Caulerpenyne from *Caulerpa taxifolia* has an antiproliferative activity on tumor cell line SK-N-SH and modifies the microtubule network. *Life Sciences* 70(4): 415–429.

Barbosa, J.P., R.C. Pereira, J.L. Abrantes, et al. 2004. In vitro antiviral diterpenes from the Brazilian brown alga *Dictyota pfaffii*. *Planta Medica* 70(9): 856–860.

Baricuatro, J.H.L. 1997. Isolation and characterization of bioactive metabolites of red alga *Halymenia durvillaei* bory. Thesis, 63. Philippines: Available online May 2011.

Barreto, M., and J.J.M. Meyer. 2006. Isolation and antimicrobial activity of a lanosol derivative from *Osmundaria serrata* (Rhodophyta) and a visual exploration of its biofilm covering. *South African Journal of Botany* 72(4): 521–528.

Barroso, E.M.A., L.S. Costa, V.P. Medeiros, et al. 2008. A non-anticoagulant heterofucan has antithrombotic activity in vivo. *Planta Medica* 74(7): 712–718.

Bazes, A., A. Silkina, et al. 2009. Investigation of the antifouling constituents from the brown alga *Sargassum muticum* (Yendo) Fensholt. *Journal of Applied Phycology* 21(4): 395–403.

Becker, B., and B. Marin. 2009. Streptophyte algae and the origin of embryophytes. *Annals of Botany* 103: 999–1004.

Benedetti-Cecchi, L., F. Pannacciulli, F. Bulleri, P.S. Moschella, L. Airoldi, G. Relini, and F. Cinelli. 2001. Predicting the consequences of anthropogenic disturbance: large-scale effects of loss of canopy algae on rocky shores. *Marine Ecology Progress Series* 214: 137–150.

Bennamara, A., A. Abourriche, M. Berrada, et al. 1999. Methoxybifurcarenone: an antifungal and antibacterial meroditerpenoid from the brown alga *Cystoseira tamariscifolia*. *Phytochemistry* 52(1): 37–40.

Bergström, L., A. Tatarenkov, K. Johannesson, R.B. Jönsson, and L. Kautsky. 2005. Genetic and morphological identification of *Fucus radicans* sp. nov. (Fucales, Phaeophyceae) in the brackish Baltic Sea. *Journal of Phycology* 41: 1025–1038.

Bespalov, V.G., N.Y. Barash, O.A. Ivanova, et al. 2005. Study of drug "mamoclam" for treatment of benign breast disease. *Voprosy Onkologii (St. Petersburg)* 51(2): 236–241.

Bhakuni, D.S., and D.S. Rawat. 2005. *Bioactive marine natural products*, 382. New Delhi: Anamaya.

Bittner, L., C.E. Payri, A. Couloux, C. Cruaud, B. de Reviers, and F. Rousseau. 2008. Molecular phylogeny of the Dictyotales and their position within the Phaeophyceae, based on nuclear, plastid and mitochondrial DNA sequence data. *Molecular Phylogenetics and Evolution* 49: 211–226.

Blondin, C., F. Chaubet, A. Nardella, et al. 1996. Relationships between chemical characteristics and anticomplementary activity of fucans. *Biomaterials* 17(6): 597–603.

Blomster, J., C.A. Maggs, and M.J. Stanhope. 1998. Molecular and morphological analysis of *Enteromorpha intestinalis* and *E. compressa* in the British Isles. *Journal of Phycology* 34: 319–340.

Blouin, N.A., J.A. Brodie, A.C. Grossman, P. Xu, and S.H. Brawley. 2011. *Porphyra*: a marine crop shaped by stress. *Trends in Plant Science* 16: 29–37.

Blunt, J.W., B.R. Copp, M.H.G. Munro, P.T. Northcote, and M.R. Prinsep. 2011. Marine natural products. *Natural Product Reports* 28: 196–268.
Boisson-Vidal, C., F. Chaubet, L. Chevolot, et al. 2000. Relationship between antithrombotic activities of fucans and their structure. *Drug Development Research* 51(4): 216–224.
Bouaïcha, N., C. Tringali, D. Pesando, M. Malléa, C. Roussakis, and J.F. Verbist. 1993. Bioactive diterpenoids isolated from *Dilophus ligulatus*. *Planta Medica* 59: 256–258.
Bouhlal, R., R. Hassane, J. Martinez, et al. 2010. The antibacterial potential of the seaweeds (Rhodophyceae) of the Strait of Gibraltar and the Mediterranean Coast of Morocco. *Africal Journal of Biotechnology* 9(38): 6365–6372.
Bourgougnon, N., M. Lahaye, B. Quemener, et al. 1996. Annual variation in composition and in vitro anti-HIV-1 activity of the sulfated glucuronogalactan from *Schizymenia dubyi* (Rhodophyta, Gigartinales). *Journal of Applied Phycology* 8(2): 155–161.
Bourgougnon, N., M. Lahaye, J.C. Chermann, et al. 1993. Composition and antiviral activities of a sulfated polysaccharide from *Schizymenia dubyi* (Rhodophyta, Gigartinales). *Bioorganic & Medicinal Chemistry Letters* 3(6): 1141–1146.
Briand, X. 1991. Seaweed harvesting in Europe. In *Seaweed resources in Europe: uses and potential*, ed. M.D. Guiry and G. Blunden, 259–308. Chichester: Wiley.
Brodie, J., R.A. Andersen, M. Kawachi, and A.J.K. Millar. 2009. Endangered algal species and how to protect them. *Phycologia* 48: 423–438.
Brodie, J., A. Mols Mortensen, M.E. Ramirez, S. Russell, and B. Rinkel. 2008. Making the links: towards a global taxonomy for the red algal genus *Porphyra* (Bangiales, Rhodophyta). *Journal of Applied Phycology* 20: 939–949.
Brodie, J., I. Bartsch, C. Neefus, S. Orfanidis, T. Bray, and A.C. Mathieson. 2007. New insights into the cryptic diversity of the North Atlantic-Mediterranean *'Porphyra leucosticta'* complex: *P. olivii* sp. nov. and *P. rosengurttii* (Bangiales, Rhodophyta). *European Journal of Phycology* 42: 3–28.
Buckle, P.J., B.A. Baldo, and K.M. Taylor. 1980. The anti-inflammatory activity of marine natural products-6-n-tridecylsalicylic acid, flexible and dendalone 3-hydroxybutyrate. *Inflammation Research* 10(4): 361–367.
Bultel-Ponce, V., S. Etahiri, and M. Guyot. 2002. New ketosteroids from the red alga *Hypnea musciformis*. *Bioorganic & Medicinal Chemistry Letters* 12(13): 1715–1718.
Caceres, P.J., M.J. Carlucci, E.B. Damonte, et al. 2000. Carrageenans from Chilean samples of *Stenogramme interrupta* (Phyllophoraceae): structural analysis and biological activity. *Phytochemistry* 53(1): 81–86.
Capon, R.J., R.A. Barrow, S. Rochfort, et al. 1998. Marine nematocides: tetrahydrofurans from a southern Australian brown alga, *Notheia anomala*. *Tetrahedron* 54(10): 2227–2242.
Carlucci, M.J., M. Ciancia, M.C. Matulewicz, et al. 1999. Antiherpetic activity and mode of action of natural carrageenans of diverse structural types. *Antiviral Research* 43(2): 93–102.
Carlucci, M.J., L.A. Scolaro, M.I. Errea, et al. 1997. Antiviral activity of natural sulphated galactans on herpes virus multiplication in cell culture. *Planta Medica* 63(5): 429–432.
Cavalier-Smith, T. 2007. Evolution and relationships of algae: major branches of the tree of life. In *Unravelling the algae: the past, present and future of algal systematics*, eds. Brodie, J. and Lewis J, pp. 21–55. CRC Press. The Systematics Association Special Volume 75.
Cavas, L., Y. Baskin, K. Yurdakoc, et al. 2006. Antiproliferative and newly attributed apoptotic activities from an invasive marine alga: *Caulerpa racemosa var. cylindracea*. *Journal of Experimental Marine Biology and Ecology* 339(1): 111–119.
Ceccherelli, G., and F. Cinelli. 1999. The role of vegetative fragmentation in dispersal of the invasive alga *Caulerpa taxifolia* in the Mediterranean. *Marine Ecology Progress Series* 182: 299–303.
Celikler, S., O. Vatan, et al. 2009. Evaluation of anti-oxidative, genotoxic and antigenotoxic potency of *Codium tomentosum* Stackhouse ethanolic extract in human lymphocytes in vitro. *Food and Chemical Toxicology* 47(4): 796–801.
Cha, S.H., K.W. Lee, and Y.J. Jeon. 2006. Screening of extracts from red algae in Jeju for potentials marine angiotensin – I converting enzyme (ACE) inhibitory activity. *Algae* 21(3): 343–348.

Chandia, N.P., and B. Matsuhiro. 2008. Characterisation of a fucoidan from *Lessonia vadosa* (Phaeophyta) and its coagulant and elicitor properties. *International Journal of Biological Macromolecules* 42(3): 235–240.
Chattopadhyay, K., C.G. Mateu, P. Mandal, et al. 2007. Galactan sulfate of *Grateloupia indica*: isolation, structural features and antiviral activity. *Phytochemistry* 68(10): 1428–1435.
Chen, Y.H., C.J. Tu, and H.T. Wu. 2004. Growth-inhibitory effects of the red alga *Gelidium amansii* on cultured cells. *Biological & Pharmaceutical Bulletin* 27(2): 180–184.
Chen, J.L., W.H. Gerwick, R. Schatzman, et al. 1994. Isorawsonol and related IMP-dehydrogenase inhibitors from the tropical green alga *Avrainvillea rawsonii*. *Journal of Natural Products* 57(7): 947–952.
Choi, J.-G., O.-H. Kang, et al. 2010. Antibacterial activity of *Ecklonia cava* against methicillin-resistant *Staphylococcus aureus* and *Salmonella* spp. *Foodborne Pathogens and Disease* 7(4): 435–441.
Choi, C.Y., and S.S. Yea. 2009. Anti-oxidative and anti-inflammatory potential of *Laminaria joponica*: Activity and mode of action. *Molecular & Cellular Toxicology* 5(3): 54–54.
Choi, J.-H., I.-S. Kim, J.-I. Kim, et al. 1992. Studies on anti-aging action of brown algae (*Undaria pinnatifida*): 2. Dose effect of alginic acid as a modulator of anti-aging action in liver membranes. *Bulletin of the Korean Fisheries Society* 25(3): 181–188.
Cock, J.M., S.M. Coelho, C. Brownlee, and A.R. Taylor. 2010. The Ectocarpus genome sequence: Insights into brown algal biology and the evolutionary diversity of the eukaryotes. *The New Phytologist* 188: 1–4.
Colliec, S., A.M. Fischer, J. Taponbretaudiere, et al. 1991. Anticoagulant properties of a fucoidan fraction. *Thrombosis Research* 64(2): 143–154.
Costa, L.S., G.P. Fidelis, et al. 2010. Biological activities of sulfated polysaccharides from tropical seaweeds. *Biomedicine & Pharmacotherapy* 64(1): 21–28.
Cumashi, A., N.A. Ushakova, M.E. Preobrazhenskaya, et al. 2007. A comparative study of the anti-inflammatory, anticoagulant, antiangiogenic, and antiadhesive activities of nine different fucoidans from brown seaweeds. *Glycobiology* 17(5): 541–552.
Damonte, E.B., C.A. Pujol, M.I. Errea, et al. 1994. Antiviral activity and mode of action of a sulfated galactan from *Pterocladia capillacea*. *Antiviral Research* 23(Suppl. 1): 84.
Dar, A., H.S. Baig, S.M. Saifullah, et al. 2007. Effect of seasonal variation on the anti-inflammatory activity of *Sargassum wightii* growing on the N. Arabian Sea coast of Pakistan. *Journal of Experimental Marine Biology and Ecology* 351: 1–9.
Davyt, D., R. Fernandez, L. Suescun, et al. 2001. New sesquiterpene derivatives from the red alga *Laurencia scoparia*. Isolation, structure determination, and anthelmintic activity. *Journal of Natural Products* 64: 1552–1555.
Deacon-Smith, R.A., J.P. Lee-Potter, and D.J. Rogers. 1985. Anticoagulant activity in extracts of British marine algae. *Botanica Marina* 28(8): 333–338.
De Clerck, O., F. Leliaert, H. Verbruggen, C.E. Lane, J.C. De Paula, D.A. Payo, and E. Coppejans. 2006. A revised classification of the Dictyoteae (Dictyotales, Phaeophyceae) based on rbcL and 26 S ribosomal DNA sequence analyses. *Journal of Phycology* 42: 1271–1288.
De Clerck, O., B. Gavio, S. Fredericq, I. Barbara, and E. Coppejans. 2005. Systematics of *Grateloupia filicina* (Halymeniaceae, Rhodophyta), based on rbcL sequence analyses and morphological evidence, including the reinstatement of *G. minima* and the description of G. capensis sp. nov. *Journal of Phycology* 41: 391–410.
de Ines, C., V.H. Argandona, J. Rovirosa, et al. 2004. Cytotoxic activity of halogenated monoterpenes from *Plocamium cartilagineum*. *Zeitschrift für Naturforschung C - A journal of Biosciences* 59(5–6): 339–344.
de Felicio, R., S. de Albuquerque, et al. 2010. Trypanocidal, leislimanicidal and antifungal potential from marine red alga *Bostrychia tenella* J. Agardli (Rhodomelaceae, Ceraniiales). *Journal of Pharmaceutical and Biomedical Analysis* 52(5): 763–769.
De Lara Isassi, G., and S. Alvarez Hernandez. 1995. Anticoagulant properties of Mexican marine algal extracts: heparin-like potency of *Halimeda discoidea* (Chlorophyta) extract. *Cryptogamie Algologie* 16(3): 199–205.

de Tischer, S.F., L.B. PC Talarico, M.D. Noseda, et al. 2006. Chemical structure and antiviral activity of carrageenans from *Meristiella gelidium* against herpes simplex and dengue virus. *Carbohydrate Polymers* 63(4): 459–465.

de Souza, E.T., D.P. de Lira, et al. 2009. The antinociceptive and anti-inflammatory activities of caulerpin, a bisindole alkaloid isolated from seaweeds of the genus *Caulerpa*. *Marine Drugs* 7(4): 689–704.

Dias, P.F., J.M. Siqueira, L.F. Vendruscolo, et al. 2005. Antiangiogenic and antitumoral properties of a polysaccharide isolated from the seaweed *Sargassum stenophyllum*. *Cancer Chemotherapy and Pharmacology* 56(4): 436–446.

Diaz-Rubio, M.E., J. Perez-Jimenez, and F. Saura-Calixto. 2009. Dietary fiber and antioxidant capacity in *Fucus vesiculosus* products. *International Journal of Food Sciences and Nutrition* 60(2): 23–34.

Diaz-Pulido, G., L. Villamil, and V. Almanza. 2007. Herbivory effects on the morphology of the brown alga *Padina boergesenii*. *Phycologia* 46: 131–136.

Dobashi, K., T. Nishino, M. Fujiara, et al. 1989. Isolation and characterization of fucose-containing sulfated polysaccharides with blood-anticoagulant activity from the brown seaweed *Hizikia fusiforme*. *Carbohydrate Research* 194: 315–320.

Domingos, T.F.S., C. Carvalho, et al. 2009. Antilonomic effects of Brazilian brown seaweed extracts. *Natural Product Communications* 4(8): 1075–1078.

Domis, L.N.D., P. Famà, A.J. Bartlett, W.F.P. van Reine, C.A. Espinosa, and G.C. Trono. 2003. Defining taxon boundaries in members of the morphologically and genetically plastic genus *Caulerpa* (Caulerpales, Chlorophyta). *Journal of Phycology* 39: 1019–1037.

Draisma, S.G.A., E. Ballesteros, F. Rousseau, and T. Thibaut. 2010. DNA sequence data demonstrate the polyphyly of the genus *Cystoseira* and other Sargassaceae genera (Phaeophyceae). *Journal of Phycology* 46: 1329–1345.

Duarte, M.E.R., J.P. Cauduro, D.G. Noseda, et al. 2004. The structure of the agaran sulfate from *Acanthophora spicifera* (Rhodomelaceae, Ceramiales) and its antiviral activity, Relation between structure and antiviral activity in agarans. *Carbohydrate Research* 339(2): 335–347.

Duarte, M.E.R., D.G. Noseda, M.D. Noseda, et al. 2001. Inhibitory effect of sulfated galactans from the marine alga *Bostrychia montagnei* on Herpes simplex virus replication in vitro. *Phytomedicine* 8(1): 53–58.

Dulger, B., N. Hacioglu, et al. 2009. Antimicrobial activity of some brown algae from Turkey. *Asian Journal of Chemistry* 21(5): 4113–4117.

Ekanayake, P.M., C. Nikapitiya, M. De Zoysa, et al. 2007. Isolation and purification of an anticoagulant from *Schizymenia dubyi* by fermentation. *Food Science and Technology International* 13(5): 355–359.

Emau, P., B.P. Tian, B.R. O'Keefe, et al. 2007. Griffithsin, a potent HIV entry inhibitor, is an excellent candidate for topical microbicide. *Journal of Medical Primatology* 36(4–5): 303–303.

Engel, S., M.P. Puglisi, P.R. Jensen, et al. 2006. Antimicrobial activities of extracts from tropical Atlantic marine plants against marine pathogens and saprophytes. *Marine Biology* 149(5): 991–1002.

Eriksson, B.K., A. Rubach, and H. Hillebrand. 2006. Community dominance by a canopy species controls the relationship between macroalgal production and species richness. *Limnology and Oceanography* 51: 1813–1818.

Fallarero, A., A. Peltoketo, J. Loikkanen, et al. 2006. Effects of the aqueous extract of *Bryothamnion triquetrum* on chemical hypoxia and aglycemia-induced damage in GT1-7 mouse hypothalamic immortalized cells. *Phytomedecine* 13(4): 240–245.

Fallarero, A., J.J. Loikkanen, P.T. Mannisto, et al. 2003. Effects of aqueous extracts of Halimeda incrassata (Ellis) Lamouroux and *Bryothamnion triquetrum* (SGGmelim) Howe on hydrogen peroxide and methyl mercury-induced oxidative stress in GT1-7 mouse hypothalamic immortalized cells. *Phytomedecine* 10(1): 39–47.

Famà, P., W.C.H.F. Kooistra, and G. Zuccarello. 2002. Molecular phylogeny of the genus *Caulerpa* (Caulerpales, Chlorophyta) inferred from chloroplast tufA gene. *Journal of Phycology* 38: 1040–1050.

Farias, W.R.L., R.A. Nazareth, and P.A.S. Mourao. 2001. Dual effects of sulfated D-galactans from the red algae *Botryocladia occidentalis* preventing thrombosis and inducing platelet aggregation. *Thrombosis and Haemostasis* 86(6): 1540–1546.

Faye, E.J., H.W. Wang, S. Kawaguchi, S. Shimada, and M. Masuda. 2004. Reinstatement of *Grateloupia subpectinata* (Rhodophyta, Halymeniaceae) based on morphology and rbcL sequences. *Phycological Research* 52: 59–67.

Feldman, S.C., S. Reynaldi, C.A. Stortz, et al. 1999. Antiviral properties of fucoidan fractions from *Leathesia difformis*. *Phytomedecine* 6(5): 335–340.

Fernandez, L.E., O.G. Valiente, V. Mainardi, et al. 1989. Isolation and characterization of an antitumor active agar-type polysaccharide of *Gracilaria dominguensis*. *Carbohydrate Research* 190(1): 77–83.

Fischel, J.L., R. Lemee, P. Formento, et al. 1995. Cell growth inhibitory effects of caulerpenyne, a sesquiterpenoid from the marine algae *Caulerpa taxifolia*. *Anticancer Research* 15(5B): 2155–2160.

Folmer, F., M. Jasmars, M. Dicato, and M. Diederich. 2010. Photosynthetic marine organisms as a source of anticancer compounds. *Phytochemistry Reviews* 9: 557–579.

Foti, M., M. Piattelli, and V. Amico. 1994. Antioxidant activity of phenolic meroditerpenoids from marine algae. *Journal of Photochmistry and Photobiology B-Biology* 26(2): 159–164.

Fowler-Walker, M.J., T. Wernberg, and S.D. Connell. 2006. Differences in kelp morphology between wave sheltered and exposed localities: Morphologically plastic or fixed traits? *Marine Biology* 148: 755–767.

Fukuda, Y., T. Sugahara, M. Ueno, et al. 2006. The anti-tumor effect of *Euchema serra* agglutinin on colon cancer cells in vitro and in vivo. *Anti-cancer Drugs* 17(8): 943–947.

Fuller, R.W., Cardellina, J.H.I., and Jurek, J., et al. (1994). Halomon analogs: additional halogenated monoterpenes from Portieria hornemannii with a novel in vitro antitumor response profile. *Abstracts of Papers American Chemical Society* 208:(1–2): Medi 92.

Gamal-Eldeen, A.M., E.F. Ahmed, et al. 2009. In vitro cancer chemopreventive properties of polysaccharide extract from the brown alga, *Sargassum latifolium*. *Food and Chemical Toxicology* 47(6): 1378–1384.

Ganesan, K., K.S. Kumar, et al. 2011. Comparative assessment of antioxidant activity in three edible species of green seaweed, *Enteromorpha* from Okha, Northwest coast of India. *Innovative Food Science & Emerging Technologies* 12(1): 73–78.

Garbary, D.J., and T. Harper. 1998. A phylogenetic analysis of the *Laurencia* complex (Rhodomelaceae) of the red algae. *Cryptogamie, Algologie* 19: 185–200.

Garg, H.S., M. Sharma, D.S. Bhakuni, et al. 1992. An antiviral sphingosine derivative from the green alga *Ulva fasciata*. *Tetrahedron Letters* 33(12): 1641–1644.

Gargiulo, G.M., M. Morabito, G. Genovese, and F. De Masi. 2006. Molecular systematics and phylogenetics of gracilariacean species from the Mediterranean Sea. *Journal of Applied Phycology* 18: 497–504.

Gavio, B., and S. Fredericq. 2002. *Grateloupia turuturu* (Halymeniaceae, Rhodophyta) is the correct name of the non-native species in the Atlantic known as *Grateloupia doryphora*. *European Journal of Phycology* 37: 349–359.

Genovese, G., L. Tedone, M.T. Hamann, and M. Morabito. 2009. The Mediterranean red alga *Asparagopsis*: A source of compounds against Leishmania. *Marine Drugs* 7: 361–366.

Ghosh, P., U. Adhikari, P.K. Ghosal, et al. 2004. In vitro anti-herpetic activity of sulfated polysaccharide fractions from *Caulerpa racemosa*. *Phytochemistry* 65(23): 3151–3157.

Given, K.C., Y. Ozsoy, and O.N. Ulutin. 1991. Anticoagulant, fibrinolytic and antiaggregant activity of carrageenans and alginic acid. *Botanica Marina* 34(5): 429–432.

Gonzalez del Val, A., G. Platas, and A. Basilio. 2001. Screening of antimicrobial activities in red, green and brown macroalgae from Gran Canaria (Canary Islands, Spain). *International Microbiology* 4: 35–40.

Govindan, M., S.A. Abbas, F.J. Schmitz, et al. 1994. New cycloartanol sulfates from the alga Tydemania expeditionis – inhibitors of the protein-tyrosine kinase PP60 (V-SRC). *Journal of Natural Products* 57(1): 74–78.

Gruenewald, N., and S. Alban. 2007. Seasonal variations of biologically active sulfated polysaccharides extracted from the red alga *Delesseria sanguinea* (Hudson) Lamouroux from the Baltic Sea. *Planta Medica* 73(9): 934.

Guiry, M.D., and G. Blunden. 1991. *Seaweed resources in Europe – Uses and potential*, 432. Chichester: Wiley.

Guiry, M.D., and G.M. Guiry. 2011. AlgaeBase. World-wide electronic publication, Galway: National University of Ireland. http://www.algaebase.org; searched on 5 April 2011.

Gunasekera, L.S., A.E. Wright, S.P. Gunasekera, et al. 1995. Antimicrobial constituent of the brown alga *Sporochnus pedunculatus*. *International Journal of Pharmacognosy* 33(3): 253–255.

Gurgel, C.F.D., and S. Fredericq. 2004. Systematics of the Gracilariaceae (Gracilariales, Rhodophyta): A critical assessment based on rbcL sequence analysis. *Journal of Phycology* 40: 138–159.

Gurgel, C.F.D., S. Fredericq, and J.N. Norris. 2004. Phylogeography of *Gracilaria tikvaihae* (Gracilariaceae, Rhodophyta): A study of genetic discontinuity in a continuously distributed species based on molecular evidence. *Journal of Phycology* 40: 748–758.

Güven, K.C., A. Percot, and E. Sezik. 2010. Alkaloids in marine drugs. *Marine Drugs* 8: 269–284.

Haneji, K., T. Matsuda, M. Tomita, et al. 2005. Fucoidan extracted from *Cladosiphon okamuranus* Tokida induces apoptosis of human T-cell leukemia virus type 1-infected T-cell lines and primary adult T-cell leukemia cells. *Nutrition and Cancer – An International Journal* 52(2): 189–201.

Harada, N., and M. Maeda. 1998. Chemical structure of antithrombin-active rhamnan sulfate from *Monostroma nitidum*. *Bioscience, Biotechnology, and Biochemistry* 62(9): 1647–1652.

Harada, H., T. Noro, and Y. Kamei. 1997. Selective antitumor activity in vitro from marine algae from Japan coasts. *Biological & Pharmaceutical Bulletin* 20(5): 541–546.

Harvey, J.B.J., and L.J. Goff. 2006. A reassessment of species boundaries in *Cystoseira* and *Halidrys* (Phaeophyceae, Fucales) along the North American West Coast. *Journal of Phycology* 42: 707–720.

Haslin, C., M. Lahaye, M. Pellegrini, et al. 2001. In vitro anti-HIV activity of sulfated cell-wall polysaccharides from gametic, carposporic and tetrasporic stages of the Mediterranean red alga *Asparagopsis armata*. *Planta Medica* 67(4): 301–305.

Hayashi, K., T. Nakano, M. Hashimoto, et al. 2008. Defensive effects of a fucoidan from brown alga *Undaria pinnatifida* against herpes simplex virus infection. *International Immunopharmacology* 8: 109–116.

Hayden, H.S., J. Blomster, C.A. Maggs, P.C. Silva, M.J. Stanhope, and J.R. Waaland. 2003. Linnaeus was right all along: *Ulva* and *Enteromorpha* are not distinct genera. *European Journal of Phycology* 38: 277–294.

Heesch, S., J.E.S. Broom, K.F. Neill, T.J. Farr, J.L. Dalen, and W.A. Nelson. 2009. *Ulva, Umbraulva* and *Gemina*: Genetic survey of New Zealand taxa reveals diversity and introduced species. *European Journal of Phycology* 44: 143–154.

Hemmingson, J.A., R. Falshaw, R.H. Furneaux, et al. 2006. Structure and antiviral activity of the galactofucan sulfates extracted from *Undaria pinnatifida* (Phaeophyta). *Journal of Applied Phycology* 18(2): 185–193.

Heo, S.J., S.H. Cha, K.W. Lee, et al. 2006. Antioxidant activities of red algae from Jeju Island. *Algae* 21(1): 149–156.

Hiraoka, M., S. Shimada, M. Uenosono, and M. Masuda. 2004. A new green-tide-forming alga, *Ulva ohnoi* Hiraoka & Shimada sp. nov. (Ulvales, Ulvophyceae) from Japan. *Phycological Research* 52: 17–29.

Hiroishi, S., K. Sugie, T. Yoshida, et al. 2001. Antitumor effects of *Marginisporum crassissimum* (Rhodophyceae), a marine red alga. *Cancer Letters* 167(2): 145–150.

Hofmann, L.C., J.C. Nettleton, C.D. Neefus, and A.C. Mathieson. 2010. Cryptic diversity of Ulva (Ulvales, Chlorophyta) in the Great Bay estuarine system (Atlantic U.S.A.): Introduced and indigenous distromatic species. *European Journal of Phycology* 45: 230–239.

Holanda, M.L., V.M.M. Melo, L.M.C.M. Silva, et al. 2005. Differential activity of a lectin from *Solieria filiformis* against human pathogenic bacteria. *Brazilian Journal of Medical and Biological Research* 38(12): 1769–1773.
Holdt, S.L., and S. Kraan. 2011. Bioactive compounds in seaweed: functional food applications and legislation. *Journal of Applied Phycology* 23(3): 543–597.
Hu, T.T., D. Liu, et al. 2010. Antioxidant activity of sulfated polysaccharide fractions extracted from *Undaria pinnitafida* in vitro. *International Journal of Biological Macromolecules* 46(2): 193–198.
Huang, H.L., S.L. Wu, H.F. Liao, et al. 2005a. Induction of apoptosis by three marine algae through generation of reactive oxygen species in human leukemic cell lines. *Journal of Agricultural and Food Chemistry* 53(5): 1776–1781.
Huang, Z.X., X.T. Mei, D.H. Xu, et al. 2005b. Protective effects of polysacchride of *Spirulina platensis* and *Sargassum thunbergii* on vascular of alloxan induced diabetic rats. *Zhongguo Zhongy ao Zazhi* 30(3): 211–215.
Huang, H.L., and B.G. Wang. 2004. Antioxidant capacity and lipophilic content of seaweeds collected from the Qingdao coastline. *Journal of Agricultural and Food Chemistry* 52(16): 4993–4997.
Hudson, J.B., J.H. Kim, M.K. Lee, et al. 1998. Antiviral compounds in extracts of Korean seaweeds: evidence for multiple activities. *Journal of Applied Phycology* 10(5): 427–434.
Ichihara, T., H. Wanibuchi, T. Taniyama, et al. 1999. Inhibition of liver glutathione S-transferase placental form-positive foci development in the rat hepatocarcinogenesis by *Porphyra tenera* (Asakusa-nori). *Cancer Letters* 141(1–2): 211–218.
Irvine, L., and M. Steentoft. 1995. Proposal to reject the name *Fucus verrucosa* Huds. (Rhodophyta). *Taxon* 44: 223–224.
Irving, A.D., D. Balata, F. Colosio, G.A. Ferrando, and L. Airoldi. 2009. Light, sediment, temperature, and the early life-history of the habitat-forming alga *Cystoseira barbata*. *Marine Biology* 156(6): 1223–1231.
Ismail-Ben Ali, A., L. Ktari, et al. 2009. Antibacterial activity of the green alga *Ulva rigida* collected from Tunisian coast: seasonal and geographical variation. *Planta Medica* 75(9): 1033–1033.
Itoh, H., H. Noda, H. Amano, et al. 1993. Antitumor-activity and immunological properties of marine algal polysaccharides, especially fucoidan, prepared from *Sargassum thunbergii* of Phaeophyceae. *Anticancer Research* 13(6A): 2045–2052.
Iwahori, Y., S. Enomoto, Y. Okada, et al. 1999. Naturally occurring substances for prevention of complications of diabetes. IV: screening of seavegetables for inhibitory effect on aldose reductase. *Natural medicines* 53(3): 138–140.
Iwashima, M., J. Mori, X. Ting, et al. 2005. Antioxidant and antiviral activities of plastoquinones from the brown alga *Sargassum micracanthum*, and a new chromene derivative converted from the plastoquinones. *Biological and Pharmaceutical Bulletin* 28(2): 374–377.
Jiao, L.L., P. Jiang, et al. 2010. Antitumor and immunomodulating activity of polysaccharides from *Enteromorpha intestinalis*. *Biotechnology and Bioprocess Engineering* 15(3): 421–428.
Jin, D.Q., C.S. Lim, J.Y. Sung, et al. 2006. *Ulva conglobata*, a marine algae has neuroprotective and anti-inflammatory effects in murine hippocampal and microglial cells. *Neuroscience Letters* 402(1–2): 154–158.
Jurd, K.M., D.J. Rogers, G. Blunden, et al. 1995. Anticoagulant properties of sulfated polysaccharides and a proteoglycan from *Codium fragile* ssp. *atlanticum*. *Journal of Applied Phycology* 7(4): 339–345.
Kajiwara, T., K. Matsui, Y. Akakabe, et al. 2006. Antimicrobial browning-inhibitory effect of flavor compounds in seaweeds. *Journal of Applied Phycology* 18(3–5): 413–422.
Kamei, Y., and A. Sagara. 2002. Neurite outgrowth promoting activity of marine algae from Japan against rat adrenal medulla pheochromocytoma cell line, PC12D. *Cytotechnology* 40(1–3): 99–106.
Kamei, Y., and M. Aoki. 2007. A chlorophyll c2 analogue from the marine brown alga *Eisenia bicyclis* inactivates the infectious hematopoietic necrosis virus, a fish rhabdovirus. *Archives of Virology* 152(5): 861–869.

Karabay-Yavasoglu, N.U., A. Sukatar, G. Ozdemir, et al. 2007. Antimicrobial activity of volatile components and various extracts of the red alga *Jania rubens*. *Phytotherapy Research* 21(2): 153–156.

Kawaguchi, S., H.W. Wang, T. Horiguchi, G. Sartoni, and M. Masuda. 2001. A comparative study of the red alga *Grateloupia filicina* (Halymeniaceae) from the northwestern Pacific and Mediterranean with the description of *Grateloupia asiatica*, sp. nov. *Journal of Phycology* 37: 433–442.

Kawamoto, H., Y. Miki, T. Kimura, et al. 2006. Effects of fucoidan from Mozuku on human stomach cell lines. *Food Science and Technology Research* 12(3): 218–222.

Kazlowska, K., T. Hsu, et al. 2010. Anti-inflammatory properties of phenolic compounds and crude extract from *Porphyra dentata*. *Journal of Ethnopharmacology* 128(1): 123–130.

Keeling, P.J. 2010. The endosymbiotic origin, diversification and fate of plastids. *Philosophical Transactions of the Royal Society B* 365: 729–748.

Khan, M.N.A., J.Y. Cho, M.C. Lee, et al. 2007. Isolation of two anti-inflammatory and one pro-inflammatory polyunsaturated fatty acids from the brown seaweed *Undaria pinnatifida*. *Journal of Agricultural and Food Chemistry* 55(17): 6984–6988.

Khanzada, A.K., W. Shaikh, T.G. Kazi, et al. 2007. Antifungal activity, elemental analysis and determination of total protein of seaweed, *Solieria robusta* (Greville) Kylin from the coast of Karachi. *Pakistan Journal of Botany* 39: 931–937.

Kikuchi, N., S. Arai, G. Yoshida, J.A. Shin, J.E. Broom, W.A. Nelson, and M. Miyata. 2010. *Porphyra migitae* sp. nov. (Bangiales, Rhodophyta) from Japan. *Phycologia* 49: 345–354.

Kim, J.H., J.B. Hudson, A.M. Huang, et al. 1997. Biological activities of seaweed extracts from British Columbia, Canada, and Korea .1. Antiviral activity. *Canadian Journal of Botany-Revue Canadienne de Botanique* 75(10): 1656–1660.

Kitamura, K., M. Matsuo, and T. Yasui. 1992. Enzymic degradation of fucoidan by fucoidanase from the hepatopancreas of *Patinopecten yessoensis*. *Bioscience, Biotechnology, and Biochemistry* 56: 490–494.

Koenig, G.M., and A.D. Wright. 1997. Sesquiterpene content of the antibacterial dichloromethane extract of the marine red alga *Laurencia obtusa*. *Planta Medica* 63(2): 186–187.

Kolender, A.A., C.A. Pujol, E.B. Damonte, et al. 1997. The system of sulfated alpha-(1>3)-linked D-mannans from the red seaweed Nothogenia fastigiata: structures, antiherpetic and anticoagulant properties. *Carbohydrate Research* 304(1): 53–60.

Kolender, A.A., M.C. Matulewicz, and A.S. Cerezo. 1995. Structural analysis of antiviral sulfated alpha-D-(1>3)-linked mannans. *Carbohydrate Research* 273(2): 179–185.

Kraan, S., and K.A. Barrington. 2005. Commercial farming of *Asparagopsis armata* (Bonnemaisoniceae, Rhodophyta) in Ireland, maintenance of an introduced species? *Journal of Applied Phycology* 17: 103–110.

Kraft, L.G.K., G.T. Kraft, and R.F. Waller. 2010. Investigations into Southern Australian *Ulva* (Ulvophyceae, Chlorophyta) taxonomy and molecular phylogeny indicate both cosmopolitanism and endemic cryptic species. *Journal of Phycology* 46: 1257–1277.

Kraft, G.T. 2009. Algae of Australia. Marine benthic algae of Lord Howe Island and the southern Great Barrier Reef, 2. Brown algae. pp. [i-iv], v-vi, 1–364, 107 figs. Erratum of fig. 73 from vol. 1. Canberra & Melbourne: Australian Biological Resources Study and CSIRO Publishing.

Ktari, L., and M. Guyot. 1999. A cytotoxic oxysterol from the marine alga *Padina pavonica* (L.) Thivy. *Journal of Applied Phycology* 11(6): 511–513.

Kubanek, J., A.C. Prusak, T.W. Snell, et al. 2005. Antineoplastic diterpene-benzoate macrolides from the Fijian red alga *Callophycus serratus*. *Organic Letters* 7(23): 5261–5264.

Kubler, J.E., and S.R. Dudgeon. 1996. Temperature dependent change in the complexity of form of *Chondrus crispus* fronds. *Journal of Experimental Marine Biology and Ecology* 207: 15–24.

Kucera, H. 2010. Species identification and discovery in common marine macroalgae: *Fucus*, *Porphyra* and *Ulva* using a DNA barcoding approach. Ph.D. Dissertation, The University of New Brunswick, 1–265.

Kuda, T., T. Kunii, H. Goto, et al. 2007. Varieties of antioxidant and antibacterial properties of *Ecklonia stolonifera* and *Ecklonia kurome* products harvested and processed in the Noto Peninsula, Japan. *Food Chemistry* 103(3): 900–905.

Kwon, H.J., S.Y. Bae, K.H. Kim, et al. 2007. Induction of apoptosis in HeLa cells by ethanolic extract of *Corallina pilulifera. Food Chemistry* 104(1): 196–201.

Lakshmi, V., R. Kumar, P. Gupta, et al. 2004. The antifilarial activity of a marine red alga, *Botryocladia leptopoda*, against experimental infections with animal and human filariae. *Parasitology Research* 93(6): 468–474.

Lane, C.E., C. Mayes, L.D. Druhel, and G.W. Saunders. 2006. A multi-gene molecular investigation of the kelp (Laminariales, Phaeophyceae) supports substantial taxonomic re-organization. *Journal of Phycology* 42: 493–512.

Lee, J.B., S. Koizumi, K. Hayashi, et al. 2010. Structure of rhamnan sulfate from the green alga *Monostroma nitidum* and its anti-herpetic effect. *Carbohydrate Polymers* 81(3): 572–577.

Lee, J.H., S.E. Park, M.A. Hossain, et al. 2007a. 2,3,6-tribromo-4,5-dihydroxybenzyl methyl ether induces growth inhibition and apoptosis in MCF-7 human breast cancer cells. *Archives of Pharmacal Research* 30: 1132–1137.

Lee, H.J., W.J. Yoon, G.J. Kang, et al. 2007b. Anti-inflammatory effects of two oxo-octadecenoic acids isolated from *Gracilaria verrucosa* in LPS-stimulated raw 264.7 cells. *Inflammation Research* 56: S443–S443.

Lee, Y.S., K.H. Shin, B.K. Kim, et al. 2004. Anti-diabetic activities of fucosterol from *Pelvetia siliquosa. Archives of Pharmacal Research* 27(11): 1120–1122.

Lee, S., Y.S. Lee, S.H. Jung, et al. 2003. Anti-oxidant activities of fucosterol from the marine algae *Pelvetia siliquosa. Archives of Pharmacal Research* 26(9): 719–722.

Lee, D.S., T.J. Nam, and J.H. Pyeun. 1998. Effect of low molecular alginates on cholesterol levels and fatty acid composition of serum and liver lipids in cholesterol-fed rats. *Journal of the Korean fisheries society* 31(3): 399–408.

Le Gall, L., and G.W. Saunders. 2010. DNA barcoding is a powerful tool to uncover algal diversity: A case study of the Phyllophoraceae (Gigartinales, Rhodophyta) in the Canadian flora. *Journal of Phycology* 46: 374–389.

Leliaert, F., H. Verbruggen, B. Wysor, and O. De Clerck. 2009a. DNA taxonomy in morphologically plastic taxa: Algorithmic species delimitation in the *Boodlea* complex (Chlorophyta: Cladophorales). *Molecular Phylogenetics and Evolution* 53: 122–133.

Leliaert, F., X. Zhang, N. Ye, E.-J. Malta, A.H. Engelen, F. Mineur, H. Verbruggen, and O. De Clerck. 2009b. Identity of the Qingdao algal bloom. *Phycological Research* 57: 147–151.

Lim, C.S., D.Q. Jin, J.Y. Sung, et al. 2006. Antioxidant and anti-inflammatory activities of the methanolic extract of *Neorhodomela aculeata* in hippocampal and microglial cells. *Biological & Pharmaceutical Bulletin* 29(6): 1212–1216.

Lima, H.C., F.H.F. Costa, A.H. Sampaio, et al. 2008. Induction and inhibition of human lymphocyte transformation by the lectin from the red marine alga *Amansia multifida. Journal of Applied Phycology* 10(2): 153–162.

Lin, S.M., H.Y. Liang, and M.H. Hommersand. 2008. Two types of auxiliary cell ampullae in Grateloupia (Halymeniacease, Rhodophyta), including *G. taiwanensis* sp. nov. and *G. orientalis* sp. nov. from Taiwan based on rbcL gene sequence analysis and cystocarp development. *Journal of Phycology* 44: 196–214.

Lin, S.M. 2006. Observations on flattened species of *Gracilaria* (Gracilariaceae, Rhodophyta) from Taiwan. *Journal of Applied Phycology* 18: 671–678.

Lindstrom, S.C. 2008. Cryptic diversity, biogeography and genetic variation in Northeast Pacific species of Porphyra sensu lato (Bangiales, Rhodophyta). *Journal of Applied Phycology* 20: 951–962.

Lins, K., D.P. Bezerra, et al. 2009. Antitumor properties of a sulfated polysaccharide from the red seaweed *Champia feldmannii* (Diaz-Pifferer). *Journal of Applied Toxicology* 29(1): 20–26.

Liu, J.N., Y. Yoshida, M.Q. Wang, et al. 1997. B cell stimulating activity of seaweed extracts. *International Journal of Immunopharmacology* 19(3): 135–142.

Lliopoulou, D., V. Roussis, C. Pannecouque, and E. De Clercq. 2002. Halogenated sesquiterpenes from the red alga *Laurencia obtusa. Tetrahedron* 58: 6749–6755.

Lopez, A., M. Rico, et al. 2011. The effects of solvents on the phenolic contents and antioxidant activity of *Stypocaulon scoparium* algae extracts. *Food Chemistry* 125(3): 1104–1109.

Lopez-Figueroa, F., N. Korbee, O. De Clerck, I. Barbara, and E. Ar Gall. 2007. Characterization of *Grateloupia lanceola* (Halymeniales, Rhodophyta), an obscure foliose *Grateloupia* from the Iberian Peninsula, based on morphology, comparative sequence analysis and mycosporine-like amino acid composition. *European Journal of Phycology* 42: 231–242.

Loughnane, C.J., L.M. McIvor, F. Rindi, D.B. Stengel, and M.D. Guiry. 2008. Morphology, rbcL phylogeny and distribution of distromatic *Ulva* (Ulvophyceae, Chlorophyta) in Ireland and southern Britain. *Phycologia* 47: 416–429.

Loya, S., M. Bakhanashvili, Y. Kashman, et al. 1995. Peyssonol-A and Peyssonal-B, 2 novel inhibitors of the reverse transcriptases of human-immunodeficiency virus type-1 and type-2. *Archives of Biochemistry and Biophysics* 316(2): 789–796.

Lüning, K., and S. Pang. 2003. Mass cultivation of seaweeds: Current aspects and approaches. *Journal of Applied Phycology* 15: 115–119.

Maggs, C.A., and J. Kelly. 2007. *Codium Stackhouse*. In *Green seaweeds of Britain and Ireland*, ed. J. Brodie, C.A. Maggs, and D.M. John, 189–201. London: British Phycological Society.

Majczak, G.A.H.,R.R.T.B. Richartz, and M.E.R. Duarte et al. (2003). Antiherpetic activity of heterofucans isolated from *Sargassum stenophyllum* (Fucales, Phaeophyta). Proceedings of the 17th International Seaweed Symposium, 169–174, January 28–February 02, 2001 Cape Town.

Mandal, P., C.G. Mateu, K. Chattopadhyay, et al. 2007. Structural features and antiviral activity of sulphated fucans from the brown seaweed *Cystoseira indica. Antiviral Chemistry and Chemotherapy* 18(3): 153–162.

Manefield, M., M. Welch, M. Givskov, et al. 2001. Halogenated furanones from the red alga, *Delisea pulchra*, inhibit carbapenem antibiotic synthesis and exoenzyme virulence factor production in the phytopathogen *Erwinia carotovora. FEMS Mircobiology Letters* 205: 131–138.

Mangialajo, L., M. Chiantore, and R. Cattaneo-Vietti. 2008. Loss of fucoid algae along a gradient of urbanisation, and structure of benthic assemblages. *Marine Ecology Progress Series* 358: 63–74.

Martin-Lescanne, J., F. Rousseau, B. De Reviers, C. Payri, A. Couloux, C. Cruaud, and L. Le Gall. 2010. Phylogenetic analyses of the *Laurencia* complex (Rhodomelaceae, Ceramiales) support recognition of five genera: *Chondrophycus, Laurencia, Osmundea, Palisada* and *Yuzurua* stat. nov. *European Journal of Phycology* 45: 51–61.

Martinez-Lozano, S.J., M.J. Verde-Star, M.T. Arizpe-Leal, et al. 2000. Extraction and characterization of kainic acid from *Digenea simplex* (Wulfen) C. Agardh, (Rhodophyta). *Phyton-International Journal of Experimental Botany* 67: 61–64.

Maruyama, H., H. Tamauchi, M. Iizuka, et al. 2006. The role of NK cells in antitumor activity of dietary fucoidan from *Undaria pinnatifida* sporophylls (Mekabu). *Planta Medica* 72(15): 1415–1417.

Matsubara, K., Y. Matsuura, A. Bacic, et al. 2001. Anticoagulant properties of a sulfated galactan preparation from a marine green alga, *Codium cylindricum. International Journal of Biological Macromolecules* 28(5): 395–399.

Matsubara, K., Y. Matsuura, K. Hori, et al. 2000. An anticoagulant proteoglycan from the marine green alga, *Codium pugniformis. Journal of Applied Phycology* 12(1): 9–14.

Matsuhiro, B., E. Zuniga, M. Jashes, et al. 1996. Sulfated polysaccharides from *Durvillaea antarctica. Hydrobiologia* 321(1): 77–81.

Mattio, L., and C.E. Payri. 2011. 190 years of *Sargassum taxonomy*, facing the advent of DNA phylogenies. *The Botanical Review* 77: 31–70.

Mattio, L., and C.E. Payri. 2009. Taxonomic revision of *Sargassum* (Fucales, Phaeophyceae) from New Caledonia based on morphological and molecular analyses. *Journal of Phycology* 45: 1374–1388.

Mattio, L., C.E. Payri, and M. Verlaque. 2009. Taxonomic revision and geographic distribution of subgenus *Sargassum* (Fucales, phaeophyceae) in the western and central Pacific islands based on morphological and molecular analyses. *Journal of Phycology* 45: 1213–1227.

Mattio, L., C.E. Payri, and V. Stiger-Pouvreau. 2008. Taxonomic revision of *Sargassum* (Fucales, Phaeophyceae) from French Polynesia based on morphological and molecular analyses. *Journal of Phycology* 44: 1541–1555.
Mayer, A.M.S., and K.R. Gustafson. 2008. Marine pharmacology in 2005–2006: Antitumour and cytotoxic compounds. *European Journal of Cancer* 44: 2357–2387.
McPhail, K.L., D. France, S. Cornell-Kennon, et al. 2004. Peyssonenynes A and B, novel enediyne oxylipins with DNA methyl transferase inhibitory activity from the red marine alga *Peyssonnelia caulifera*. *Journal of Natural Products* 67(6): 1010–1013.
Meinesz, A., T. Belsher, T. Thibault, B. Antolic, B. Ben Mustapha, C. Chiavarini, F. Cinelli, J. Cottalorda, A. Djellouli, A. El Abed, C. Orestano, A. Grau, L. Ivesa, A. Jaklin, H. Langar, E. Masutti-Pascual, A. Peirano, L. Tunesi, J. De Vaugelas, N. Zavodnik, and A. Zuljevic. 2001. The introduced green alga *Caulerpa taxifolia* continues to spread in the Mediterranean. *Biological Invasions* 3: 201–210.
Mendonca, P., and J.C. Freitas. 2000. Topical antiedematous activity of the organic extract of *Liagora farinosa* algae (Rhodophyta, Nemaliales). *Ciência e Cultura (São Paulo)* 52(3): 175–178.
Merechal, J.P., G. Culioli, C. Hellio, et al. 2004. Seasonal variation in antifouling activity of crude extracts of the brown alga *Bifurcaria bifurcata* (Cystoseiraceae) against cyprids of Balarus amphytrite and the marine bacteria *Cobetia marina* and *Pseudoalteromonas haloplanktis*. *Journal of Experimental Marine Biology and Ecology* 313(1): 47–62.
Meusnier, I., M. Valero, C. Destombe, J.L. Olsen, and W.T. Stam. 2004. Invasive algal species splits: Successive deletions in the rDNA ITS1 suggests nascent speciation in *Caulerpa taxifolia*. *European Journal of Phycology* 39: 83–92.
Millar, A.J.K. 2003. The world's first recorded extinction of a seaweed. In *Proceedings of the XVIIth International Seaweed Symposium*, ed. A.R.O. Chapman, R.J. Anderson, V.J. Vreeland, and I.R. Davison, 313–318. Oxford: Oxford University Press.
Millar, A.J.K., and D.W. Freshwater. 2005. Morphology and molecular phylogeny of the marine algal order Gelidiales (Rhodophyta) from New South Wales, including Lord Howe and Norfolk Islands. *Australian Systematic Botany* 18: 215–263.
Mine, I., D. Menzel, and K. Okuda. 2008. Morphogenesis in giant-celled algae. *International Review of Cell and Molecular Biology* 266: 37–83.
Moghaddam, M.F., W.H. Gerwick, and D.L. Ballantine. 1991. Discovery of the mammalian insulin release modulator, hepoxilin-B3, from the tropical red algae *Platysiphonia miniata* and *Cottoniella filamentosa*. *Journal of Biological Chemistry* 265(11): 6126–6130.
Morales, J.L., Z.O. Cantillo-Ciau, I. Sanchez-Molina, et al. 2006. Screening of antibacterial and antifungal activities of six marine macroalgae from coasts of Yucatan peninsula. *Pharmaceutical Biology* 44(8): 632–635.
Moreau, D., H. Thomas-Guyon, C. Jacquot, et al. 2006. An extract from the brown alga *Bifurcaria bifurcata* induces irreversible arrest of cell proliferation in a non-small-cell bronchopulmonary carcinoma line. *Journal of Applied Phycology* 18(1): 87–93.
Moreau, J., D. Pesando, and B. Caram. 1984. Antifungal and antibacterial screening of Dictyotales from the French Mediterranean coast. *Hydrobiologia* 116–117(1): 521–524.
Mori, T., B.R. O'Keefe, R.C. Sowder, et al. 2005. Isolation and characterization of griffithsin, a novel HIV-inactivating protein, from the red alga *Griffithsia* sp. *Journal of Biological Chemistry* 280(10): 9345–9353.
Na, H.J., P.D. Moon, H.J. Lee, et al. 2005. Regulatory effect of atopic allergic reaction by *Carpopeltis affinis*. *Journal of Ethnopharmacology* 101(1–3): 43–48.
Nakamura, H., S. Yamaguchi, T. Hayashi, et al. 1997. Studies on the biological activities of marine algae (III) anti-tumor promoting activity and inhibitory effect on aldose reductase. *Natural Medicines* 51(2): 162–169.
Nakamura, T., K. Nakayama, K. Uchida, et al. 1996. Antioxidant activity of phlorotannins isolated from the brown alga *Eisenia bicyclis*. *Fisheries Science* 62(6): 923–926.
Nakamura, H., N. Ohnuki, K. Sadamasu, et al. 1994. Anti-human immunodeficiency virus (HIV) activities of aqueous extracts from marine algae. *Natural Medicines* 48(3): 173–179.

Nam, K.W. 2007. Validation of the generic name *Palisada* (Rhodomelaceae, Rhodophyta). *Algae* 22: 53–55.
Nam, K.W., C.A. Maggs, and D.J. Garbary. 1994. Resurrection of the genus *Osmundea* with an emendation of the generic delineation of Laurencia (Ceramiales, Rhodophyta). *Phycologia* 33: 384–395.
Navarro, D.A., A.M. Ricci, M.C. Rodríguez, and C.A. Stortz. 2011. Xylogalactans from *Lithothamnion heterocladum*, a crustose member of the Corallinales (Rhodophyta). *Carbohydrate Polymers* 84(3): 944–951.
Nazarova, I.V., N.M. Shevchenko, B.M. Kovalev, et al. 1998. Immunomodulatory properties of polysaccharides from red algae: influence on the complement system. *Biologiya Morya (Vladivostok)* 24(1): 49–52.
Nelson, W.A., T.J. Farr, and J.E.S. Broom. 2006. Phylogenetic relationships and generic concepts in the red order Bangiales: Challenges ahead. *Phycologia* 45: 249–259.
Neves, S.A., A.L.P. Freitas, B.W. Sousa, et al. 2007. Antinociceptive properties in mice of a lectin isolated from the marine alga *Amansia multifida* Lamouroux. *Brazilian Journal of Medical and Biological Research* 40(1): 127–134.
Ní Chualáin, F., C.A. Maggs, G.W. Saunders, and M.D. Guiry. 2004. The invasive genus *Asparagopsis* (Bonnemaisoniaceae, Rhodophyta): Molecular systematics, morphology, and ecophysiology of Falkenbergia isolates. *Journal of Phycology* 40: 1112–1126.
Nicoletti, E., F. Della Pieta', V. Calderone, et al. 1999. Antiviral properties of a crude extract from a green alga *Caulerpa taxifolia* (Vahl) C-Agardh. *Phytotherapy Research* 13(3): 245–247.
Nishino, T., and T. Nagumo. 1991. The sulfate content dependence of the anticoagulant activity of a fucan sulfate from the brown seaweed *Ecklonia kurome*. *Carbohydrate Research* 214(1): 193–197.
Norte, M., J.J. Fernandez, M.L. Souto, et al. 1996. Two new antitumoral polyether squalene derivatives. *Tetrahedron Letters* 37(15): 2671–2674.
Nylund, G.M., G. Cervin, F. Persson, M. Hermansson, P.D. Steinberg, and H. Pavia. 2008. Seaweed defence against bacteria: a poly-brominated 2-heptanone from the red alga *Bonnemaisonia hamifera* inhibits bacterial colonization. *Marine Ecology Progress Series* 369: 39–50.
Oh, K.B., J.H. Lee, S.C. Chung, et al. 2008. Antimicrobial activities of the bromophenols from the red alga *Odonthalia corymbifera* and some synthetic derivatives. *Bioorganic & Medicinal Chemistry Letters* 18(1): 104–108.
Okai, Y., and K. Higashi-Okai. 1997. Potent anti-inflammatory activity of pheophytin a derived from edible green alga, *Enteromorpha prolifera* (Sujiao-nori). *International Journal of Immunopharmacology* 19(6): 355–358.
O'Kelly, C.J., A. Kurihara, T.C. Shipley, and A.R. Sherwood. 2010. Molecular assessment of *Ulva* spp. (Ulvophyceae, Chlorophyta) in the Hawaiian Islands. *Journal of Phycology* 46: 728–735.
Olsen, J.L., M. Valero, I. Meusnier, S.A. Boele-Bos, and W.T. Stam. 1998. Mediterranean *Caulerpa taxifolia* and *C. mexicana* (Chlorophyta) are not conspecific. *Journal of Phycology* 34: 850–856.
Olsen, J.L., F.W. Zechman, G. Hoarau, J.A. Coyer, W.T. Stam, M. Valero, and P. Aberg. 2010. The phylogeographic architecture of the fucoid seaweed *Ascophyllum nodosum*: An intertidal 'marine tree' and survivor of more than one glacial-interglacial cycle. *Journal of Biogeography* 37: 842–856.
Orhan, I., B. Sener, T. Atici, et al. 2006. Turkish freshwater and marine macrophyte extracts show in vitro antiprotozoal activity and inhibit FabI, a key enzyme of *Plasmodium falciparum* fatty acid biosynthesis. *Phytomedicine* 13(6): 388–393.
Ozdemir, G., Z. Horzum, A. Sukatar, et al. 2006. Antimicrobial activities of volatile components and various extracts of *Dictyopteris membranacea* and *Cystoseira barbata* from the Coast of Izmir, Turkey. *Pharmaceutical Biology* 44(3): 183–188.
Pardee, K.I., P. Ellis, M. Bouthillier, et al. 2004. Plant virus inhibitors from marine algae. *Canadian Journal of Botany-Revue Canadienne de Botanique* 82(3): 304–309.
Park, H.J., M. Kurokawa, K. Shiraki, et al. 2005. Antiviral activity of the marine alga *Symphyocladia latiuscula* against Herpes simplex virus (HSV-1) in vitro and its therapeutic efficacy against HSV-1 infection in mice. *Biological & Pharmaceutical Bulletin* 28(12): 2258–2262.

Park, Y.-B., I.-S. Kim, S.-J. Yoo, et al. 1998. Elucidation of anti-tumor initiator and promoter derived from seaweed-3: anti-tumor promoters of *Ecklonia stolonifera* extracts. *Journal of the Korean Fisheries Society* 31(4): 587–593.

Partschefeld, J., and S. Alban. 2007. Sulfated polysaccharides from *Delesseria sanguinea* (Hudson) Lamouroux inhibit the release of inflammatory cytokines in vitro. Conference Information: 55th Annual Congress of the Society-for-Medicinal-Plant-Research, Date: Sep 02–06, 2007 Karl Franzens Univ Graz Austria. *Planta Medica* 73(9): 832–833.

Paul, N.A., R. de Nys, and P.D. Steinberg. 2006. Chemical defence against bacteria in the red alga *Asparagopsis armata*: Linking structure with function. *Marine Ecology Progress Series* 306: 87–101.

Pengzhan, Y., L. Ning, L. Xiguang, et al. 2003. Antihyperlipidemic effects of different molecular weight sulfated polysaccharides from *Ulva pertusa* (Chlorophyta). *Pharmacological Research* 48(6): 543–549.

Perkol-Finkel, S., and L. Airoldi. 2010. Loss and recovery potential of marine habitats: An experimental study of factors maintaining resilience in subtidal algal forests at the Adriatic Sea. *PloS One* 5: e10791.

Perry, N.B., J.W. Blunt, and M.H.G. Munro. 1991. A cytotoxic and antifungal 1,4-naphtoquinone and related compounds from a New Zealand brown alga, *Landsburgia quercifolia*. *Journal of Natural Products* 54(4): 978–985.

Piazzi, L., A. Meinesz, M. Verlaque, B. Akcali, B. Antolic, M. Argyrou, D. Balata, E. Ballesteros, S. Calvo, F. Cinelli, S. Cirik, A. Cossu, R. D'Archino, A.S. Djellouli, F. Javel, E. Lanfranco, C. Mifsud, D. Pala, P. Panayotidis, A. Peirano, G. Pergent, A. Petrocelli, S. Ruitton, A. Zuljevic, and G. Ceccherelli. 2005. Invasion of *Caulerpa racemosa* var. *cylindracea* (Caulerpales, Chlorophyta) in the Mediterranean Sea: An assessment of the spread. *Cryptogamie, Algologie* 26: 189–202.

Ponce, N.M.A., C.A. Pujol, E.B. Damonte, et al. 2003. Fucoidans from the brown seaweed *Adenocystis utricularis*: extraction methods, antiviral activity and structural studies. *Carbohydrate Research* 338(2): 153–165.

Potin, P., P. Patier, J.Y. Floch, et al. 1992. Chemical characterization of cell-wall polysaccharides from tank-cultivated and wild plants of *Delesseria sanguinea* (Hudson) Lamouroux (Ceramiales, Delesseriaceae) – culture patterns and potent anticoagulant activity. *Journal of Applied Phycology* 4(2): 119–128.

Premakumara, G.A.S., W.D. Ratnasooriya, and L.M.V. Tillekeratne. 1996. Isolation of a non-steroidal contragestative agent from Sri Lankan marine red alga, *Gelidiella acerosa*. *Contraception* 54(6): 379–383.

Puglisi, M.P., S. Engel, P.R. Jensen, et al. 2007. Antimicrobial activities of extracts from Indo-Pacific marine plants against marine pathogens and saprophytes. *Marine Biology* 150(4): 531–540.

Pujol, C.A., M.I. Errea, M.C. Matulewicz, et al. 1996. Antiherpetic activity of S1, an algal derived sulphated galactan. *Phytotherapy Research* 10(5): 410–413.

Pushpamali, W.A., C. Nikapitiya, M. De Zoysa, et al. 2008. Isolation and purification of an anticoagulant from fermented red seaweed *Lomentaria catenata*. *Carbohydrate Polymers* 73(2): 274–279.

Qi, H.M., D. Li, J.J. Zhang, et al. 2008. Study on extraction of agaropectin from *Gelidium amansii* and its anticoagulant activity. *Chinese Journal of Oceanology and Limnology* 26(2): 186–189.

Ren, D.L., J.Z. Wang, H. Noda, et al. 1995. The effects of an algal polysaccharide from *Gloiopeltis tenax* on transplantable tumors and immune activities in mice. *Planta Medica* 61(2): 120–125.

Reunov, A., V. Nagorskaya, L. Lapshina, et al. 2004. Effect of kappa/beta-carrageenan from red alga *Tichocarpus crinitus* (Tichocarpaceae) on infection of detached tobacco leaves with Tobacco mosaic virus. *Zeitschrift fur Pflanzenkrankheiten und Pflanzenschutz-Journal of Plant Diseases and Protection* 111(2): 165–172.

Rhimou, B., R. Hassane, et al. 2010. Antiviral activity of the extracts of Rhodophyceae from Morocco. *African Journal of Biotechnology* 9(46): 7968–7975.

Rindi, F. 2008. *A review of the recent literature on medical and pharmaceutical uses of seaweeds*. Unpublished technical report, Marine Institute of Ireland, pp. 98.

Robba, L., S.J. Russell, G.L. Barker, and J. Brodie. 2006. Assessing the use of the mitochondrial cox1 marker for use in DNA barcoding of red algae (Rhodophyta). *American Journal of Botany* 93: 1101–1108.

Robles Centeno, P.O., and D.L. Ballantine. 1999. Effects of culture conditions on production of antibiotically active metabolites by the marine alga *Spyridia filamentosa* (Ceramiaceae, Rhodophyta). I. Light. *Journal of Applied Phycology* 10(5): 453–460.

Rocha de Souza, M.C., C.T. Marques, C.M. Guerra Dore, et al. 2007. Antioxidant activities of sulfated polysaccharides from brown and red seaweeds. *Journal of Applied Phycology* 19(2): 153–160.

Rogers, D.J., K.M. Jurd, G. Blundsen, S. Paoletti, and F. Zanetti. 1990. Anticoagulant activity of a proteoglycan in extracts of *Codium fragile* spp. *atlanticum*. *Journal of Applied Phycology* 2: 357–361.

Rovirosa, J., and A. San-Martin. 1997. Antimicrobial activity of the brown alga *Stypopodium flabelliforme* constituents. *Fitoterapia* 68(5): 473–475.

Rozas, E., and J.C. Freitas. 2007. Anti-inflammatory activity of the apolar extract from the seaweed *Galaxaura marginata* (Rhodophyta, Nemaliales). *Journal of Venomous Animals and Toxins including Tropical Diseases* 13(2): 544–548.

Sabina, H., S. Tasneem, Sambreen, et al. 2005. Antileishmanial activity in the crude extract of various seaweed from the coast of Karachi, Pakistan. *Pakistan Journal of Botany* 37(1): 163–168.

Salvador, N., A. Gomez-Garreta, and M.A. Ribera Siguan. 2008. Characterization of two frequently confused species, *Bonnemaisonia asparagoides* and *Bonnemaisonia clavata* (Bonnemaisoniales, Rhodophyta), on the basis of morphological and molecular evidence. *Phycologia* 47: 177–190.

Sampaio Assreuy, A.M., D.M. Gomes, M.S. Josino da Silva, et al. 2008. Biological effects of a sulfated-polysaccharide isolated from the marine red algae *Champia feldmannii*. *Biological & Pharmaceutical Bulletin* 31(4): 691–695.

Sasidharan, S., I. Darah, and K. Jain. 2008. In Vivo. and In Vitro. Toxicity Study of *Gracilaria changii*. *Pharmaceutical Biology* 46(6): 413–417.

Saunders, G.W. 2005. Applying DNA barcoding to red macroalgae: a preliminary appraisal holds promise for future applications. *Philosophical Transactions of the Royal Society B* 360: 1879–1888.

Saunders, G.W., and M.H. Hommersand. 2004. Assessing red algal supraordinal diversity and taxonomy in the context of contemporary systematic data. *American Journal of Botany* 91: 1494–1507.

Saunders, G.W., and K.V. Lehmkuhl. 2005. Molecular divergence and morphological diversity among four cryptic species of *Plocamium* (Plocamiales, Florideophyceae) in northern Europe. *European Journal of Phycology* 40: 293–312.

Sekine, H., N. Ohonuki, and K. Sadamasu. 1995. The inhibitory effect of the crude extract from a seaweed of Digenea-simplex Agardh, C on the in-vitro cytopathic activity of HIV-1 and its antigen production. *Chemical & Pharmaceutical Bulletin* 43(9): 1580–1584.

Serkedjieva, J. 2000. Antiherpes virus effect of the red marine alga *Polysiphonia denudata*. *Zeitschrift für Naturforschung C - A Journal of Biosciences* 55(9–10): 830–835.

Shanmugam, M., B.K. Ramavat, K.H. Mody, R.M. Oza, and A. Tewari. 2001. Distribution of heparinoid-active sulphated polysaccharides in some Indian marine green algae. *Indian Journal of Marine Sciences* 30(4): 222–227.

Shanmughapriya, S., A. Manilal, A. Sujith, et al. 2008. Antimicrobial activity of seaweed extracts against multiresistant pathogens. *Annals of Microbiology* 58(3): 535–541.

Sherwood, A.R. 2008. Phylogeography of *Asparagopsis taxiformis* (Bonnemaisoniales, Rhodophyta) in the Hawaiian Islands: two MtDNA markers support three separate introductions. *Phycologia* 47: 79–88.

Sheu, J.H., G.H. Wang, P.J. Sung, et al. 1997. Cytotoxic sterols from the Formosan brown alga *Turbinaria ornata*. *Planta Medica* 63(6): 571–572.

Shi, D.Y., L.J. Han, J. Sun, et al. 2005. Studies on chemical constitutes of green alga *Chaetomorpha basiretorsa* and their bioactivity. *China Journal of Chinese Material Medica*. doi:CNKI:SUN:ZGZY.0.2005-15-006.

Shibata, T., K. Ishimaru, S. Kawaguchi, et al. 2007. Antioxidant activities of phlorotannins isolated from Japanese Laminariaceae. *Journal of Applied Phycology* 20(5): 705–711.

Shibata, H., I. Kimura-Takagi, M. Nagaoka, et al. 2000. Properties of fucoidan from *Cladosiphon okamuranus* Tokida in gastric mucosal protection. *Biofactors* 11(4): 235–245.

Shoeib, N.A., M.C. Bibby, G. Blunden, et al. 2003. In-vitro cytotoxic activities of the major bromophenols present in the British alga, *Polysiphonia lanosa*. *Journal of Pharmacy and Pharmacology* 55(Supplement): S.1.

Silberfeld, T., J.W. Leigh, H. Verbruggen, C. Cruaud, B. de Reviers, and F. Rousseau. 2010. A multi-locus time-calibrated phylogeny of the brown algae (Heterokonta, Ochrophyta, Phaeophyceae): investigating the evolutionary nature of the "brown algal crown radiation". *Molecular Phylogenetics and Evolution* 56: 659–674.

Silva, T.M.A., L.G. Alves, K.C.S. Queiroz, et al. 2005. Partial characterization and anticoagulant activity of a heterofucan from the brown seaweed *Padina gymnospora*. *Brazilian Journal of Medical and Biological Research* 38(4): 523–533.

Siqueira, R.C.L., M.S.J. da Silva, et al. 2011. *In vivo* anti-inflammatory effect of a sulfated polysaccharide isolated from the marine brown algae *Lobophora variegata*. *Pharmaceutical Biology* 49(2): 167–174.

Spavieri, J., A. Allmendinger, et al. 2010. Antimycobacterial, antiprotozoal and cytotoxic potential of twenty-one brown algae (Phaeophyceae) from British and Irish waters. *Phytotherapy Research* 24(11): 1724–1729.

Smit, A.J. 2004. Medicinal and pharmaceutical uses of seaweed natural products: A review. *Journal of Applied Phycology* 16: 245–262.

Stam, W.T., J.L. Olsen, S.F. Zaleski, S.N. Murray, K.R. Brown, and L.J. Walters. 2006. A forensic and phylogenetic survey of *Caulerpa* species (Caulerpales, Chlorophyta) from the Florida coast, local aquarium shops and e-commerce: Establishing a proactive baseline for early detection. *Journal of Phycology* 42: 1113–1124.

Steentoft, M., Irvine, L.M., Farnham, W.F. (1995). Two terete species of Gracilaria and Gracilariopsis (Gracilariales, Rhodophyta) in Britain. Phycologia 34: 113–127.

Stout, E.P., A.P. Hasemeyer, et al. 2009. Antibacterial Neurymenolides from the Fijian Red Alga *Neurymenia fraxinifolia*. *Organic Letters* 11(1): 225–228.

Sugiura, Y., K. Matsuda, Y. Yamada, et al. 2006. Isolation of a New Anti-allergic Phlorotannin, Phlorofucofuroeckol-B, from *Eisenia* Brown Alga, *Eisenia arborea*. *Bioscience, Biotechnology, and Biochemistry* 70(11): 2807–2811.

Sukatar, A., N.U. Karabay-Yavasoglu, G. Ozdemir, et al. 2006. Antimicrobial activity of volatile component and various extracts of *Enteromorpha linza* (Linnaeus) J. Agardh from the coast of Izmir, Turkey. *Annals of Microbiology* 56(3): 275–279.

Sunilson, J.A.J., R. Suraj, and K. Anandarajagopal. 2009. Preliminary phytochemical analysis, elemental determination and antibacterial screening of *Codium decorticatum* – a marine green algae. *International Journal of Biological Chemistry* 3: 84–89.

Takahashi, S., T. Matsunaga, C. Hasegawa, et al. 1998. Martefragin A, a novel indole alkaloid isolated from red alga, inhibits lipid peroxidation. *Chemical & Pharmaceutical Bulletin* 46(10): 1527–1529.

Talarico, L.B., C.A. Pujol, R.G.M. Zibetti, et al. 2005. The antiviral activity of sulfated polysaccharides against dengue virus is dependent on virus serotype and host cell. *Antiviral Research* 66(2–3): 103–110.

Tan, I.H., J. Blomster, G. Hansen, E. Leskinen, C.A. Maggs, D.G. Mann, H.J. Sluiman, and M.J. Stanhope. 1999. Molecular phylogenetic evidence for a reversible morphogenetic switch controlling the gross morphology of two common genera of green seaweeds, *Ulva* and *Enteromorpha*. *Molecular Biology and Evolution* 16: 1011–1018.

Tanaka, K., A. Yamada, K. Noda, et al. 1998. A novel glycoprotein obtained from *Chlorella vulgaris* strain CK22 shows antimetastatic immunopotentiation. *Cancer Immunology, Immunotherapy* 45(6): 313–320.

Tariq, V.N. 1991. Antifungal activity in crude extracts of marine red algae. *Mycological Research* 95: 1433–1440.

Teixeira, V.L., F.D. Rocha, P.J. Houghton, et al. 2007. Alpha-amylase inhibitors from Brazilian seaweeds and their hypoglycemic potential. *Fitoterapia* 78(1): 35–36.

Tian, X.H., J.F. Gu, C.P. Sun, et al. 1997. An EPR study of the neuroprotective effect of the sulfated polysaccharide from *Laminaria japonica*. *Society for Neuroscience Abstracts* 23(1–2): 2187.

Tierney, M.S., A.K. Croft, and M. Hayes. 2010. A review of antihypertensive and antioxidant activities in macroalgae. *Botanica Marina* 53: 387–408.

Todd, J.S., P.J. Proteau, and W.H. Gerwick. 1994. The absolute configuration of *Ecklonia* lactones A, B and E, novel oxylipins, from brown algae of the genera *Ecklonia* and *Egregia*. *Journal of Natural Products* 57(1): 171–174.

Tsuda, M., Y. Okamoto, and S. Shimada. 2007. Test for antiviral activity against Epstein-Barr virus and Kaposi sarcoma related herpes virus by seaweeds distributed in Hokkaido, Japan. *Bulletin of Marine Sciences and Fisheries Kochi University* 25: 1–4.

Tronholm, A., M. Sansòn, J. Afonso-Carrillo, H. Verbruggen, and O. De Clerck. 2010a. Niche partitioning and the coexistence of two cryptic *Dictyota* (Dictyotales, Phaeophyceae) species from the Canary Islands. *Journal of Phycology* 46: 1075–1087.

Tronholm, A., F. Steen, L. Tyberghein, F. Leliaert, H. Verbruggen, M.A. Ribera Siguan, and O. De Clerck. 2010b. Species delimitation, taxonomy and biogeography of *Dictyota* in Europe. *Journal of Phycology* 46: 1301–1321.

Trono, G.C., Jr 1999. Diversity of the seaweed flora of the Philippines and its utilization. *Proceedings of the International Seaweed Symposium* 16: 1–6.

Trowbridge, C.D. 1998. Ecology of the green macroalga *Codium fragile* (Suringar) Hariot: Invasive and noninvasive subspecies. *Oceanography and Marine Biology Annual Reviews* 36: 1–64.

Tuney, I., B.H. Cadirci, D. Unal, et al. 2006. Antimicrobial activities of the extracts of marine algae from the coast of Urla (Izmir, Turkey). *Turkish Journal of Biology* 30(3): 171–175.

Tziveleka, L.A., C. Vaglas, and V. Roussis. 2003. Natural products with anti-HIV activity from marine organisms. *Current Topics in Medicinal Chemistry* 3(13): 1512–1535.

Valls, R., and L. Piovetti. 1995. The chemistry of the Cystoseiraceae (Fucales, Phaeophyceae): Chemotaxonomic relationships. *Biochemical Systematics and Ecology* 23: 723–745.

Valls, R., B. Banaigs, L. Piovetti, et al. 1993. Linear diterpene with antimitotic activity from the brown alga *Bifurcaria bifurcata*. *Phytochemistry* 34(6): 1585–1588.

Verbruggen, H., F. Leliaert, C.A. Maggs, S. Shimada, T. Schils, J. Provan, D. Booth, S. Murphy, O. De Clerck, D.S. Littler, M.M. Littler, and E. Coppejans. 2007. Species boundaries and phylogenetic relationships within the green algal genus *Codium* (Bryopsidales) based on plastid DNA sequences. *Molecular Phylogenetics and Evolution* 44: 240–254.

Verbruggen, H., C.A. Maggs, G.W. Saunders, L. Le Gall, H.S. Yoon, and O. De Clerck et al. 2006. Data mining approach identifies research priorities and data requirements for resolving the red algal tree of life. *BMC Evolutionary Biology* 10: 16.

Verlaque, M., C.F. Boudouresque, A. Meinesz, and V. Gravez. 2000. The *Caulerpa racemosa* complex (Caulerpales, Ulvophyceae) in the Mediterranean Sea. *Botanica Marina* 43: 49–68.

Verlaque, M., C. Durand, J.M. Huisman, C.F. Boudouresque, and Y. Parco. 2003. On the identity and origin of the Mediterranean invasive *Caulerpa racemosa* (Caulerpales, Chlorophyta). *European Journal of Phycology* 38: 325–339.

Viana, G.S., A.L.P. Freitas, M.M.L. Lima, et al. 2002. Antinociceptive activity of sulfated carbohydrates from the red algae *Bryothamnion seaforthii* (Turner) Kutz. and *B. triquetrum* (S.G. Gmel.) M. Howe. *Brazilian Journal of Medical and Biological Research* 35(6): 713–722.

Villarreal-Gomez, L.J., I.E. Soria-Mercado, et al. 2010. Antibacterial and anticancer activity of seaweeds and bacteria associated with their surface. *Revista de Biología Marina y Oceanografía* 45(2): 267–275.

Vlachos, V., A.T. Critchley, and A. vonHoly. 1997. Antimicrobial activity of extracts from selected southern African marine macroalgae. *South African Journal of Science* 93(7): 328–332.

Wall, M.E., M.C. Wani, G. Manikumar, et al. 1989. Plant antimutagenic agents. 7. Structure and antimutagenic properties of cymobarbatol and 4-isocymobarbatol, new cymopols from a green alga (*Cymopolia barbata*). *Journal of Natural Products* 52(5): 1092–1099.

Wang, J., L. Liu, Q. Zhang, et al. 2009. Synthesized oversulphated, acetylated and benzoylated derivatives of fucoidan extracted from *Laminaria japonica* and their potential antioxidant activity in vitro. *Food Chemistry* 114(4): 1285–1290.
Wang, H., L.C.M. Chiu, V.E.C. Ooi, et al. 2008. Seaweed polysaccharides with anticancer potential. *Botanica Marina* 51(4): 313–319.
Wang, H., V.E.C. Ooi, and P.O. Ang. 2007. Antiviral polysaccharides isolated from Hong Kong brown seaweed *Hydroclathrus clathratus*. *Science in China. Series C, Life Sciences* 50: 611–618.
Wilkes, R.J., L.M. McIvor, and M.D. Guiry. 2005. Using rbcL sequence data to reassess the taxonomic position of some *Grateloupia* and *Dermocorynus* species (Halymeniaceae, Rhodophyta) from the north-eastern Atlantic. *European Journal of Phycology* 40: 53–60.
Witvrouw, M., J.A. Este, M.Q. Mateu, et al. 1994. Activity of a sulfated polysaccharides extracted from the red seaweed *Agardhiella tenera* against human immunodeficiency-virus and other enveloped viruses. *Antiviral Chemistry & Chemotherapy* 5(5): 297–303.
Wong, S.L., C.H. Liu, W. Cheng, et al. 2008. Improved survival under Vibrio challenge and enhanced immunogene transcription of the Pacific white shrimp, *Litopenaeus vannamei*, which were oral administrated with brown algae extract, *Endarachne binghamiae*, via enriched *Artemia*. *Journal of the Fisheries Society of Taiwan* 35(1): 17–34.
Wong, C.K., V.E.C. Ooi, and P.O. Ang. 2004. Hepatoprotective effect of seaweeds' methanol extract against carbon tetrachloride-induced poisoning in rats. *Hydrobiologia* 512(1–3): 267–270.
Wright, J.T. 2005. Differences between native and invasive *Caulerpa taxifolia*: A link between asexual fragmentation and abundance in invasive populations. *Marine Biology* 147: 559–569.
Xu, N.J., X. Fan, X.J. Yan, et al. 2004. Screening marine algae from China for their antitumor activities. 5th Asia-Pacific Conference on Algal Biotechnology, Date: Oct 18–21, 2003 Qingdao. *Journal of Applied Phycology* 16(6): 451–456.
Xu, N.J., X. Fan, X.J. Yan, et al. 2003. Antibacterial bromophenols from the marine red alga *Rhodomela confervoides*. *Phytochemistry* 62(8): 1221–1224.
Yano, T., M. Kamiya, S. Arai, and H. Kawai. 2004. Morphological homoplasy in Japanese *Plocamium* species (Plocamiales, Rhodophyta) inferred from the Rubisco spacer sequence and intracellular acidity. *Phycologia* 43: 383–393.
Yano, T., M. Kamiya, A. Murakami, H. Sasaki, and H. Kawai. 2006. Biochemical phenotypes corresponding to molecular phylogeny of the red algae *Plocamium* (Plocamiales, Rhodophyta): Implications of incongruence with the conventional taxonomy. *Journal of Phycology* 42: 155–169.
Yoon, H.S., K.M. Müller, R.G. Sheath, F.D. Ott, and D. Bhattacharya. 2006. Defining the major lineages of red algae. *Journal of Phycology* 42: 482–492.
Yoshizawa, Y., A. Ametani, J. Tsunehiro, et al. 1995. Macrophage stimulation activity of the polysaccharide fraction from a marine alga (*Porphyra yezoensis*) – structure, function, relationships and improved solubility. *Bioscience, Biotechnology, and Biochemistry* 59(10): 1933–1937.
Yoshizawa, Y., A. Enomoto, H. Todho, A. Ametani, and S. Kaminogawa. 1993. Activation of murine macrophages by polysaccharide fractions from marine alga (*Porphyra yezoensis*). *Bioscience, Biotechnology, and Biochemistry* 57: 1862–1866.
Youngwan, S., S.H. Kang, H.J. Lee, et al. 2006. In vitro screaning of seaweed extract on the proliferation of mouse spleen and thymus cell. *Biotechnology and Bioprocess Engineering* 11(2): 160–163.
Yuan, Y.V., and N.A. Walsh. 2006. Antioxidant and antiproliferative activities of extracts from a variety of edible seaweeds. *Food and Chemical Toxicology* 44(7): 1144–1150.
Yuan, H.M., and J.M. Song. 2005. Preparation, structural characterization and in vitro antitumor activity of kappa-carrageenan oligosaccharide fraction from *Kappaphycus striatum*. *Journal of Applied Phycology* 17(1): 7–13.
Yuan, Y.V., D.E. Bone, and Y.V. Yuan. 2005. Antioxidant activity of dulse (*Palmaria palmata*) extract evaluated in vitro. *Food Chemistry* 91(3): 485–494.

Zamora Tovar, C., and D.L. Ballantine. 2000. Multiple antimicrobial activities of the marine alga *Spyridia filamentosa* (Ceramiaceae, Rhodophyta). *Botanica Marina* 43(3): 233–238.

Zardi, G.I., K.R. Nicastro, F. Canovas, J.F. Costa, E.A. Serrão, and G.A. Pearson. 2011. Adaptive traits are maintained on steep selective gradients despite gene flow and hybridization in the intertidal zone. *PlosOne* 6(6): e19402.

Zhao, T., Q. Zhang, H. Qi, et al. 2006. Degradation of porphyran from *Porphyra haitanensis* and the antioxidant activities of the degraded porphyrans with different molecular weight. *International Journal of Biological Macromolecules* 38(1): 45–50.

Zhang, H.J., W.J. Mao, F. Fang, et al. 2008. Chemical characteristics and anticoagulant activities of a sulfated polysaccharide and its fragments from *Monostroma latissimum*. *Carbohydrate Polymers* 71(3): 428–434.

Zhou, G.F., W.X. Sheng, W.H. Yao, et al. 2006. Effect of low molecular lambda-carrageenan from *Chondrus ocellatus* on antitumor H-22 activity of 5-Fu. *Pharmacological Research* 53(2): 129–134.

Ziółkowska, N.E., and A. Wlodawer. 2006. Structural studies of algal lectins with anti-HIV activity. *Acta Biochimica Polonica* 53: 617–626.

Zubia, M., M.S. Fabre, et al. 2009. Antioxidant and cytotoxic activities of some red algae (Rhodophyta) from Brittany coasts (France). *Botanica Marina* 52(3): 268–277.

Chapter 2
Extraction and Characterization of Bioactive Compounds with Health Benefits from Marine Resources: Macro and Micro Algae, Cyanobacteria, and Invertebrates

Elena Ibañez, Miguel Herrero, Jose A. Mendiola, and María Castro-Puyana

2.1 Bioactive Compounds from Marine Sources and Functional Foods

The occurrence and incidence of different diseases such as cancer, cardiovascular diseases, obesity, and diabetes may be related to the consumption of high caloric diets combined with sedentary lifestyles. The concept of functional foods first appeared in Japan where it was considered to be a tool to promote health and well-being. In 1992, the Japanese government established a policy of "Foods of Specific Health Uses" (FOSHU). This concept was further developed in Europe within the "Functional Food Science in Europe" (FUFOSE) project supported by the European Commission (EC) and co-ordinated by the International Life Sciences Institute (ILSI). Several interesting points were observed at the end of this project (Bellisle et al. 1998; Diplock et al. 1999), including a definition of a functional food as "a food which is demonstrated to positively affect one or more physiological functions, so that it is able to increase the well-being and/or to reduce the risk to suffer a disease" (Diplock et al. 1999). This definition implies that a functional food must maintain the shape of the food (thereby excluding pills and capsules) and that the functional food must impart a physiological effect following consumption that is above and beyond any observed nutritional effects.

One of the ways most often employed by food manufacturers in the production of novel functional foods is the addition of one or more interesting bioactive

E. Ibañez (✉) • M. Herrero • J.A. Mendiola • M. Castro-Puyana
Bioactivity and Food Analysis Department, Institute of Food Science Research (CIAL-CSIC), Nicolás Cabrera 9 Campus UAM Cantoblanco, Madrid 28049, Spain
e-mail: elena@ifi.csic.es

M. Hayes (ed.), *Marine Bioactive Compounds: Sources, Characterization and Applications*, DOI 10.1007/978-1-4614-1247-2_2,

compounds to a traditional food. The added bioactive compounds are usually referred to as functional ingredients and they are responsible for the functional bioactivities that the new product might present. By using this strategy, several functional foods have already been developed and commercialized. For instance, products claiming antihypertensive activity, hypo-cholesterolemic effect, antioxidant properties, probiotic or prebiotic effects, or regulatory effects on the appetite, among others, are available on the market.

The marine environment, which contains a vast array of organisms with unique biological properties, is one of the most underutilized biological resources. To date, algae and microalgae are referenced in the literature as sources of bioactive compounds for use as functional food ingredients (Plaza et al. 2009; Plaza et al. 2008). The huge diversity in terms of the number of different species of macro- and microalgae that exist (as discussed in Chapter 1), coupled with the hostile environments in which these organisms live, make macro- and microalgae key targets for bioactive compound screening projects.

Algae comprise a complex and heterogeneous group of organisms characterized by their photosynthetic nature and their simple reproductive structures. According to their size, algae can be roughly divided into unicellular organisms, known as *microalgae* and multicellular organisms referred to as *macroalgae*. Algae frequently live in extreme environments of light, salinity, and temperature. In order to adapt to these extreme conditions, most algae produce a high variety of secondary metabolites that often have potent biological activities. Most algae are relatively easy to cultivate or produce at industrial scale. Thus, the production of algal-derived biologically active compounds may be tuned by the selection of appropriate cultivation conditions, making these algae true natural bioreactors. Bioactive compounds have also been isolated previously from other marine organisms including crustaceans, fish, and their by-products and this is reviewed in later chapters of this book. In some of these matrices, interesting functional compounds were isolated previously (Kadam and Prabhasankar 2010; Kim and Wijesekara 2010).

It is important to consider how functional ingredients are obtained from new matrices, such as micro- and macroalgae. In this regard, there is a need to combine appropriate, selective, cost-effective, and environmentally friendly extraction procedures with the legal requirements regarding the use of food-grade solvents and processes. Traditional extraction techniques such as soxhlet, solid–liquid extraction (SLE), or liquid–liquid extraction (LLE) are characterized by using high volumes of solvents and long extraction times. These techniques often produce low extraction yields of bioactives and present low selectivity. Furthermore, traditional extraction techniques are usually not automated procedures and their reproducibility can therefore be compromised. As described later in this chapter, the use of new state-of-the-art extraction techniques, such as supercritical fluid extraction (SFE), pressurized liquid extraction (PLE), accelerated solvent extraction (ASE®), pressurized hot water extraction (PHWE), ultrasound-assisted extraction (UAE), and microwave-assisted extraction (MAE) techniques, among others, may provide an effective alternative to the problems encountered with the use of traditional extraction procedures.

2.2 Screening of Bioactives from Marine Sources: An Integrated Approach

Considering the great biodiversity of marine species, the use of appropriate methodologies that can rapidly screen different marine sources for bioactive compounds is of great interest. To design this screening methodology, different parameters have to be considered. These parameters include the possible nature of the sought-after bioactive compounds (in terms of solubility, heat resistance, or molecular weight) and the bioactivity that is sought. Figure 2.1 proposes a screening methodology for the extraction and identification of bioactive compounds from marine sources. Initially, a suitable extraction technique should be selected. This selection must be carried out in accordance with the predicted nature of the expected/target bioactive compound(s). However, several extraction techniques could also be used to fully characterize the potential of the different natural sources, introducing different extraction selectivity.

The use of environmentally clean advanced extraction techniques allows for the attainment of the target compound(s) of interest with more efficient extraction procedures, while, at the same time, minimizing the use of organic toxic solvents. Depending on the extraction techniques selected, diverse extraction parameters should be tested in order to study the influence of solvents, temperatures, pressures, and other important parameters that might have a significant influence on the outcome of the extraction process employed. The different extracts, obtained using diverse conditions, must then be tested for biological bioactivities by performing the appropriate functional activity assay(s). The main aim of this step is to confirm that the obtained extracts from step one possess the sought-after bioactivity.

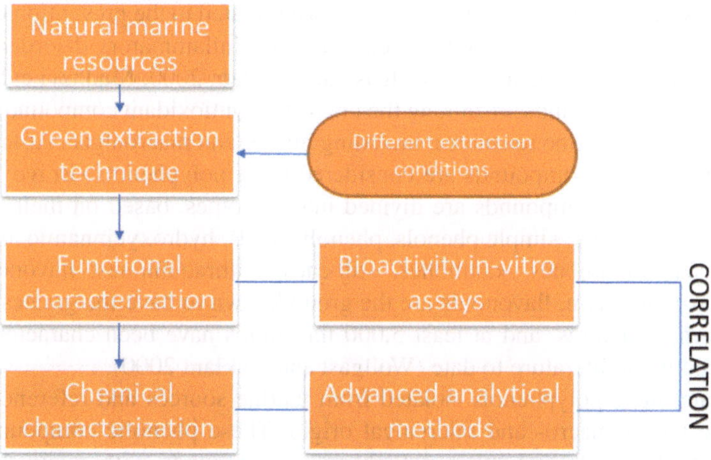

Fig. 2.1 Basic scheme showing the proposed workflow for the screening of bioactive compounds from marine sources

Functional characterization should be assessed through the application of fast in vitro assays directed to the confirmation of the sought biological properties, for instance, antioxidant capacity assays, antimicrobial activity assays, or antihypertensive activity assays.

Once the target biological activities have been confirmed, the next step involves chemical characterization of the bioactive components present in the initial extract, which may often be referred to as lead functional components (LFCs). Again, the analytical technique employed at this stage of the characterization process will depend on the nature of the initial extract or LFC in terms of its solubility, stability at different pH conditions, and heat stability, as well as the nature of the suspected bioactive compounds. In general, advanced analytical techniques are employed, even coupled, in order to maximize the identification potential. The final aim of this stage of the characterization process is the correlation between the chemical composition of the LFC and the bioactivities observed. Ideally, it will be possible to identify which compounds are responsible for the observed bioactivities. It is often necessary to return to the initial extraction method employed and to perform a fine-tuning of this extraction procedure in order to enrich the initial extract(s) with the target compounds, which has demonstrated biological activities and has a defined chemical structure. Further use of the information gathered using this integrated approach can be directed towards the design of upscale procedures for this extraction and characterization process and this is described later in this chapter (Sect. 2.4).

In the following section, a brief description of the main potential functional food ingredients that may be isolated from different marine sources is presented.

2.2.1 Antioxidants

The antioxidant capacity of compounds has been related to the prevention of several diseases including cancer, coronary heart diseases, inflammatory disorders, neurological degeneration, and aging (Wollgast and Anklam 2000; Madhavi et al. 1996). Polyphenolic compounds are among the interesting antioxidant compounds isolated previously from marine resources, including micro- and macroalgae. At least 8,000 different bioactive compounds are considered to be polyphenols (Bravo 1998). In general, phenolic compounds are divided into ten types, based on their structure. These ten groups are: simple phenols, phenolic acids, hydroxycinnamic acids, coumarins, naphtoquinones, xanthones, stilbenes, anthraquinones, flavonoids, and lignins. Among them, flavonoids are the group known to have the greatest number of different structures, and at least 5,000 flavonoids have been characterized and referenced in the literature to date (Wollgast and Anklam 2000).

To date, most polyphenols isolated from marine sources and referenced in the literature are of macro- and microalgal origin. These phenolic compounds range from phenolic acids and other polyphenolic compounds with relatively simple chemical structures, to the more complex structures of phlorotannins, typically isolated from brown algae belonging to the Phaeophyceae class, and which consist of polymeric structures made up of units of phloroglucinol (1,3,5-trihydroxybenzene). It is understood that the intensity of the antioxidant activity of these complex polyphenols is related to the degree of polymerization of the polyphenol. In general, lower degrees of polymerization result in greater antioxidant activities. Nevertheless, the main activity related to phenolic compounds is antioxidant activity (Li et al. 2009). In addition to their strong antioxidant activities, phlorotannins are known to possess other activities (Wijesekara et al. 2010) including antibacterial (Nagayama et al. 2002), chemopreventive (Kang et al. 2003), UV- protective (Artan et al. 2008), and antiproliferative effects (Kong et al. 2009). They are also known to act as detoxifying agents against heavy metals (Eide et al. 1980), and have myriad other bioactivities that could potentially be exploited for use in functional foods (Lee et al. 2009; Jung et al. 2006; Yoon et al. 2009) For example, the total antioxidant capacity of a *Polysiphoma urceolata* red algal extract is directly related to the total phenolic content (Duan et al. 2006). However, as occurs with other chemical constituents, the composition of phenolic compounds both qualitatively and quantitatively might vary depending on the species as well as on the environmental conditions and location of the seaweed.

In addition to polyphenolic compounds, other interesting antioxidants, such as carotenoids, can be found in marine resources. Carotenoids are a family of natural pigments that are widely distributed in nature. There are more than 600 known carotenoid structures described in the literature, and these natural pigments are often responsible for the color of the different natural matrices. The basic carotenoid structure is formed by eight isoprenoid units that constitute the symmetrical skeleton of the compound, along with a long chain with conjugated double bonds. This characteristic arrangement is responsible for the antioxidant activities of carotenoids. Carotenoids are synthesized by plants, algae, fungi, and other microorganisms but are not made by animals and need to be ingested in the diet if required. Carotenoids possess other bioactivities and are thought to be active agents for the prevention of cancer (De Spirt et al. 2010; Silberstein and Parsons 2010), cardiovascular diseases (Riccioni et al. 2008), and macular degeneration (Snodderly 1995).

Certain algal species in addition to crustaceans are known reservoirs for carotenoids. For instance, the microalga *Dunaliella salina* is able to accumulate large amounts of β-carotene when cultivated under certain conditions (Zhu and Jiang 2008). Other carotenoids are more typical from these organisms such as fucoxanthin or astaxanthin. Fucoxanthin is the main pigment found in brown algae. It has been observed that this oxygenated carotenoid is a very effective inhibitor of cellular growth and promotes apoptosis in human cancer cell lines (Kotate-Nara et al. 2001; Hosokawa et al. 1999). Moreover, this pigment possesses anti-inflammatory (Shiratori et al. 2005), antidiabetic (Maeda et al. 2007), and antioxidant activities (Sachindra et al. 2007). Astaxanthin has been found in different marine organisms but *Haematococcus pluvialis* is the major producer, being able to selectively accumulate this carotenoid in quantities that amount to approximately 5% of its dry weight (Yuan and Chen 2000). As a result of its color, this carotenoid is usually employed as an additive in aquaculture feed for salmon, trout, and shrimp. As occurs with other carotenoids, the antioxidant activity associated with astaxanthin also influences other bioactivities and astaxanthin is thought to be active in the prevention of several diseases associated with oxidation (Higuera-Ciapara et al. 2006).

2.2.2 Lipids

Marine sources are widely regarded as possessing interesting lipid compositions, which make them attractive as a source for lipid extraction. The main polar lipids found in these substrates include monogalactosyl diacylglycerols (MGDG), digalactosyl diacylglycerols (DGDG), and phosphatidylglycerols (PG). These polar lipids possess several functional activities, but are mainly referenced in the literature for their anti-inflammatory activities (Bruno et al. 2005; Larsen et al. 2003). The lipid fraction consists primarily of the polyunsaturated fatty acids (PUFAs) that are well documented as essential for good human health. For instance, it has been widely studied how long chain ω-3 fatty acids, such as eicosapentaenoic (EPA, $C_{20:5}$) or docosahexaenoic acids (DHA, $C_{22:6}$) are useful in the prevention of cardiovascular diseases (Calzolari et al. 2009; Schuchardt et al. 2010; Zuliani et al. 2009). These fatty acids can be presented bound to polar lipids or even as triacylglycerols.

The composition and extraction of PUFAs from algae (Plaza et al. 2009), fish (Sahena et al. 2009), fish by-products (Wu and Bechtel 2008), and other marine sources (Juárez et al. 2010) have been studied previously. However, PUFAs are prone to lipid oxidation and this represents one drawback for their inclusion in functional foods. For this reason, PUFAs are often used in combination with other antioxidants in order to enhance their preservation.

Other attractive lipid bioactive compounds from marine sources include the group known as sterols. The composition of sterols isolated from macro- and microalgae (Cardozo et al. 2007) and other marine invertebrates (Kanazawa 2001) has been extensively studied. Sterols and some of their derivatives were found previously to play an important role in lowering LDL cholesterol levels in vivo (Francavilla et al. 2010). Other bioactivities are associated with sterols, include anti-inflammatory and antiaterogenic activity (Francavilla et al. 2010). In addition, phytosterols (C_{28} and C_{29} sterols) are important precursors of compounds including vitamins. For example, ergosterol is a precursor of vitamin D_2 and cortisone.

2.2.3 Carbohydrates

Macroalgae are regarded as a rich source of sulfated polysaccharides and the particular type of polysaccharide is different depending on the taxonomic group. Different carbohydrates including agar, carrageenan, or alginates are extracted from macroalgae and these carbohydrates are used widely in the food and pharmaceutical industries as functional ingredients such as stabilizers. For example, the alga

Chondrus crispus is traditionally employed for the extraction of carrageenan (also known as Irish moss), a highly sulfated polysaccharide. However, macroalgal polysaccharides also have potential for use as prebiotics as they are not digested in the human gut and, in addition, can be considered as a rich source of dietetic fiber. Among their associated bioactive properties, immunomodulating, anticancer, anti-inflammatory, antiviral, or antioxidant activities have been pointed out (Li et al. 2008). Other minor sulfated polysaccharides, typically from particular species, are also known, such as porphyrans produced by *Porphyra*. In general, these polysaccharides can significantly vary their composition, and therefore their related properties. For instance, the bioactivity might differ depending on the degree of sulfation, molecular weight, type of sugar found predominantly, and/or glycosidic branching (Qi et al. 2005).

Another carbohydrate-type product, abundant in different marine sources, is chitin. This compound is a polymer composed of B(1→4)-linked N-acetylglucosamine units. Chitin is one of the most extensive biopolymers in nature, and is found in different marine sources such as crustaceans where it is part of their exoskeleton. Chitosan is produced by the alkaline deacetylation of chitin, and is employed in a wide range of applications. Some interesting effects have also been associated with this polymer, including dietary fiber, lipid absorption reduction, and hypocholesterolemic or antidiabetic effects, among others (Ngo et al. 2011; Liu et al. 2007; Liao et al. 2007; Muzzarelli et al. 2007).

2.2.4 Peptides and Proteins

As discussed in the next chapter, proteins of marine origin have received attention recently due to their potential bioactive and functional properties. Phycobiliproteins are one of the most important groups of proteins from seaweed. These water-soluble proteins, mainly found in some blue-green and red algae, are characterized by possessing a tetrapirrolic ring covalently attached to their structure. This pigment can be either phycocyanobilin (blue-green algae) or phycoerythrobilin (red algae) and it is partially responsible for the functional properties associated with these proteins, mainly hepatoprotective, anti-inflammatory, and antioxidant activities (Bhat et al. 1998; Romay et al. 2003; Bhat and Madyastha 2000). In fact, the extraction of these proteins from different algae has been extensively studied, considering their economic importance. Usually, the extraction method is the key factor for enhanced recovery of phycobiliproteins. The extraction protocols usually involve the selection of a suitable source followed by the disruption of the algae cells and release of proteins (Moraes et al. 2010; Patil et al. 2008).

However, interest in marine proteins might not be only directly correlated to the intact protein, but also to the possibility of generating bioactive peptides. In this sense, different peptides derived from marine proteins have been identified as having antioxidant activity (Byun et al. 2009) as well as antihypertensive, anticoagulant, or antimicrobial activities (Ngo et al. 2011). These peptides have been isolated from diverse marine sources such as algae, crustaceans, and also different fish species. In general, bioactive peptides comprise relatively short chains of amino acids (from 3 to 20), which do not present any bioactivity prior to release from the intact parent protein. Nevertheless, upon digestion or hydrolysis in vivo or due to technological processes such as high pressure processing, these peptides are released demonstrating their bioactivities.

As shown in this section, marine-related products can be a good source of potential functional compounds that may be used as ingredients in the food industry. In the following sections, the application of different, environmentally clean extraction techniques to obtain these kinds of bioactive compounds from different marine sources are described.

2.3 Green Extraction Techniques for Bioactive Compound Isolation from Marine Resources

Since the early 1990s, the Green Chemistry movement has been exploring ways to reduce the risk of chemical exposure to humans and the environment. Simply stated, Green Chemistry reduces or eliminates the use or generation of hazardous substances from chemical products and processes and improves all types of chemical products and processes by reducing the impacts on human health and the environment (Fig. 2.2a). Green Chemistry technologies encompass all types of chemical processes, including synthesis, catalysis, reaction conditions, separations, analysis and monitoring. A Green Chemistry technology can involve implementing incremental improvements at any stage (Majors and Raynie 2011). In the 12 principles that rule Green Chemistry (Table 2.1) there are three main aspects that dominate: waste, hazard (health, environmental, and safety), and energy (Anastas and Warner 1998).

Green Engineering is the development and commercialization of industrial processes that are economically feasible and which reduce risk to human health and the environment. Both of these "Green concepts" are intimately related to sustainability (Fig. 2.2b), which means using methods, systems, and materials that will not deplete natural resources or harm natural cycles. The principles of Green Chemistry and Green Engineering (Table 2.1) provide a framework for scientists and engineers to use when designing new materials, products, processes, and systems (Anastas and Zimmerman 2003).

The discovery and development of marine bioactive compounds is a relatively new area when compared to the discovery of bioactive compounds from terrestrial sources. Therefore, in the development of this area, new, environmentally friendly, and sustainable trends should, if possible, be followed. All of the above-mentioned rules enable scientists and engineers to develop a compendium of novel protocols for the discovery of bioactive compounds for use in functional foods and pharmaceuticals. These new protocols should comply, as much as possible, with the principles of Green Chemistry and Green Engineering. Among the different advanced

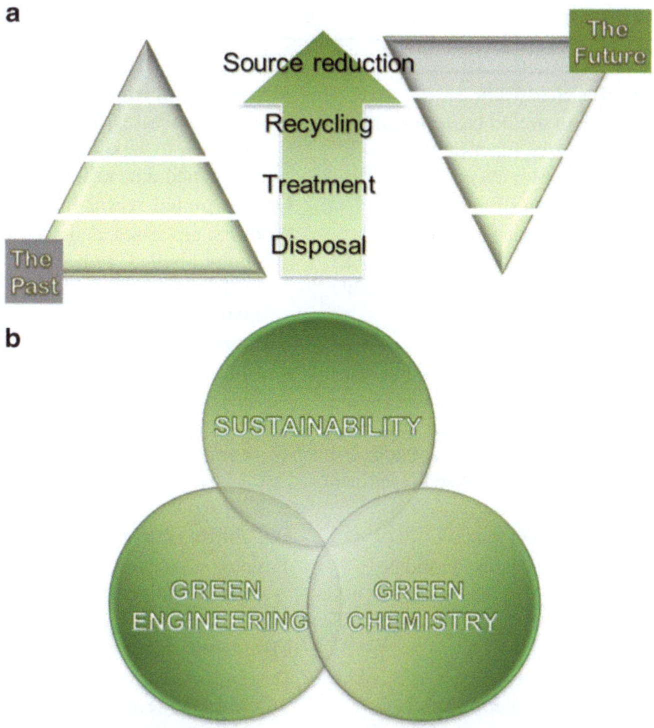

Fig. 2.2 Changes in waste prevention hierarchy (**a**) and environmental stewardship perspectives: green chemistry, green engineering and sustainability (**b**) (Adapted from "Pollution Prevention: The Basics" http://www.epa.state.oh.us/portals/41/P2basics_web.pdf (last accessed April 2011))

Table 2.1 Principles of green chemistry Anastas and Warner 1998 and green engineering Anastas and Zimmerman 2003

#	Green chemistry	Green engineering
1	**Prevention** It is better to prevent waste than to treat or clean up waste after it has been created	**Inherent rather than circumstantial** Designers need to strive to ensure that all materials and energy inputs and outputs are as inherently nonhazardous as possible
2	**Atom economy** Synthetic methods should be designed to maximize the incorporation of all materials used in the process into the final product	**Prevention instead of treatment** It is better to prevent waste than to treat or clean up waste after it is formed
3	**Less hazardous chemical syntheses** Wherever practicable, synthetic methods should be designed to use and generate substances that possess little or no toxicity to human health and the environment	**Design for separation** Separation and purification operations should be designed to minimize energy consumption and materials use

(continued)

Table 2.1 (continued)

#	Green chemistry	Green engineering
4	**Designing safer chemicals** Chemical products should be designed to effect their desired function while minimizing their toxicity	**Maximize efficiency** Products, processes, and systems should be designed to maximize mass, energy, space, and time efficiency
5	**Safer solvents and auxiliaries** The use of auxiliary substances (e.g., solvents, separation agents, etc.) should be made unnecessary wherever possible and innocuous when used	**Output-pulled versus input-pushed** Products, processes, and systems should be "output pulled" rather than "input pushed" through the use of energy and materials
6	**Design for Energy efficiency** Energy requirements of chemical processes should be recognized for their environmental and economic impacts and should be minimized. If possible, synthetic methods should be conducted at ambient temperature and pressure	**Conserve complexity** Embedded entropy and complexity must be viewed as an investment when making design choices on recycle, reuse, or beneficial disposition
7	**Use of renewable feedstocks** A raw material or feedstock should be renewable rather than depleting whenever technically and economically practicable	**Durability rather than immortality** Targeted durability, not immortality, should be a design goal
8	**Reduce derivatives** Unnecessary derivatization (use of blocking groups, protection/ deprotection, temporary modification of physical/ chemical processes) should be minimized or avoided if possible, because such steps require additional reagents and can generate waste	**Meet need, minimize excess** Design for unnecessary capacity or capability (e.g., "one size fits all") solutions should be considered a design flaw
9	**Catalysis** Catalytic reagents (as selective as possible) are superior to stoichiometric reagents	**Minimize material diversity** Material diversity in multicomponent products should be minimized to promote disassembly and value retention
10	**Design for degradation** Chemical products should be designed so that at the end of their function they break down into innocuous degradation products and do not persist in the environment	**Integrate material and energy flows** Design of products, processes, and systems must include integration and interconnectivity with available energy and materials flows
11	**Real-time analysis for pollution prevention** Analytical methodologies need to be further developed to allow for real-time, in-process monitoring and control prior to the formation of hazardous substances	**Design for commercial "Afterlife"** Products, processes, and systems should be designed for performance in a commercial "afterlife"
12	**Inherently safer chemistry for accident prevention** Substances and the form of a substance used in a chemical process should be chosen to minimize the potential for chemical accidents, including releases, explosions, and fires	**Renewable rather than depleting** Material and energy inputs should be renewable rather than depleting

extraction techniques that fit within the remit of the fore-mentioned rules, supercritical fluid extraction, pressurized liquid extraction, ultrasound-assisted extraction and microwave-assisted extraction are the most promising for the isolation of marine-sourced bioactive compounds. In the following sections, detailed information about these techniques together with the necessary instrumentation and their application to different marine sources for the isolation of bioactive compounds is provided.

2.3.1 Supercritical Fluid Extraction (SFE)

Hannay and Hogarth first introduced supercritical fluid extraction as an alternative extraction method in 1879. However, it was not until circa 1960 that this extraction method started to be thoroughly investigated (Hosikian et al. 2010) as an alternative to conventional extraction methods, such as SLE and LLE, which utilize large amounts of hazardous chemicals such as chlorinated solvents. Supercritical fluid extraction is based on the use of solvents at temperatures and pressures above their critical points. This technique has been employed previously to extract a wide variety of interesting compounds from different food-related materials (Mendiola et al. 2007a), and algae are no exception (Herrero et al. 2006a).

One of the most valuable characteristics of SFE is the highly reduced (often to zero) employment of toxic organic solvents. Carbon dioxide (CO_2) is the solvent most commonly used to extract bioactive compounds from natural sources using SFE. In fact, CO_2 has a series of interesting properties for bioactive extraction: it is cost efficient, its critical conditions are easily attainable (30.9°C and 73.8 bars), and it is an environmentally friendly solvent that is Generally Recognized As Safe (GRAS) for use in the food industry. The main drawback of supercritical CO_2 is its low polarity, a problem that can be overcome by employing polar modifiers or cosolvents to change the polarity of the supercritical fluid and to increase its solvating power towards the analyte of interest. For example, the addition of relatively small percentages (1–10%) of methanol to carbon dioxide expands its extraction range to include more polar analytes. The modifiers can also reduce the analyte–matrix interactions, improving their quantitative extraction (Björklund et al. 2005). Other solvents have been proposed for SFE and these include propane, butane, and dimethyl ether but none of these fulfill all the principles of Green Chemistry and Green Engineering as well as CO_2. Advantages of using CO_2 under supercritical conditions for the extraction of marine bioactive compounds include its high diffusivity and the relative ease in tuning the temperature and pressures applied, so that solvent strength and density can be modified easily. Another advantage of SFE when using CO_2 is the possibility of attaining solvent-free extracts. Once the extraction procedure is finished, the depressurization of the system converts CO_2 from a liquid to a gas making it easier to recover the extract. These properties are responsible for the extended use of supercritical CO_2 for extraction of bioactive compounds.

2.3.1.1 Instrumentation

Several parameters are involved in the extraction of bioactive compounds from marine resources by SFE. It is necessary to precisely control the effect of the extraction temperature, pressure, percentage, and type of modifier addition, amount of sample to be extracted, as well as its particle size and the use of dispersing agents. The first parameters are related to the solubility of the target compounds in the supercritical fluid, inasmuch as changes to the extraction temperature and pressure will have a strong influence on solvent properties, such as density. Although supercritical solvents have diffusivity in the matrix beyond those of liquids, a decrease in the sample particle size generally produces an increase in the extraction yield obtained. This is due mainly to the increase that results in the surface contact between sample and solvent, which enables an increase in mass transfer. In some applications, the use of dispersing agents such as sand, glass beads, and diatomaceous earth or a hydromatrix to absorb liquid from the sample can be useful.

A basic supercritical-fluid extractor (Fig. 2.3a) consists of a tank of the mobile phase, usually CO_2, a pump to pressurize the gas, an oven containing the extraction vessel, a restrictor to maintain the high pressure inside the system, and a trapping vessel. Extracts are trapped when the solute-containing supercritical fluid decompresses into an empty vial, through a solvent, or onto a solid or liquid material.

Extractions are done in dynamic mode, static mode, or a combination of both. In a dynamic extraction, the supercritical fluid continuously flows through the sample in the extraction vessel and out of the restrictor to the trapping vessel. In static mode, the supercritical fluid circulates in a loop contained in the extraction vessel for some period of time before being released through the restrictor to the trapping vessel. In combination mode, a static extraction is performed for some period of time, followed by a dynamic extraction.

Use of supercritical CO_2 also results in a "clean" extract when compared to other conventional extraction techniques. Indeed, the selectivity obtained through the use of supercritical CO_2 often results in the attainment of more purified extracts, reducing to a great extent the amount of interfering compounds extracted from the matrix. However, if the extraction of more polar compounds is targeted, other strategies have to be conceived. As mentioned, the main alternative in this case is the use of a given percentage of a modifier together with the supercritical fluid, which is normally CO_2 (Fig. 2.3b). When the modifier, usually a polar solvent, is added to the supercritical fluid, a change in the properties of the extracting mixture is produced, allowing the collection of more polar compounds, increasing the polarity of the solvent used for the extraction and also the range of applications of the SFE method.

With some simple engineering modifications, such as multiple extraction vessels connected in series that can be taken offline at any time, greater processing and economic efficiencies are produced (Fig. 2.3b). For example, at the end of the extraction period (for one of the vessels), carbon dioxide flow can be routed to another vessel (that has previously been filled with material to be extracted), and while the extraction process is continuing in the second vessel, the first vessel (containing the spent matrix) can be unloaded and loaded again with fresh sample.

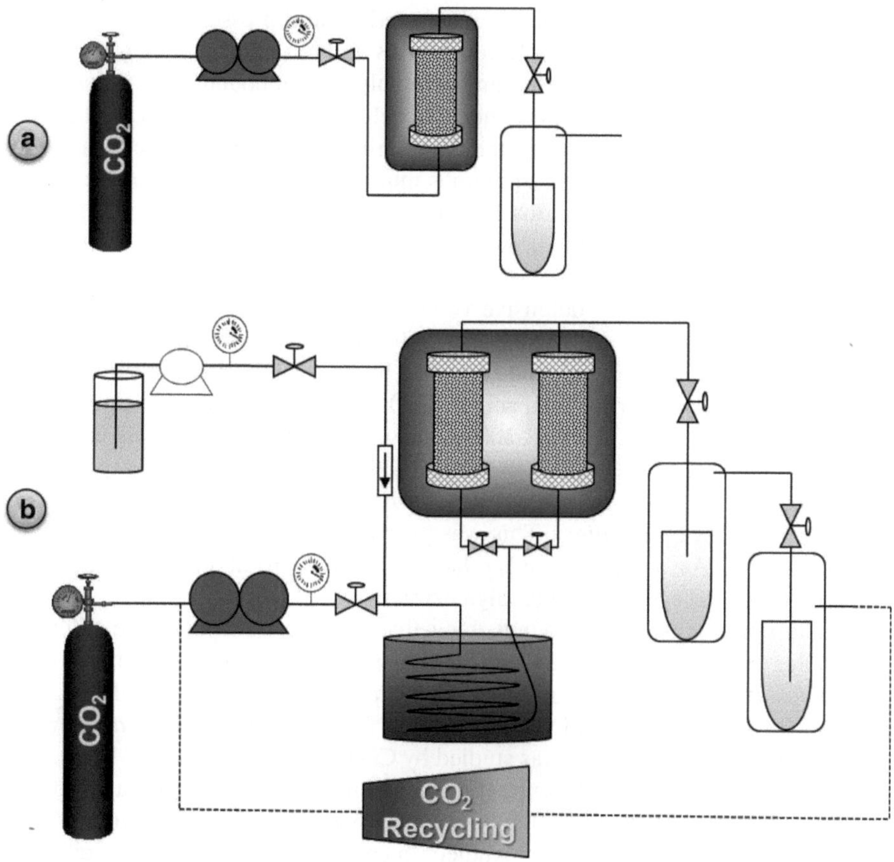

Fig. 2.3 Basic Scheme of supercritical fluid extractor (**a**) and typical scheme of industrial-scale supercritical fluid extractor including modifier circuit, CO_2 recycling circuit, extract fractionation, and dual extraction vessel and dual extraction vessel (**b**)

Instead of being vented to the atmosphere, the carbon dioxide present in the offline extractor at the end of the extraction period can be sent to the supply vessel and subsequently reused. The use of cascade depressurization allows fractionation of the extracts in different collectors depending on the solubility of the compounds at the selected depressurization conditions (Fig. 2.3b).

2.3.1.2 Application of SFE to Macroalgae, Microalgae, and Cyanobacteria

In spite of the potential of this technique, its usefulness will be related to the type of compounds to be extracted from the algae. Considering the low polarity of supercritical CO_2, SFE is more suitable for the extraction of compounds with low polarity.

However, it is also notable that CO_2, when used at mild pressure and temperature conditions, allows obtaining volatile compounds without affecting its properties. In general, naturally produced volatile and semivolatile compounds play an essential role in the survival of organisms for chemical defense mechanisms and food gathering. When attacked by herbivores, land and marine plants can produce a variety of volatile compounds that attract carnivorous mutualists. One example of the extraction of volatile metabolites from marine sources includes the isolation of terpenoids and sulfur compounds from the brown alga *Dictyopteris membranacea* (El Hattab et al. 2007).

Microalgae also produce defensive volatile compounds. It is common in many microalgae to share their ecological niche with bacteria and other micro-organisms. Therefore, the defensive compounds secreted by microalgae possess antibacterial, antifungal, and often antiprotozoal activities. Our findings with the green microalga *Dunaliella salina* (Mendiola et al. 2008a) showed that an extract of this microalga isolated using SFE with CO_2 at 314 bar and 9.8°C displayed potent antimicrobial activity against the pathogens *Escherichia coli*, *Staphylococcus aureus*, *Candida albicans*, and *Aspergillus niger.* This was probably due to the presence of indolic compounds, PUFAs, and compounds related to carotene metabolism, such as β-ionone and neophytadiene in the microalga extract.

In addition to volatile compounds, bioactive lipids, both saponifiable and unsaponifiable, are also preferentially extracted using SFE. Among saponifiable compounds, essential fatty acids can be easily extracted using supercritical CO_2. The effect of the extraction conditions in obtaining fatty acids from *Hypnea charoides* algae, using supercritical CO_2, was studied by Cheung in 1999 (Cheung 1999), who suggested the usefulness of this extraction technique for converting this alga species to a new source of omega-3 (ω-3) fatty acids. Temperatures from 40°C to 50°C and pressures from 241 to 379 bar were studied. In general, lipid recovery, as well as the ratio of unsaturated fatty acids, increased with extraction pressure and temperature. Extraction of ω-3 fatty acids was shown to depend on their chain length. The cyanobacterium *Spirulina platensis* was studied to determine the amount of lipids and GLA present in the microalgae (Qiuhui 1999). The maximum extraction yield was obtained at 350 bar, 40°C and a flow rate of 24 kg CO_2/h. Bioactive compounds from algal matter have been studied previously (vitamin E and carotenoids). For example, Mendiola et al. (2008b) optimized, by means of a central composite circumscribed design (CCCD), a process based on SFE at pilot scale, to obtain fractions highly enriched in vitamin E from *Spirulina platensis* (Mendiola et al. 2008c). The authors achieved a tocopherol enrichment of more than 12 times the initial concentration of tocopherol in the raw material by extracting with neat CO_2 at 361 bar and 83.3°C. In addition, SFE has previously been used to extract carotenoids from *Chlorella vulgaris* (Mendes et al. 1995), and *Spirulina plantesis* (Mendiola et al. 2008). In general, high pressures allowed high extraction yields and carotenoid extraction improved slightly by crushing the raw material. The addition of polar modifiers such as ethanol in the supercritical CO_2 allowed the extraction of more polar carotenoids, but also chlorophylls, thus decreasing the selectivity of the extraction process.

Other target bioactive compounds, such as diolefines, have been extracted from *Botrycoccus braunii* cells by SFE (Mendes et al. 2003). This organism can store high amounts of long-chain hydrocarbons (i.e., 25–31 carbon atoms) that can be used as substitutes of paraffinic and natural waxes. Authors proved that the solubility of these types of compounds in CO_2 increased with pressure and found that 300 bar provided the optimum in terms of yield and extraction speed (Mendes et al. 2003).

One interesting area of research, which, is less studied, is the supercritical fluid extraction of phenolic compounds (simple phenols, flavonoids) from marine sources. Klejdus et al. (2010) developed a new hyphenated technique for the extraction and determination of isoflavones from sea macroalgae (*Sargassum muticum, Sargassum vulgare, Hypnea spinella, Porphyra* sp., *Undaria pinnatifida, Chondrus crispus, and Halopytis incurvus*), freshwater algae (*Spongiochloris spongiosa*), and cyanobacteria (Scenedesmus and Nostoc 17; Klejdus et al. 2010). The method involved sample pretreatment using sonication, followed by extraction using supercritical CO_2 modified with 3% (v/v) of MeOH/H_2O mixture (9:1, v/v) at 350 bar and 40°C for 60 min. This was followed by fast chromatography analysis and MS/MS determination. By using this approach, eight isoflavones were reported for the first time in different algae and cyanobacteria. Furthermore, Wang et al. (2010) reported the use of SC–CO_2 extraction to obtain antioxidants (flavonoids) from a novel microalga, *Chlorella vulgaris C-C* (Wang et al. 2010). Authors compared SC–CO_2 extraction at 310 bar, 50°C, using 50% aqueous ethanol as a modifier, to ultrasonic extraction with 50% aqueous ethanol and reported that flavonoid content obtained under SFE conditions was significantly higher than those obtained using UAE. This resulted in a higher antioxidant activity but also in better inhibition of lung cancer metastasis (Wang et al. 2010).

2.3.1.3 Applications to Invertebrates

Supercritical CO_2 extraction has not only been employed to extract bioactive compounds from macroalgae, microalgae, and cyanobacteria but also from marine invertebrates, such as crustacean (krill, crawfish, crab, or shrimp), squid, urchin, or starfish.

Crustacean shell wastes are a rich source of astaxanthin, which is the pigment responsible for their orange-pink coloration. Crustaceans are able to modify some carotenoids such as β-carotene and transform them into astaxanthin (Félix-Valenzuela et al. 2001). Yamaguchi et al. (1986) reported for the first time the application of SFE to crustacean waste. These authors extracted nonpolar lipids, mainly triglycerides, and astaxanthin (which is unstable at high pressures and temperatures around 80°C) from krill samples using a one-step extraction utilizing SC–CO_2 at 60°C and 245 bar (Yamaguchi et al. 1986). More recent work was focused on the extraction of astaxanthin from different crustacean shells using SC–CO_2 with 10% (w/w) ethanol as a cosolvent. After studying the experimental extraction conditions (in terms of temperature, pressure, and sample load), extraction efficiency of astaxanthin was 57% from blue crab shells and was achieved using 45°C, 340 bar, and 17.5 g of sample load (Félix-Valenzuela et al. 2001). Charest et al. (2001) employed temperatures ranging from 50°C to 70°C and pressures from 138 to 310 bar to extract astaxanthin from crawfish shells (Charest et al. 2001). Results showed that 157.1 mg/kg of astaxanthin could be extracted at 60°C and 224 bar. Another interesting study compared the extraction efficiencies of *cis*- and *trans*-astaxanthin by supercritical CO_2 and solvent extraction from spear shrimp shells. The formation of *cis*- isomers or its derivatives may possess different biological activities to *trans*-astaxanthin (Lin et al. 2005). A SC–CO_2 extraction was carried out with pressures ranging from 250 to 400 bars, temperatures between 45°C and 75°C and with or without 10% ethanol as a modifier. Higher contents (μg/g) of *cis*- and *trans*-astaxanthin, as well as their esters, were achieved with ethanol, low temperature (45°C), and high pressure (400 bar). Compared to solvent extraction, which generates a greater content of *trans*-astaxanthin and astaxanthin ester, SC–CO_2 extraction could be used to produce slightly higher levels of 9-*cis*- and 13-*cis* astaxanthin.

Squid viscera, which is normally discarded during processing, and sea urchin (*Strongylocentrotus nudus*) gonads are rich in polyunsaturated fatty acids, which are essential nutrients for humans. Therefore, marine invertebrates can also be used as a potential source of high quality lipids. For instance, palmitic, oleic, eicosapentaenoic acid, and docosahexaenoic acid were the major fatty acids found in the oil extracted from waste squid viscera using SC–CO_2 with 1.5% (w/w) ethanol at temperatures between 25°C and 50°C and pressures ranging from 80 to 170 bar. Cholesterol (less than 54%) was also coextracted with the lipids (Kang et al. 2005a). From sea urchin gonad, lipids were extracted and compared to an enzyme-assisted aqueous method (papain, neutral protease, alkaline protease, and trypsin were tested) and a SC–CO_2 method (at 50°C and 280 bar). Lipid yields were similar for samples extracted using both methods (53.7% and from 61.5% to 82.0% for SC–CO2 and the enzyme-assisted method, respectively). PUFAs were the main components, corresponding to approximately 35% of the total fatty acids extracted (Zhu et al. 2010). Recently, Chun et al. (2010) studied SFE with CO_2 for defatting, as an alternative to organic solvent use in defatting samples. The main goal of this work was to purify the phospholipase A_2 (PLA_2) from the defatted pyloric ceca of starfish (defatting conditions: SC–CO_2 at 40°C and 250 bar) to be used in the food industry (Chun et al. 2010).

2.3.1.4 Applications to Marine By-Products

Waste and by-products discharged by fisheries, such as fish heads, viscera, skin, tails, and blood often possess the highest concentration of bioactive compounds suitable for human health (Ferraro et al. 2010).

Marine organisms (fish, macroalgae, and microalgae) are an important natural source of Omega-3 PUFA. In fact, fish oil usually presents higher amounts of PUFA than seed oils or microalgae. This makes fish oil a valuable product as a raw material for the generation of Omega-3 concentrates. SFE using CO_2 as the extracting solvent has been widely employed for fish oil extraction due to the good solubility of fish oil and its individual components in CO_2 (Rubio-Rodríguez et al. 2010). Dunford et al. (1997, 1998) investigated the effect of moisture content on mackerel oil extracted by SC–CO_2 at 345 bar and 35°C, concluding that the extraction yield increased when fish muscle moisture content decreased (Dunford et al. 1997). They also developed a mathematical model based on the interaction between oil and water in the fluid phase of SFE to explain the extraction behavior (Dunford et al. 1998). Esquível et al. (1997) studied the SC–CO_2 extraction of highly polyunsaturated oil from sardine muscle. According to this study, at 40°C and 180 bar, it was possible to recover the oil (92% containing 209.7 and 136.7 g/kg oil EPA and DHA, respectively) without degradation of Omega-3 PUFA (Esquível et al. 1997).

Fish by-products have also been employed as raw materials for Omega-3 rich oil SC–CO_2 extraction, demonstrating the applicability of SFE technology for valorizing waste products of the fish industry. For instance, oil rich in EPA and DHA was obtained by SC–CO_2 from freeze-dried sardine heads. A yield of 10.36% extracted oil containing 10.95% EPA and 13.01% DHA was achieved at 300 bar and 75°C (Létisse et al. 2006). From hake skin, more than 96% of the total oil contained in the raw material was extracted at 40°C and 250 bar in a semipilot SFE plant. The oil extracted contained a ratio of Omega-3/Omega-6 around 7 with a high content of EPA and DHA (6% and 14%, respectively, of the total fatty acid in the hake skin; Rubio-Rodríguez et al. 2008). Another study reported by Sahena et al. (2010) employed various SFC extraction techniques including continuous use of cosolvent (ethanol), soaking, and pressure swing of SC–CO_2 to extract oil from the skin of Indian mackerel. Soaking and pressure swing techniques (at 350 bar and 75°C) gave the highest yields of PUFA, especially arachidonic acid, EPA, and DHA. Considering extractability, CO_2 consumption, and recovery of EPA and DHA, pressure swing was deemed the more attractive technique for isolation of PUFA from Indian mackerel skins (Sahena et al. 2010). However, SC–CO_2 was also used previously for removal and deodorizing of lipids in tuna viscera before protein recovery. At optimum conditions (CO_2 + 3% ethanol, 124 bar and 35°C), 97% of the lipids (palmitic acid, heptadecanoic acid, and DHA) were extracted and the protein content found in the residue was about 50%, and consisted of L-Proline, taurine, and L-aminoadipic acid (Kang et al. 2005b). Létisse and Comeau (2008) developed a fractionation process for fatty acid methyl esters (FAME) using SC–CO_2 to obtain extracts enriched in EPA and DHA from sardine and tails. Under optimized temperature (60°C) and pressure (264 bar) conditions, efficient fractionation according to chain length and degree of saturation was achieved, with high purity (up to 95% of FAME) and good yields (45% of EPA and DHA), (Létisse and Comeau 2008). Chang et al. (2008) carried out the concentration enhancement of Omega-3 fatty acids extracted from soft-shelled turtle fish oil bags using SC–CO_2. They obtained the crude oil by top-down pressurized CO_2 at 60 bar and 55°C; then long-chain PUFAs were enriched from the esterified oil mixture using a countercurrent SC–CO_2 extractive distillation at 200 bar and 80°C (Chang et al. 2008).

Although fish oils are a natural source of Omega-3 fatty acids, the use of EPA and DHA as functional ingredients requires a concentration step into a chemical form that can be easily metabolized by humans. Hence, different processes to fractionate or concentrate PUFA from fish oil have been developed. Many of these methods refer to Omega-3 ethyl esters formed by the esterification or saponification of triglycerides with ethanol (Rubio-Rodríguez et al. 2010). Also, these methods often involve supercritical fluid fractionation to achieve higher yield and purity at lower cost. Thus, Fleck et al. (1998) carried out the fractionation of fatty acid ethyl ester (FAEE) from fish oil using SC–CO_2 (60°C, 145 bar) and an automated countercurrent column to obtain high separation efficiency (Fleck et al. 1998). Perretti et al. (2007) studied the fractionation of fish oil FAEE to concentrate the Omega-3 FAEEs with a suitable EPA/DHA ratio. Experiments were carried out using a pilot plant with three columns at 40°C, 50°C and 60°C, respectively, starting from the bottom and applying pressure changes from 100 to 150 bar and CO_2 flow rates of between 2.5 and 5 kg/h. Higher pressure (150 bar) and flow rates (5 kg/h) resulted in a higher EPA and DHA concentration and decreased the EPA/DHA ratio (Perretti et al. 2007). Moreover, Davarnejad et al. (2008) studied the solubility of menhaden fish oil in SC–CO_2 at temperatures from 40°C to 70°C and pressures of between 136 and 272 bar. The fractionated samples were converted to methyl esters by chemical esterification with methanol. The maximum solubility was achieved at 40°C and 272 bar (Davarnejad et al. 2008).

Following the fractionation of FAEE derived from fish oil according to their C-atom number using a phase equilibrium study with SC–CO_2 (Riha and Brunner 1999), a method to separate FAEE by countercurrent multistage extraction with SC–CO_2 in a pilot scale column was developed (Riha and Brunner 2000). Temperatures ranged from 40°C to 80°C and pressure from 65 to 195 bar. Low-molecular-weight components (C_{14}–C_{18}) were extracted as the top products and high-molecular-weight components (EPA- and DHA-rich esters) constituted the bottom products with 95% purity and 95% yield. Moreover, using a thermodynamic model, it was possible to determine the optimal conditions (145 bars and 60°C) for separation of FAEE by countercurrent SFE. This process, applied on an industrial scale, could produce EPA, DHA, and DPA (DHA and DPA in a mixture) at concentrations higher than 90% (Brunner 2000).

Other thermodynamic and mathematical models have been employed to optimize the fractionation of fish oil FAEE by SC–CO_2. For instance, Espinosa et al. (2002) developed a thermodynamic modeling, simulation, and optimization process for the supercritical fractionation of FAEE from fish oil based on Group Contribution Equation of State (GC-EOS), which provides reliable phase equilibrium predictions. They proposed a method for obtaining the maximum EPA and DHA recovery and purity from fish oil using a countercurrent system with three columns operating at similar temperatures (around 60°C) and pressures from 145 to 158 bar (Espinosa et al. 2002). Although recoveries of 80.34% of DHA (purity of 80.09 mol%) and 98.18% EPA (purity of 60.46 mol%) were obtained, the process needed a complex scheme to achieve complete fractionation, implying that high purity Omega-3 PUFA are associated with high extraction and purification costs. For that reason, the authors extended the thermodynamic model to include ethane–ester interaction parameters in order to use ethane as an alternative supercritical solvent (Espinosa et al. 2008). Concentrations of 60% wt of EPA + DHA FAEE were obtained at optimal conditions (60°C and 84 bar), showing that the production of food additives containing EPA and DHA ester mixtures with moderate concentrations was possible at a relatively low cost.

Alternatively, a previous step based on urea complexation to increase the purity was also proposed. Gironi and Maschietti (2006) developed a thermodynamic model (assuming that the oil was composed of five ethyl esters) to represent high-pressure phase equilibria for the system. The method was validated with experimental data (temperature and pressure ranged from 42°C to 70°C and 101–172 bar, respectively) obtained in a semicontinuous single-stage fractionation process carried out on fish oil ethyl ester mixtures (Gironi and Maschietti 2006). Then the model was used to develop a simulation program of a continuous multistage fractionation process demonstrating that it is possible to produce a raffinate with 95% weight of EE-C_{20} and EE-C_{22}, together with 95% recovery of these compounds. Martín and Cocero (2007) proposed a mathematical method, based on the differential mass and energy balances over the height of the extractor, to study theoretically the effect of operating conditions and extractor configuration in the fractionation of liquids with CO_2. They validated the model for the fractionation of fish oil ethyl ester mixtures and found that it was possible to predict the variation of the composition of the extract at the different operating parameters (Martín and Cocero 2007).

Catchpole et al. (2000) reported the countercurrent extraction and fractionation of different crude fish oils by SC–CO_2 and SC–CO_2 with ethanol. Vitamin A palmitate was recovered from cod liver oil ethyl ester/vitamin mixtures using pure CO_2 at 60°C and 200–300 bar. Under these conditions, vitamin A was preferentially recovered in the raffinate; then the cod liver oil ethyl esters virtually free from vitamin A could be used to obtain concentrates of DHA and EPA (Catchpole et al. 2007). In the same work, authors also investigated the use of mixtures CO_2 + ethanol to fractionate orange roughy (*Hoplostethus atlanticus*) and deep sea shark liver oils at a pilot scale. Peroxides, fatty acids, and odor compounds were removed from orange roughy oil and squalene was obtained from deep sea shark liver oil, at higher throughputs but lower selectivity using CO_2 with ethanol than using pure CO_2. An extraction process with pure CO_2 at a demonstration scale to produce squalene-rich extract and a diacylglycerylethers-rich raffinate (both compounds are used as health food supplements) was also developed.

Although fatty acids are usually converted into FAEE for oil fractionation, there are some references where Omega-3 PUFAs were obtained in their natural forms. For instance, Sarrade et al. (1998) combined SC–CO_2 extraction (310 bar and 40°C) with nanofiltration (using a multilayer composite nanofilter) for the fractionation and purification of fatty acids from fish oil, obtaining a significant concentration of EPA and DHA in the retentate stream and short-chained fatty acids in the permeate (Sarrade et al. 1998). Antunes Correa et al. (2008) also studied the fractionation of fish oil and PUFA concentration with SC–CO_2. Phase equilibrium was measured at temperatures from 28°C to 50°C and pressures from 78 to 294 bar. They found that the best conditions to fractionate the fish oil were 78 bar and 28°C (Antunes Corrêa et al. 2008).

Finally, it is important to highlight that several methods based on supercritical fluid chromatography to obtain Omega-3 PUFA with a high purity have also been reported in the literature. Most of these methods, as for supercritical fluid fractionation, deal with ethyl ester separation (Alkio et al. 2000; Petinello et al. 2000).

2.3.2 Pressurized Liquid Extraction (PLE)

PLE is also known as pressurized fluid extraction (PFE), enhanced solvent extraction (ESE), high-pressure solvent extraction (HPSE), or accelerated solvent extraction (ASE; Nieto et al. 2010). This technique was described for the first time in 1996 (Richter et al. 1996). In PLE, pressure is applied to allow the use of liquids at temperatures higher than their normal boiling point. Among them, ASE (which can be considered a new version of the Soxhlet apparatus but operating at high pressures and temperatures), pressurized hot water extraction (PHWE or SWE, subcritical water extraction), near-critical fluid extraction, and enhanced fluidity extraction are the most promising techniques in bioactive extraction from different raw materials (Mendiola et al. 2007; Herrero et al. 2006).

The combined use of high pressures and temperatures provides faster extraction processes that require small amounts of solvents (e.g., 20 min using 10–50 ml of solvent in PLE can be compared with a traditional extraction procedure in which 10–48 h and up to 300 ml are required). The increase on the extraction temperature can promote higher analyte solubility by increasing both solubility and mass transfer rate. In addition, high temperatures decrease the viscosity and the surface tension of the solvents, helping to reach areas of the matrices more easily, thus improving the extraction rate. On the other hand, PLE is broadly recognized as a green extraction technique, mainly due to its low organic solvent consumption, which meets perfectly the basics of Pollution Prevention addressed in Fig. 2.2 and with Green Chemistry and Engineering principles (Table 2.1).

2.3.2.1 Instrumentation

In general terms it could be said that instrumentation for PLE is quite simple, but nowadays there is little commercial equipment available. Basically, the instrumentation consists of a solvent reservoir coupled to a high-pressure pump to introduce the solvent into the system, an oven where the extraction cell is placed and extraction occurs, and a restrictor or valve to keep the pressure inside the system (Fig. 2.4). Extracts are collected in a vial placed at the end of the extraction system. In addition, the system can be equipped with a coolant device for rapid cooling of the resultant extract.

After preheating, the extraction cell is filled with solvent and kept in the oven at constant temperature and pressure for a user-set static time. The solvent, which contains the extracted analytes, is collected in a vial and the cell is then flushed and purged with nitrogen gas. Together these steps constitute a cycle that can be repeated

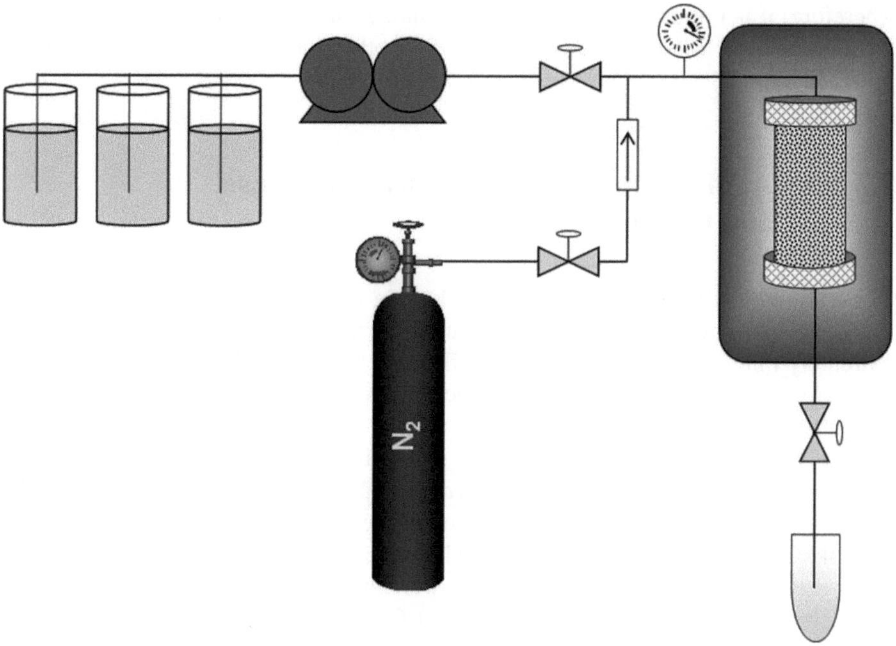

Fig 2.4 Basic scheme of a pressurized liquid extractor

several times, if necessary. The total extraction time is normally 15–45 min, although sometimes a longer extraction time is necessary (Nieto et al. 2010).

The PLE processes can be carried out in both dynamic and static mode, as SFE. The static mode has been the most utilized and is the more frequent when using commercial instruments. The dynamic mode, presumably, could improve the extraction rate by allowing a better contact between the matrix and fresh solvent pumped in a continuous way through the extraction cell and is used mainly with lab-made devices.

Different parameters can be optimized in order to obtain the highest recoveries; the most important are the extraction solvent, temperature, pressure, static time, and number of cycles. Other parameters (e.g., purge time and flush volume) have shown little influence on the final recoveries, so these are usually fixed. Each parameter can be optimized separately or using experimental designs (Nieto et al. 2010). Other aspects that have to be considered are the arrangement of the sample inside the extraction vessel and the collection of the analytes. In PLE the analyte recovery is not as critical as in SFE because, in most of the automatic systems, the solute is recovered in solution in a closed vial minimizing solute losses. Of course, for highly volatile compound recovery, a cooling step should be included.

Compared to SFE, PLE is more versatile in terms of the extraction solvents that can be used and therefore, is more flexible in terms of bioactive compounds to be extracted. In PLE the solvent will be selected depending on the polarity of the target compounds. However, this technique is considered by far less selective than SFE.

Therefore, it is important to keep in mind that, even if the extraction of the bioactives is attained, it would be possible to find other interfering compounds in the obtained extract.

2.3.2.2 Applications to Macroalgae, Microalgae, and Cyanobacteria

Since the early times of PLE, there has been a lot of effort focused in obtaining bioactive compounds from marine sources. Probably the main reason for the extremely important development of PLE-based techniques is the possibility of its automation along with the reduced extraction time and solvents required, as previously mentioned.

PLE has been used, for example, to concentrate carotenoids from *Dunaliella salina* microalgae (Breithaupt 2004; Herrero et al. 2006b). Results showed that the extraction temperature was the factor having the strongest influence (positive) on both the extraction yield and the antioxidant activity of the extracts (determined using the TEAC method). The best yields were obtained with ethanol at the highest extraction temperature and time tested (160°C and 17.5 min). Results of chemical characterization by high-performance liquid chromatography coupled with diode array detection (HPLC-DAD) pointed out that the extracts contained, besides all-trans-β-carotene and isomers, several different minor carotenoids that seemed to contribute strongly to the antioxidant activity of the extracts.

As seen for Dunaliella, ethanol was selected as the optimum solvent to extract antioxidants from *Synechocystis sp.* and *Himanthalia elongata* (Plaza et al. 2010a) and antimicrobial compounds from *Haematococcus pluvialis* (Santoyo et al. 2009). Generally, the best extraction conditions in these applications have been obtained at mild temperatures, around 100°C. The different species produced active extracts in terms of both antioxidant and antimicrobial activities. The obtained pressurized liquid extracts have been chemically characterized by GC–MS (gas chromatography–mass spectrometry) and HPLC-DAD in order to find the compounds responsible for their associated bioactivity. Several fatty acids and volatile compounds with antimicrobial activity have been identified in the different extracts, such as phytol, fucosterol, neophytadiene, or palmitic, palmitoleic, and oleic acids. On the other hand, antioxidant activity has been mainly attributed to the presence of carotenoids and phenolic compounds, among others.

Among cyanobacteria, several species have been screened for their content in bioactive compounds. For example, *Spirulina platensis* has been investigated as a natural source of different functional compounds together with PLE (Herrero et al. 2005a; Herrero et al. 2007). Different factors were optimized such as solvent polarity (hexane, petroleum ether, ethanol, and water), extraction temperature, and time and distribution of the sample inside the extraction cell. By combining PLE with a multistep chemical analysis including preparative thin layer chromatography (TLC), DPPH staining to test the antioxidant activity of the separated bands, and HPLC-DAD to identify the separated compounds in each band, it was possible to determine different chlorophylls and carotenoids in the *S. platensis* PLE extracts, among them, β- and α-carotene, β-cryptoxanthin, asthaxanthin, zeaxanthin, lutein, echinenone, oscillaxanthin, and myxoxanthophyll (Jaime et al. 2005). Among the different solvents tested to maximize the attainment of antioxidant compounds, ethanol was found to be the most appropriate, considering both the high yields and the good antioxidant activities achieved.

Extraction of bioactive compounds from macroalgae has great potential and applications will continue to grow in the following years Plaza et al. 2008. PLE has been used, for example, to isolate carotenoids from brown macroalgae, such as *Eisenia bicyclis* (Shang et al. 2011), *Cystoseira abies-marina,* and *Himanthalia elongata* Plaza 2010. Results showed that ethanol at high temperatures provides high recoveries of fucoxanthin and other oxygenated carotenoids. Besides ethanol, Plaza et al. (2010) tested other solvents including water and hexane, and analyzed both the antioxidant and the antimicrobial activity of extracts, finding the highest bioactivities (antimicrobial and antioxidant) in ethanol extracts. In terms of antimicrobial activity, this seems to be related to the presence of several fatty acids such as linoleic or α-linoleic and certain volatile compounds such as phytol or furanones.

Depending on the selected solvent, PLE could be an ideal technique for the separation of bioactive phenolic compounds. In this sense, marine algae (*Stypocaulon scoparium*, *Porphyra tenera* -nori-, and *Undaria pinnatifida* -wakame-) and freshwater algae (*Spongiochloris spongiosa* and *Anabaena doliolum*) have been studied (López et al. 2011; Onofrejová et al. 2010). In general terms it could be stated that pressurized methanol/water mixtures provided the highest recoveries of phenolic compounds and, therefore, with the highest antioxidant activity. In the work carried out by Onofrejová et al. (2010), a combination of PLE and solid phase extraction (SPE) was used concluding that PLE–SPE extraction and HPLC-ESI-MS (HPLC-electrospray ionization mass spectrometry) analysis of phenolic compounds in selected algal species can detect bioactive phenolic compounds in subnanomolar concentrations (Onofrejová et al. 2010). Using optimized extraction conditions, the average recovery for studied phenols was 96%. In addition, the antioxidant activity analysis indicated that algae PLE–SPE extracts could be used as a source of antioxidants. This extraction procedure could be useful for the rapid extraction of bioactive phenols in various cyanobacteria or algae materials and their food products.

2.3.2.3 Applications to Invertebrates

PLE has also been used as an extraction technique for enrichment of high-value compounds from low-value raw materials. For instance, ethanol at elevated temperature and pressure has been used as a Green solvent for the extraction of astaxanthin from shrimp wastes (consisting of head, shell, and tail of shrimp). Parameters as such temperature and extraction time were shown to have a significant influence on the extraction yield whereas pressure only had a minor effect. Under optimal conditions, astaxanthin yields of around 24 mg/kg shrimp waste were obtained (Quan and Turner 2009).

In addition, PLE has been used as a first step in the isolation of novel natural products from marine sponges. The PLE extraction has been carried out using three independent runs of three solvents, hexane, dichloromethane, and methanol, obtaining the highest percentage yield of the total organic extract in the first extraction. Meroditerpenes, such as isojaspic acid, cacospongin, and jaspquinol, with antimicrobial properties against *Staphylococcus epidermidis* have been isolated from *Cacospongia* marine sponge (Rubio et al. 2007) whereas sesquiterpenes, such as aignopsanoic acid A and methyl aignopsanoate A, which are moderately active against the parasite responsible for sleeping sickness, along with isoaignopsanoic acid have been obtained from *Cacospongia mudofijiensis* (Johnson et al. 2009). In a recent work, the utility of methanol partitioning and PLE to extract 12 bioactive natural products (including isojaspic acid and aignopsanpic acid A) from five marine sponges has been investigated. In this work, an aqueous extraction prior to extractions with hexane, dichloromethane, and methanol, was carried out to remove residual inorganic salts. In all cases, a pressure of 117.2 bar and temperatures equal to 22°C and 100°C were tested. Results obtained in this study showed that the extraction efficiency of PLE was three times higher than the solvent partitioning method (Johnson et al. 2010), demonstrating that it could be an effective high-throughput methodology to process marine sponges for the rapid discovery of novel and bioactive marine natural products.

2.3.2.4 Applications to Marine By-Products

PLE has been used to recover the total lipid content from fish tissue. In addition to the known advantages, less tissue than in conventional methods is required because high extraction efficiency is attainable by PLE. Different solvents have been tested for the extraction of total lipids from fish tissues (salmon and halibut muscle; Dodds et al. 2004). Each sample was extracted at 100°C and 138 bar with two static extraction cycles of 5 min each using chloroform/methanol, hexane/isopropanol, and methylene chloride. The performance of each solvent system was evaluated using gravimetry and GC analysis of FAME prepared from the extracts. Extraction using chloroform/methanol gave the highest recovery of FAs. On the other hand, Isaac et al. (2005) evaluated the reliability and efficiency of PLE for the extraction of lipid content from cod and the effect of sample treatment on the extraction efficiency. Under the best extraction conditions (using 2-propanol:hexane (65:35, v/v) as the first extracting solvent and hexane:diethyl ether (90:10, v/v) as second solvent, at 115°C, 100 bar, and two static cycles of 10 min each) the total lipid yields from homogenized cod (lean fish) and herring (fat fish) muscle exceeded those of conventional batch extractions (modified Jensen methods) by a factor of 10% (Isaac et al. 2005).

Lately, Spiric et al. (2010) carried out statistical studies to compare the fat content, the fatty acid profiles and cholesterol in carp fish muscles obtained by two extraction procedures, modified Soxhlet and PLE. PLE was performed using hexane:isopropanol (60:40 v/v) at 100°C and 103 bar in two static cycles of 10 min each. Results obtained showed that modified Soxhlet gave higher Omega-6-fatty acid content than PLE, although there was no statistically significant difference in the Omega-3-fatty acid between both methods. They concluded that PLE could be considered as an alternative to the modified Soxhlet extraction (Spiric et al. 2010).

2.3.3 Pressurized Hot Water Extraction (PHWE)

Pressurized hot water extraction (PHWE), also known as subcritical water extraction, pressurized low polarity water (PLPW) extraction, or superheated water extraction (SHWE) is a particular use of PLE with water as extracting solvent. Water has many advantages in terms of versatility and environmental impact and can be used in extraction processes to isolate functional ingredients from different raw materials including plants and food wastes. PHWE is based in the use of water at temperatures above its atmospheric boiling point, while keeping it as liquid by applying pressure; under these conditions, physical and chemical properties of water change dramatically, for instance, the dielectric constant of water decreases from around 80 at room temperature (25°C) to around 33 at 200°C, that is, close to a polar organic solvent such as methanol. Moreover, the viscosity and surface tension are both reduced with increasing temperature, whereas diffusivity is increased, altogether enhancing the extraction process in terms of efficiency and speed. In addition, water's solubility parameter is also modified by temperature, thus favoring the solubility a of different type of compound and modifying its selectivity. Water is also the greenest solvent that can be used, perfectly complying with the rules of Green Chemistry and Green Engineering. For more information on the extraction process and the most important parameters to control (temperature, pressure, extraction time, flow rate, selectivity, etc.), readers are referred to a number of book chapters and review articles published covering general (Turner and Ibañez 2011; Teo et al. 2010), food and drug (Herrero et al. 2006), and medicinal plant (Glomohamad et al. 2008; ONg et al. 2006; Wang and Weller 2006) applications.

The conditions at which PHWE will be used will mainly depend on the target compound and the target application. In this sense, pressurized hot water is a very versatile medium in both sub- and supercritical conditions. For instance, supercritical or near-critical conditions can be used for supercritical water oxidation, hydrolysis, and molecular transformations, such as biomass cònversion, whereas in the subcritical region, extraction of health-beneficial compounds can be performed.

2.3.3.1 Instrumentation

The instrumentation for PHWE is essentially the same as for PLE (Fig. 2.4), but basic instrumentation may differ depending on whether a static or dynamic process is used and whether a commercial or homemade setup is employed. Static PHWE consists in a batch process with one or several extraction cycles with replacement of solvent in between. In dynamic PHWE, the extraction solvent is continuously pumped at a selected flow rate through the extraction vessel containing the sample; thus, it requires a slightly more sophisticated high-pressure or HPLC pump to control the water flow rate as well as a pressure restrictor or a micrometering valve rather than a static open/close valve. Among other advantages, dynamic operation may avoid, to some extent, thermal degradation of bioactive compounds inasmuch as water is continuously flowing through the matrix at a certain fluid velocity that improves the efficiency of the extraction avoiding the excessive heating of the sample.

The main advantages of homemade systems compared to commercial setups are the possibility to perform both dynamic and static extractions with fewer operating restrictions, the working temperature range, and the possibility to carry out different processes (reaction, drying, and extraction) by modifying the basic setup. For a complete description of how to build your own system, refer to Turner and Ibañez (2011).

2.3.3.2 Applications to Macroalgae, Microalgae, and Cyanobacteria

As seen for PLE, the use of solvents above their boiling points provides faster extractions; moreover the use of water instead of organic solvents makes the PHWE a greener extraction process. However, optimization of this type of extraction is often empirical and fails to take advantage of the experimental findings in associated fields or theoretical correlations (King 2006). Although PHWE is a promising technique, it needs more research in the marine field; only few applications can be found dealing with the extraction of bioactives from marine sources.

The extraction and characterization of compounds with antioxidant and antimicrobial activity from *Haematococcus pluvialis* microalga in red phase have been studied using PHWE (Rodriguez-Meizoso et al. 2010). Extraction yields achieved were as high as 30% (dry weight) using 200°C as the extraction temperature. Moreover, the extract obtained at 200°C presented the highest antioxidant activity, and temperature did not significantly affect the antimicrobial activity. Chemical composition was determined by HPLC–DAD, HPLC–QqQ-MS, and GC–MS. Short-chain fatty acids turned out to be responsible for the antimicrobial activity, whereas the antioxidant activity was correlated with vitamin E (present exclusively in the 200°C extract), together with simple phenols, carmelization products, and neoformed antioxidants produced by Maillard reaction during the extraction at high temperatures (Plaza et al. 2010b).

Recent evidence showed that PHWE at high temperatures may generate new bioactive (antioxidant) compounds, as seen, for instance in *Haematococcus pluvialis* microalga. Plaza et al. (2010b) studied both antioxidants naturally found in raw samples and those formed during PHWE via Maillard reaction and other chemical events. Samples of different natures such as microalgae (*Chlorella vulgaris*), macroalgae (*Sargassum vulgare*, *Porphyra* spp., *Cystoseira abies-marina*, *Sargassum muticum*, *Undaria pinnatifida,* and *Halopitys incurvus*), and plants (rosemary, thyme, and verbena) were studied (Plaza et al. 2010b). Results demonstrated that the extent of each reaction depends on the chemical composition and nature of the particular sample being extracted; for example, it has been shown how in samples richer in phenolic compounds (i.e., rosemary and thyme), the occurrence of these reactions is more limited. *Undaria pinnatifida* was the sample in which these reactions progressed to a higher extent. These observations were also in agreement with the tremendous difference between the antioxidant capacity shown by the two extracts obtained at low and high temperatures from this alga. Data supported the formation of new antioxidants during PHWE processing.

PLE has been tested to extract phycobiliproteins from the cyanobacteria *Spirulina platensis*. Capillary electrophoresis coupled with mass spectrometry (CE-MS) was used to monitor and optimize the pressurized liquid extraction of proteins from *Spirulina platensis* (Herrero et al. 2005b). The combined use of PLE and CE-MS allowed the attainment of extracts rich in phycobiliproteins in short extraction times (namely, yields of 20% can be obtained in less than 2 h under the optimum process in an automatic way). Different extraction conditions were tested, including time, extraction temperature and pressure, nature of pressurized solvent, distribution of microalga inside the extraction cell, type of packing, and so on.

2.3.3.3 Applications to Invertebrates

Water at high temperatures has been suggested to hydrolyze several marine sources, such as invertebrate viscera wastes (squid or scallop), to produce valuable compounds such as amino acids, fatty acids, organic acids, and so on.

A method for recovery of valuable compounds (amino acids, organic acids, fat and oil phases, soluble proteins, and peptides) from scallop viscera has been proposed by Tavakoli and Yoshida. In this work, reactions were conducted in the temperature range of 170°C (subcritical region)–400°C (supercritical region) and pressure range from 7.92 to 300 bar. Maximum yield of amino acids (0.15 kg/kg dry scallop waste) and organic acids (0.08 kg/kg dry scallop waste) were obtained at 240°C (reaction time of 50 min) and 280°C (reaction time of 40 min), respectively, with glycine the predominant amino acid and pyroglutamic acid the most abundant organic acid (Tavakoli et al. 2006a). These authors have also described the use of subcritical water hydrolysis for the treatment of squid wastes; they described a two-step process in which the first step was optimized to recover amino acids (maximum yield 0.1031 kg/kg dry entrails) and extracting fat and oil (using 200°C, 5 min) and the second step (performed at 240°C, 50 min) was used for processing the solid phase, achieving the highest amount of organic acids (maximum yield 0.055 kg/kg dry entrails; Yoshida and Tavakoli 2004). Later, they applied subcritical water hydrolysis to produce oil and fat phases and hydrolyzed the triglycerides to free fatty acids. Among fatty acids, DHA (0.103 kg/kg oil) and EPA (0.062 kg/kg oil) were obtained at 240°C (20 min) and 200°C (40 min), respectively. From the oil and fat phases, it was possible to carry out the transesterification of fatty acid to the corresponding fatty acid methyl ester, which are important compounds in biodiesel fuel production (Tavakoli et al. 2006b).

In another interesting work, Uddin et al. (2010) compared the production of valuable compounds, mainly amino acids, by subcritical water hydrolysis from raw and deoiled squid viscera (Uddin et al. 2010). Deoiled viscera were obtained using SC–CO_2 at 45°C and 250 bar (Uddin et al. 2009). The highest yield of amino acids was found at low (180°C) and high (280°C) temperatures from raw and deoiled squid viscera, respectively, being about 1.5 times lower from raw viscera than from deoiled viscera. These results demonstrate the possibility to integrate green technologies based on supercritical and subcritical technologies to produce valuable compounds from low-cost raw materials.

2.3.3.4 Applications to Marine By-Products

Sub- and supercritical water technologies have been used to recover useful substances from fish wastes. For instance, Yoshida et al. (1999) studied the production of useful compounds by subcritical water hydrolysis from fish meats. Amino acids such as cysteine, alanine, glycine, and leucine (from 0.004 to 0.024 kg/kg dry meat) were produced at 270°C and 55.1 bar. The production of other useful organic acids (lactic acid, pyroglutamic acid) was also observed. In addition, the oil phase extracted with hexane contained EPA and DHA (Yoshida et al. 1999). Following this work, the authors proposed a simplified kinetic model to explain the hydrolysis reaction under subcritical and supercritical water conditions to generate major products such as oil, amino acids, or organic acids from fish meat (Yoshida et al. 2003). Kang et al. (2001) investigated the recovery of amino acids from white croaker fish entrails by sub- and supercritical water reaction using semibatch and batch reactors (Kang et al. 2001). They observed two consecutive reactions in the treatment: hydrolysis of proteins to amino acids, and decomposition of amino acids to other products such as organic acids. Therefore, a proper control of sub- and supercritical conditions was necessary to obtain the target compounds. The maximum yield of total amino acids (137 mg/g dry fish) was obtained under subcritical conditions (250°C, 40 bar) at 60 min reaction time by using a batch reactor. Under supercritical conditions (380°C, 450 bar) the yield decreased due to the rapid decomposition compared to the production rate of amino acids.

2.3.4 Ultrasound Assisted Extraction (UAE) and Microwave-Assisted Extraction (MAE)

The last two green techniques discussed in this chapter are ultrasound-assisted extraction and microwave-assisted extraction. UAE uses acoustic cavitation to cause disruption of cell walls, reduction of particle size, and enhancement of contact between the solvent and the target compounds. MAE uses microwave radiation that causes motion of polar molecules and rotation of dipoles to heat solvents and to promote transfer of target compounds from the sample matrix into the solvent (Ying et al. 2011).

Both methods are very versatile due to the possibility of using several solvents of different polarities; in fact, both can couple extraction and reaction at the same time. Moreover, both techniques allow fast extractions, which is a key point to avoid degradation of labile compounds. Both techniques are fast, use low amounts of solvents, and are cost effective, therefore the development of both methods could represent a key point in Sustainable Development.

2.3.4.1 Instrumentation for Ultrasound-Assisted Extraction

In UAE only a small portion of the ultrasound spectrum is used, namely power ultrasound. Power ultrasound, having frequencies between 20 kHz and 100 MHz, are now well known to have significant effects on the rate of various physical and chemical processes such as cleaning, degassing, solubilization, homogenization, emulsification, sieving, filtration, and crystallization. Power ultrasound involves the mechanical and chemical effects of cavitation. When a liquid is irradiated by ultrasound, microbubbles form, grow, and oscillate extremely quickly, and eventually collapse powerfully if the acoustic pressure is high enough. These collapses, occurring near a solid surface, generate microjets and shock waves that result in cleaning, erosion, and fragmentation of the surface. Microdischarges due to high electrical fields generated by deformation and fragmentation of the bubbles and the formation of radicals could be responsible for the observed chemical effects (Ötles 2009).

The two ultrasound apparatus most commonly used for extraction are the ultrasonic cleaning bath and the more powerful probe system. For small extraction volumes, an ultrasound horn with the tip submerged in the fluid can be sufficient. Large volumes of fluids have to be sonicated in an ultrasound bath or in continuous or recycled-flow sonoreactors (Ötles 2009). Although most of the research effort in UAE has concentrated on ultrasound itself, some studies have also examined the coupling between ultrasound and other techniques. When combined with supercritical fluid extraction, UAE enhances mass transfer of the species of interest from the solid phase to the extraction solvent (Hu et al. 2007).

UAE of bioactive compounds is increasingly efficient at directly transferring knowledge into technology for commercial development. This novel process can extract analytes under a concentrated form (low volumes of solvent) and free from any contaminants or artifacts. The new systems developed so far clearly demonstrated the advantages of UAE in terms of yield, selectivity, operating time, energy input, and even preservation of thermolabile compounds (Ötles 2009).

2.3.4.1b Instrumentation for Microwave-Assisted Extraction

Microwave heating for sample digestion in elemental analysis (atomic absorbance spectroscopy, AAS; inductively coupled plasma-mass spectrometry, ICP-MS) has been a recognized technique for decades. The application of microwave irradiation to the extraction of compounds from biological samples has been a more recent development. MAE was first described in 1986 (Ganzler et al. 1986). In this description, a domestic microwave was used as a sample preparation technique prior to chromatography. A variety of compounds from seeds, foods, and feeds were extracted; the process was found to be more efficient than classical Soxhlet or LLE.

In general, samples for MAE are homogenized and mixed with a solvent and the suspension irradiated at greater than 2,000 MHz for short periods of time. Heating is usually repeated several times with periods of cooling in between to prevent boiling. Efficiencies seen with this technique approach those of classical Soxhlet extraction but can be performed much more rapidly. Further modifications of the technique mimic sweep codistillation or steam distillation with air, sweeping the surface of the heated sample and being collected by a condenser that protrudes through the oven housing (Worsfold et al. 2005). Techniques have also been developed using closed-system microwave heating, the most used nowadays, which lets the mixture of sample and solvent increase the pressure due to reaching the boiling point; therefore, the extraction takes place in a similar condition as PLE (see above).

There are two approaches of applying microwave energy, namely, the mere bulk heating of a mixture through the use of absorbing containers and solvents (oven-type apparatus with samples in closed-vessel conditions), and the more refined, albeit demanding, selective heating of the target materials. These two approaches led to the development of two main types of laboratory extraction instruments.

By using selective heating it is possible to operate safely and much more efficiently under open-vessel conditions (i.e., atmospheric pressure); under these conditions, operating temperatures remain low. Once fitted with an appropriate reflux column, the system is autocontrolled in terms of temperature Ötles 2009. A technology combining microwave and Soxhlet extraction was designed in 1998 (Garcia-Ayuso et al. 1998). This extraction technique, called microwave-assisted Soxhlet extraction, uses two sources of energy, namely microwaves, applied on the extraction chamber of a modified Soxhlet, and electrical heating applied on the distillation flask. The combination of both techniques, UAE and MAE, is also possible (see Fig. 2.5) and it was developed as a modification of microwave-assisted Soxhlet extraction.

2.3.4.2 Applications to Macroalgae, Microalgae, and Cyanobacteria

Despite the possibilities that MAE and UAE techniques offer, there are not many applications dealing with the isolation of bioactive compounds from marine resources to this date.

Fig. 2.5 Apparatus for simultaneous US/MW irradiation and detail of immersion horn (25 kHz) inserted in the cavitating tube (19 kHz); the inset shows the latter as seen from the top (Reprinted from Cravotto et al. (2008) with permission from Elsevier)

Different species of the microalgal genus Dunaliella have been treated with these techniques, mainly for carotenoid extraction. The performance of MAE in *Dunaliella tertiolecta* (chlorophyte) and *Cylindrotheca closterium* (diatom bacillariophyte) was studied in terms of pigment extraction (Pasquet et al. 2011). The process performed on *Dunaliella tertiolecta* led to rapid pigment extraction, mainly because of the absence of frustule in microalgae cells thus allowing immediate solvent penetration into the matrix. In contrast, presence of the frustule in the diatom *Cylindrotheca closterium* constituted a mechanical barrier to pigment extraction by MAE. On the other hand, MAE was identified as the best extraction process for Cylindrotheca pigments as it combined rapidity, reproducibility, homogeneous heating, and high extraction yields. Compared to conventional processes, MAE offers the advantage of an homogeneous thermoregulation of the medium, as no heat transfer is required to heat cells located in the center of the flask. Microalgal organites are also homogeneously heated, allowing synchronous pigment extraction, whatever their subcellular localization (cytoplasmic or chloroplastic). It was also observed that the extraction process had no effect on cell integrity and shape.

In order to compare two green extraction methods, SFE and UAE, *Dunaliella salina* was subjected to both and the pigment extraction was compared (Macías-Sánchez et al. 2009). Results indicated that the supercritical CO_2 extraction process was comparable to the ultrasound-assisted extraction when methanol was used as solvent, but not when using N,N′-dimethylformamide (DMF). Extraction yields obtained for carotenoids and chlorophylls using DMF were higher than those obtained with methanol. Results demonstrated that DMF was more selective than methanol in the recovery of carotenoids from chlorophylls. This behavior can be attributed to the fact that chlorophyll is heterogenically bound to other compounds in the chloroplast and at least two or even three fractions of chlorophyll exist in the chloroplast.

Oil recovery from different microalgae (*Chaetoceros gracilis, Chaetoceros mulleri, Chlorella vulgaris, Dunaliella* sp., *Isochrysis* sp., *Nannochloropsis oculata, Tetraselmis* sp., *Tetraselmis chui, Tetraselmis tetrathele,* and *Thalassiosira weissflogii*) has been recently studied using UAE (Araujo et al. 2011). Regarding oil production, the choice of salinity in the culture media was an important parameter and reflected directly on the suitability of the microalgae towards oil production. The diatoms *C. vulgaris* and *C. gracilis* were the most suitable microalgae for large-scale oil production. Comparing these results with the extraction of lipids of Chaetoceros by SFE (Mendiola et al. 2007b) revealed that the use of UAE is more suitable than SFE to obtain extracts from diatoms; this fact can be attributed to the better penetration of the solvent in the cells with frustule because of cavitation.

Cravotto et al. (2008) developed a method to work simultaneously with MAE and UAE in order to extract oil from soy and from the microalga *Crypthecodinium cohnii* (Cravotto et al. 2008). This microalga is very rich in DHA but presents a very tough cell wall. Extraction times were reduced up to tenfold and yields increased by 50–500% in comparison with conventional extraction methods. GC analyses showed only slight or negligible differences in methyl ester profiles of oils extracted under high-intensity MAE/UAE and Soxhlet. It should also be pointed out that techniques using US or combined US/MW irradiation should be well suited for other processes, such as two-step extraction and transesterification for the production of biofuels.

As already mentioned in Sect. 2.3.1.2, the combination of SFE and UAE has been studied to isolate phenolic compounds from algae (Klejdus et al. 2010).

2.3.4.3 Applications to Marine By-Products

Only one work in which MAE has been employed as an extraction technique from fish wastes has been reported in the literature. Batista et al. (2001) used this technique to extract lipids from mackerel fillet and cod liver for the determination of the fatty acid profile composition (Batista et al. 2001). MAE extraction was carried out using a solvent mixture of ethyl acetate:cyclohexane, 1:1 v/v. After lipid extraction, the esterification was performed using trimethylsulfonium hydroxide and the FAME were determined by GC-FID. The lipid content extracted from mackerel fillet and cod liver was 5.6% ± 0.4% and 62.6% ± 3.1%, respectively. Results obtained in this work demonstrated that MAE can be a good alternative to conventional methods for lipid extraction.

2.4 From Analytical to Industrial Scale: Multiple Integrated Processes as Future Trends

As previously mentioned, there has been an enormous interest in the field of pressurized fluids, mainly when using safer, less harmful, and green solvents. The versatility of pressurized solvents is due to their very different physicochemical properties (density, diffusivity, viscosity, dielectric constant) and the possibility of controlling these parameters and, therefore, their solvating power and selectivity, by changing the pressure and temperature of the extraction system.

Most applications of the green extraction processes reported in this chapter have been carried out at analytical scale, although some of them have also been tested at pilot scale and, to a lesser extent, at industrial scale. Readers may have realized that the processes that have been studied at pilot or industrial scale refer only to the use of SFE; the reason is simple: SFE is a more mature process and industrial plants can be found worldwide for multiple applications. Nevertheless, it is important to realize that, when talking about pressurized fluid systems, the basic design of the extraction equipment is quite similar, although the solvents are basically different (and with different physicochemical properties). Systems used to carry out processes such as SFE, PLE, and PHWE consist of a solvent supply, a pump for transporting the solvent, a heater for heating the solvent, a pressure vessel where the extraction occurs, a pressure control in the system, and a collection device for the extract (see Figs. 2.3 and 2.4).

Lab scale studies can be used to determine factors required for scale-up. Moreover, knowledge of phase equilibria, mass transfer rate, and solubility data may also be necessary to scale up the extraction process and equipment. For a more in-depth discussion on the design and scale-up of pressurized fluid extractors, readers are referred to Pronyk and Mazza (2009); in this review, aspects such as modeling of the extraction process (del Valle and de la Fuente 2006), mass transfer effects (internal and external), and effects of processing parameters (temperature, pressure, flow rate, material properties, and length to diameter ratio) on the extraction process, that is, in mass transfer and solubility of the target compounds, and how this affects the design of the extraction system, are carefully considered. Readers are also referred to other publications related to the scale-up of SFE (Meireles 2003; Berna et al. 2002) and PHWE (Lagadec et al. 2000) for further information.

As for MAE and UAE, few reports can be found considering pilot scale units (Terigar et al. 2011; Boonkird et al. 2008) undoubtedly, more research is needed in this field to demonstrate the usefulness of these techniques at larger scale.

Therefore, considering the interesting applications discussed in the present chapter for bioactive compound extraction from marine resources, it is expected that once both viability and feasibility of the process at large scale are studied, not only SFE but also other pressurized fluids such as ethanol and water, could be used for both pilot and industrial process development. We should consider that, as for CO_2, water and ethanol have important advantages in terms of the low environmental impact of the solvent and, therefore, their use might be an alternative to more conventional extraction processes, using less green solvents. Undoubtedly, other aspects of the process should also be considered such as the total amount of sample, other chemicals needed, energy usage, and the need for solvent recycling and storage. However, in terms of sustainability it is a good point to be able to switch to a more Greener solvent such as water or ethanol.

As for future trends, the idea of developing multiple integrated processes, able to face some of the challenges in our society such as environmental impact, sustainability, energy preservation, and health is suggested. Previously, King and Srinivas (2009) and Turner and Ibañez (2011) discussed this proposal. The idea of building a multiunit operations system with the possibility of using different fluids can provide unique characteristics and advantages to develop ad hoc platforms tuned to the different processes that want to be optimized. Main characteristics of this "green" processing platform are: it should work with environmentally benign solvents such as liquefied or supercritical CO_2, for nonpolar to moderately polar solutes, and with pressurized hot water (between its boiling and critical points) for a wider range of polarities, considering also the use of ethanol as cosolvent together with water or carbon dioxide. As mentioned previously, the use of pressurized hot water could be exploited to include extraction of bioactive compounds from natural raw materials such as marine resources, reaction of targeted substrates (oxidation, hydrolysis, etc.), and biomass conversion for renewable fuels.

Several examples can be found in the literature about integrated processes that may favor the extraction and purification of bioactives; some of them already deal with some green processes to extract bioactive compounds from marine resources (Liau et al. 2010) whereas others can be used as a base for converting the reported processes to more green, sustainable, and efficient ones (Siriwardhana et al. 2008; Athukorala et al. 2006). In the first approach (Liau et al. 2010), a process was studied considering SFE of lipids and carotenoids from the microalgal species of *Nannochloropsis oculata* and supercritical antisolvent precipitation of a carotenoid-rich solution. In this approach, both processes were considered independently but it can be easily inferred that the development of a multiple integrated process can allow purification of carotenoids in a cleaner and efficient way. Other processes that can be easily included in the green platform are those dealing with enzymatic hydrolysis and extraction; the pioneer work of Turner et al. (2006) demonstrated the viability of a process combining enzymatic hydrolysis in hot water, using a thermostable β-glucosidase to catalyze hydrolysis of quercetin glucosides in onion waste, plus extraction with water at high temperatures. The developed process was preferred over more conventional extraction/hydrolysis processes based on methanol extraction and hydrochloric acid hydrolysis at 80°C, regarding primary energy consumption and global warming potential (Lindahl et al. 2010). Following a similar approach, integrated treatments considering enzymatic hydrolysis plus extraction carried out on different algae, such as *Hizikia fusiformis* (164), *Ecklonia cava,* and other brown algae (Athukorala et al. 2006; Moreda-Piñeiro et al. 2007; Turner et al. 2006; Heo et al. 2005) could be optimized in a more efficient and Green way using the above-mentioned concepts and ideas. Moreover, studies have demonstrated that processes such as enzymatic hydrolysis are accelerated under pressure conditions, thus giving even stronger support to the possibility of improving the processes through the use of integrated pressurized fluid technologies (Moreda-Piñeiro et al. 2007).

2.5 Conclusions

Different aspects have been addressed in this chapter. First of all, we attempted to demonstrate the important possibilities offered by marine resources (such as micro- and macroalgae, cyanobacteria, invertebrates, and marine by-products) as a source of natural bioactive compounds with health benefits, with potential use in the food industry. The new integrated approach consisting of a screening of extraction conditions and an in vitro measuring of functional activities together with an exhaustive chemical characterization will provide us with a new tool to discover new bioactive compounds and to help the further design of processes to obtain such products in the most Green, sustainable, and efficient way, complying with the rules of green chemistry and green engineering. In this chapter we discuss and present some selected applications of the extraction of target compounds with, among others, antioxidant, antimicrobial, and antiproliferative activities from different sustainable marine sources. Either the use of microalgae as bioreactors to produce and enrich target compounds, the gathering and reuse of low-cost by-products from the fish industry, or the collection or growing of macroalgae in nonprofitable lands, is a good start for a process tending to recover new and valuable compounds from these sources in a sustainable, economical, Green, and efficient way. We have shown many possibilities at laboratory scale, which are the basis of the knowledge, and that can be used as the first step for a bigger production scale.

Undoubtedly, the replacement of environmentally burdensome solvents such as acetonitrile, methanol, dichloromethane, and toluene, long used in such extraction processes, for more green solvents such as CO_2, ethanol, and water is one of the goals and one of the proposals that we are offering to the reader in this chapter. Some technologies are mature enough to be used at large scale; others require more study and development but, in any case, it is advisable that new steps be taken to help build a more rational use of our natural resources. The possibility, mentioned in this chapter as a future trend, to build new platforms able, in a sustainable way, to run integrated processes including pretreatments, extractions, reactions, and transformations in a more integrated way is one of our main goals and might help all of us to build a better future.

References

Bellisle, F., A.T. Diplock, G. Hornstra, B. Koletzko, M. Roberfroid, S. Salminen, et al. 1998. Functional food science in Europe. *British Journal of Nutrition* 80: 1–193.

Diplock, A.T., P.J. Agget, M. Ashwell, F. Bornet, E.B. Fern, and M.B. Roberfroid. 1999. Scientific concepts of functional foods in Europe: consensus document. *British Journal of Nutrition* 81: S1–S27.

Plaza, M., M. Herrero, A. Cifuentes, and E. Ibañez. 2009. Innovative natural functional ingredients from Microalgae. *Journal of Agricultural and Food Chemistry* 57: 7159–7170.

Plaza, M., A. Cifuentes, and E. Ibañez. 2008. In the search of new functional food ingredients from algae. *Trends in Food Science & Technology* 19: 31–39.

Kadam, S.U., and P. Prabhasankar. 2010. Marine foods as functional ingredients in bakery and pasta products. *Food Research International* 43: 1975–1980.
Kim, S.K., and I. Wijesekara. 2010. Development and biological activities of marine-derived bioactive peptides: a review. *Journal of Functional Foods* 2: 1–9.
Wollgast, J., and E. Anklam. 2000. Review on polyphenols in Theobroma cacao: changes in composition during the manufacture of chocolate and methodology for identification and quantification. *Food Research International* 33: 423–447.
Madhavi, D.V., S.S. Despande, and D.K. Salunkhe. 1996. *Food antioxidants*. New York: Marcel Dekker.
Bravo, L. 1998. Polyphenols: chemistry, dietary sources, metabolism, and nutritional significance. *Nutrition Reviews* 56: 317–333.
Wijesekara, I., N.Y. Yoon, and S.K. Kim. 2010. Phlorotannins from *Ecklonia cava* (Phaeophyceae): biological activities and potential health benefits. *BioFactors* 36: 408–414.
Nagayama, K., Y. Iwamura, T. Shibata, I. Hirayama, and T. Nakamura. 2002. Bactericidal activity of phlorotannins from the brown alga *Ecklonia kurome*. *Journal of Antimicrobial Chemotherapy* 50: 889–893.
Kang, K., Y. Park, H.J. Hwang, S.H. Kim, J.G. Lee, and H.C. Shin. 2003. Antioxidative properties of brown algae polyphenolics and their perspectives as chemopreventive agent against vascular risk factors. *Archives of Pharmacal Research* 26: 286–293.
Artan, M., Y. Li, F. Karadeniz, S.H. Lee, M.M. Kim, and S.K. Kim. 2008. Anti-HIV-1 activity of phloroglucinol derivative, 6, 6′-bieckol, from *Ecklonia cava*. *Bioorganic & Medicinal Chemistry* 16: 7921–7926.
Kong, C.S., J.A. Kim, N.Y. Yoon, and S.K. Kim. 2009. Induction of apoptosis by phloroglucinol derivative from *Ecklonia cava* in MCF-7 human breast cancer cells. *Food and Chemical Toxicology* 47: 1653–1658.
Eide, I., S. Myklestad, and S. Melson. 1980. Longterm uptake and release of heavy metals by *Ascophyllum nodosum* (L.). *Environmental Pollution* 23: 19–28.
Lee, S.H., Y. Li, F. Karadeniz, M.M. Kim, and S.K. Kim. 2009. α-Glycosidase and α-amylase inhibitory activities of phloroglucinal derivatives from edible marine brown alga, *Ecklonia cava*. *Journal of the Science of Food and Agriculture* 89: 1552–1558.
Jung, H.A., S.K. Hyun, H.R. Kim, and J.S. Choi. 2006. Angiotensin-converting enzyme I inhibitory activity of phlorotannins from *Ecklonia stolonifera*. *Fisheries Science* 72: 1292–1299.
Yoon, N.Y., S.H. Lee, Y. Li, and S.K. Kim. 2009. Phlorotannins from *Ishige okamurae* and their acetyl- and butyry-lcholinesterase inhibitory effects. *Journal of Functional Foods* 1: 331–335.
Li, Y., Z.J. Qian, B.M. Ryu, S.H. Lee, M.M. Kim, and S.K. Kim. 2009. Chemical components and its antioxidant properties in vitro: an edible marine brown alga, *Ecklonia cava*. *Bioorganic & Medicinal Chemistry* 17: 1963–1973.
Duan, X.J., W.W. Zhang, X.M. Li, and B.G. Wang. 2006. Evaluation of antioxidant property of extract and fractions obtained from a red alga, Polysiphonia urceolata. *Food Chemistry* 95: 37–43.
De Spirt, S., K. Lutter, and W. Stahl. 2010. Carotenoids in photooxidative stress. *Current Nutrition & Food Science* 6: 36–43.
Silberstein, J.L., and J.K. Parsons. 2010. Evidence-based principles of bladder cancer and diet. *Current Nutrition & Food Science* 6: 2–12.
Riccioni, G., B. Mancini, E. Di Ilio, T. Bucciarelli, and N. D'Orazio. 2008. Protective effect of lycopene in cardiovascular disease. *European Review for Medical and Pharmacological Sciences* 12: 183–190.
Snodderly, M.D. 1995. Evidence for protection against age-related macular degeneration by carotenoids and antioxidant vitamins. *American Journal of Clinical Nutrition* 62: S1448–S1461.
Zhu, Y.H., and J.G. Jiang. 2008. Continuous cultivation of *Dunaliella salina* in photobioreactor for the production of β-carotene. *European Food Research and Technology* 227: 953–959.
Kotate-Nara, E., M. Kushiro, H. Zhang, T. Sagawara, K. Miyashita, and A. Nagao. 2001. Carotenoids affect proliferation of human prostate cancer cells. *Journal of Nutrition* 131: 3303–3306.

Hosokawa, M., S. Wanezaki, K. Miyauchi, H. Kurihara, H. Kohno, J. Kawabata, et al. 1999. Apoptosis inducing effect of fucoxanthin on human leukemia cell HL-60. *Food Science and Technology Research* 5: 243–246.

Shiratori, K., K. Ohgami, I. Ilieva, X.H. Jin, Y. Koyama, K. Miyashita, et al. 2005. Effects of fucoxanthin on lipopolysaccaride-induced inflammation in vitro and in vivo. *Experimental Eye Research* 81: 422–428.

Maeda, H., M. Hosokawa, T. Sashima, and K. Miyashita. 2007. Dietary combination of fucoxanthin and fish oil attenuates the weight gain of white adipose tissue and decreases blood glucose in obese/diabetic KK-Ay mice. *Journal of Agricultural and Food Chemistry* 55: 7701–7706.

Sachindra, N.M., E. Sato, H. Maeda, M. Hosokawa, Y. Niwano, M. Kohno, et al. 2007. Radical scavenging and singlet oxygen quenching activity of marine carotenoid fucoxanthin and its metabolites. *Journal of Agricultural and Food Chemistry* 55: 8516–8522.

Yuan, J.P., and F. Chen. 2000. Purification of trans-astaxanthin from a high-yielding astaxanthin ester-producing strain of the alga *Haematococcus pluvialis*. *Food Chemistry* 68: 443–448.

Higuera-Ciapara, I., L. Felix-Valenzuela, and F.M. Goycoolea. 2006. Astaxanthin: A review of its chemistry and applications. *Critical Reviews in Food Science and Nutrition* 46: 185–196.

Bruno, A., C. Rossi, G. Marcolongo, A. Di Lena, A. Venzo, C.P. Berrie, et al. 2005. Selective in vivo anti-inflammatory action of the galactolipid monogalactosyldiacylglycerol. *European Journal of Pharmacology* 524: 159–168.

Larsen, E., A. Kharazmi, L.P. Christensen, and S.B. Christensen. 2003. An antiinflammatory galactolipid from rose hip (*Rosa canina*) that inhibits chemotaxis of human peripheral blood neutrophils in vitro. *Journal of Natural Products* 66: 994–995.

Calzolari, I., S. Fumagalli, N. Marchionni, and M. Di Bari. 2009. Polyunsaturated fatty acids and cardiovascular disease. *Current Pharmaceutical Design* 15: 4149–4156.

Schuchardt, J.P., M. Huss, M. Stauss-Grabo, and A. Hahn. 2010. Significance of long-chain polyunsaturated fatty acids (PUFAs) for the development and behaviour of children. *European Journal of Pediatrics* 169: 149–164.

Zuliani, G., M. Galvani, E. Leitersdorf, S. Volpato, M. Cavelieri, and R. Fellin. 2009. The role of polyunsaturated fatty acids (PUFA) in the treatment of dyslipidemias. *Current Pharmaceutical Design* 15: 4173–4185.

Sahena, F., I.S.M. Zaidul, S. Jinap, N. Saari, H.A. Jahurul, K.A. Abbas, et al. 2009. PUFAs in fish: extraction, fractionation, importance in health. *Comprehensive Reviews in Food Science and Food Safety* 8: 59–74.

Wu, T.H., and P.J. Bechtel. 2008. Salmon by-product storage and oil extraction. *Food Chemistry* 111: 868–871.

Juárez, M., A. Juárez, N. Aldai, C. Avilés, and O. Polvillo. 2010. Validation of a gas-liquid chromatographic method for analysing samples rich in long chain n-3 polyunsaturated fatty acids: application to seafood. *Journal of Food Composition and Analysis* 23: 665–670.

Francavilla, M., P. Trotta, and R. Luque. 2010. Phytosterols from *Dunaliella tertiolecta* and *Dunaliella salina*: a potentially novel industrial application. *Bioresource Technology* 101: 4144–4150.

Cardozo, K.H.M., T. Guaratini, M.P. Barros, V.R. Falcão, A.P. Tonon, N.P. Lopes, et al. 2007. Metabolites from algae with economical impact. *Comparative Biochemistry and Physiology Part C: Toxicology and Pharmacology* 146: 60–78.

Kanazawa, A. 2001. Sterols in marine invertebrates. *Fisheries Science* 67: 997–1007.

Li, B., F. Lu, X. Wei, and R. Zhao. 2008. Fucoidan: Structure and bioactivity. *Molecules* 13: 1671–1695.

Qi, H., Q. Zhang, T. Zhao, R. Chen, H. Zhang, X. Niu, et al. 2005. Antioxidant activity of different sulfate content derivatives of polysaccharide extracted from *Ulva pertusa* (Chlorophyta) invitro. *International Journal of Biological Macromolecules* 37: 195–199.

Ngo, D.H., I. Wijesekara, T.S. Vo, Q.V. Ta, and S.V. Kim. 2011. Marine food-derived functional ingredients as potential antioxidants in the food industry: an overview. *Food Research International* 44: 523–529.

Liu, B., W.S. Liu, B.Q. Han, and Y.Y. Sun. 2007. Antidiabetic effects of chitooligosaccharides on pancreatic islet cells in streptozotocin-induced diabetic rats. *World Journal of Gastroenterology* 13: 725–731.

Liao, F.H., M.J. Shieh, N.C. Chang, and Y.C. Chien. 2007. Chitosan supplementation lowers serum lipids and maintains normal calcium, magnesium, and iron status in hyperlipidemic patients. *Nutrition Research* 27: 146–151.

Muzzarelli, R.A.A., P. Morganti, G. Morganti, P. Palombo, M. Palombo, G. Biagini, et al. 2007. Chitin nanofibrils/chitosan glycolate composites as wound medicaments. *Carbohydrate Polymers* 70: 274–284.

Bhat, B.V., N.W. Gaikwad, and K.M. Madyastha. 1998. Hepatoprotective effect of C-phycocyanin: protection for carbon tetrachloride and R-(+)-pulegone-mediated hepatotoxicty in rats. *Biochemical and Biophysical Research Communications* 249: 428–431.

Romay, C.H., R. Gonzalez, N. Ledón, D. Remirez, and V. Rimbau. 2003. C-Phycocyanin; Abiliprotein with antioxidante, anti-inflammatory and neuroprotective effects. *Current Protein & Peptide Science* 4: 207–216.

Bhat, B.V., and K.M. Madyastha. 2000. C-Phycocyanin: a potent peroxyl radical scavenger in vivo and in vitro. *Biochemical and Biophysical Research Communications* 275: 20–25.

Moraes, C.C., J.F. De Medeiros Burkert, and S.J. Kalil. 2010. C-phycocyanin extraction process for large-scale use. *Journal of Food Biochemistry* 34: 133–148.

Patil, G., S. Chethana, M.C. Madhusudhan, and K.S.M.S. Raghavarao. 2008. Fractionation and purification of the phycobiliproteins from *Spirulina platensis*. *Bioresource Technology* 99: 7393–7396.

Byun, H.G., J.K. Lee, H.G. Park, J.K. Jeon, and S.K. Kim. 2009. Antioxidant peptides isolated from the marine rotifer, *Brachionus rotundiformis*. *Process Biochemistry* 44: 842–846.

Majors, R., and D. Raynie. 2011. The greening of the chromatography laboratory. *LCGC Europe.* 24: 72–78.

Anastas, P.T., and J.C. Warner. 1998. *Green chemistry: theory and practice*. New York: Oxford University Press.

Anastas, P.T., and J.B. Zimmerman. 2003. Design through the twelve principles of green engineering. *Environmental Science and Technology* 37: 94A–101A.

Hosikian, A., Lim, S., Halim, R., and M.K. Danquah. 2010. Chlorophyll extraction from Microalgae: a Review on the process engineering aspects. International Journal of Chemical Engineering. Article ID 391632, 11 pages. doi:10.1155/2010/391632.

Mendiola, J.A., M. Herrero, A. Cifuentes, and E. Ibáñez. 2007a. Use of compressed fluids for sample preparation: food applications. *Journal of Chromatography. A* 1152: 234–246.

Herrero, M., A. Cifuentes, and E. Ibáñez. 2006a. Sub- and supercritical fluid extraction of functional ingredients from different natural sources: plants, food-by-products, algae and microalgae: a review. *Food Chemistry* 98: 136–148.

Björklund, E., C. Sparr-Eskilsson, W. Paul, T. Alan, and P. Colin. 2005. EXTRACTION: supercritical Fluid Extraction. In *Encyclopedia of Analytical Science*, ed. P. Worsfold, A. Townshend, and C. Poole, 597–604. Oxford: Elsevier.

El Hattab, M., G. Culioli, L. Piovetti, S.E. Chitour, and R.J. Valls. 2007. Comparison of various extraction methods for identification and determination of volatile metabolites from the brown alga *Dictyopteris membranacea*. *Journal of Chromatography. A* 1143: 1–7.

Mendiola, J.A., S. Santoyo, A. Cifuentes, G. Reglero, E. Ibáñez, and F.J. Señoráns. 2008a. Antimicrobial activity of sub- and supercritical CO2 extracts of the green alga *Dunaliella salina*. *Journal of Food Protection* 71: 2138–2143.

Cheung, P.C.K. 1999. Temperature and pressure effects on supercritical carbon dioxide extraction of n – 3 fatty acids from red seaweed. *Food Chemistry* 65: 399–403.

Qiuhui, H. 1999. Supercritical carbon dioxide extraction of *Spirulina platensis* component and removing the stench. *Journal of Agricultural and Food Chemistry* 47: 2705–2706.

Mendiola, J.A., D. García-Martínez, F.J. Rupérez, P.J. Martín-Álvarez, G. Reglero, A. Cifuentes, et al. 2008b. Enrichment of vitamin E from *Spirulina platensis* microalga by SFE. *Journal of Supercritical Fluids* 43: 484–489.

Mendes, R.L., H.L. Fernandes, J.P. Coelho, E.C. Reis, J.M.S. Cabral, J.M. Novais, et al. 1995. Supercritical CO_2 extraction of carotenoids and other lipids from *Chlorella vulgaris*. *Food Chemistry* 53: 99–103.

Mendiola, J.A., F.R. Marín, S.F. Hernández, B.O. Arredondo, F.J. Señoráns, E. Ibañez, et al. 2008c. Characterization via liquid chromatography coupled to diode array detector and tandem mass spectrometry of supercritical fluid antioxidant extracts of *Spirulina platensis* microalga. *Journal of Separation Science* 28: 1031–1038.

Mendes, R.L., B.P. Nobre, M.T. Cardoso, A.P. Pereira, and A.F. Palabra. 2003. Supercritical carbon dioxide extraction of compounds with pharmaceutical importance from microalgae. *Inorganica Chimica Acta* 356: 328–334.

Klejdus, B., L. Lojková, M. Plaza, M. Šnóblová, and D. Štěrbová. 2010. Hyphenated technique for the extraction and determination of isoflavones in algae: ultrasound-assisted supercritical fluid extraction followed by fast chromatography with tandem mass spectrometry. *Journal of Chromatography. A* 1217(51): 7956–7965.

Wang, H.M., J.L. Pan, C.Y. Chen, C.C. Chiu, M.H. Yang, H.W. Chang, et al. 2010. Identification of anti-lung cancer extract from *Chlorella vulgaris C-C* by antioxidant property using supercritical carbon dioxide extraction. *Process Biochemistry* 45: 1865–1872.

Félix-Valenzuela, L., I. Higuera-Ciaparai, and F. Goycoolea-Valencia. 2001. Supercritical CO_2/ethanol extraction of astaxanthin from blue crab (*callinectes sapidus*) shell waste. *Journal of Food Process Engineering* 24: 101–112.

Yamaguchi, K., M. Murakami, H. Nakano, S. Konosu, T. Kokura, H. Yamamoto, et al. 1986. Supercritical carbon dioxide extraction of oils from Antarctic Krill. *Journal of Agricultural and Food Chemistry* 34: 904–907.

Charest, D.J., M.O. Balaban, M.R. Marshall, and J.A. Cornell. 2001. Astaxanthin extraction from crawfish shells by supercritical CO_2 with ethanol as cosolvent. *Journal of Aquatic Food Product Technology* 10: 79–93.

Lin, W.-C., J.-T. Chien, and B.-H. Chen. 2005. Determination of carotenoids in spear shrimp shells (*Parapenaeopsis hardwickii*) by liquid chromatography. *Journal of Agricultural and Food Chemistry* 53: 5144–5149.

Kang, K.-Y., D.-H. Ahn, G.T. Wilkinson, and B.-S. Chun. 2005a. Extraction of lipids and cholesterol from squid oil with supercritical carbon dioxide. *Korean Journal of Chemical Engineering* 22: 399–405.

Zhu, B.-W., L. Qin, D.-Y. Zhou, H.-T. Wu, J. Wu, J.-F. Yang, et al. 2010. Extraction of lipid from sea urchin (*Strongylocentrotus nudus*) gonad by enzyme-assisted aqueous and supercritical carbon dioxide methods. *European Food Research and Technology* 230: 737–743.

Chun, B.-H., H. Kishimura, H. Kanzawa, S. Klomklao, S. Nalinanon, S. Benjakul, et al. 2010. Application of supercritical carbon dioxide for preparation of starfish phospholipase A2. *Process Biochemistry* 45: 689–693.

Ferraro, V., I.B. Cruz, R. Ferreira Jorge, F.X. Malcata, M.E. Pintado, and P.M.L. Castro. 2010. Valorization of natural extracts from marine source focused on marine by-products: A review. *Food Research International* 43: 2221–2233.

Rubio-Rodríguez, N., S. Beltrán, I. Jaime, S.M. de Diego, M.T. Sanz, and J. Rovilla Carballido. 2010. Production of omega-3 polyunsaturated fatty acid concentrates: a review. *Innovative Food Science & Emerging Technologies* 11: 1–12.

Dunford, N.T., F. Temelli, and E. LeBlanc. 1997. Supercritical CO_2 extraction of oil and residual proteins from atlantic mackerel (*Scomber scombrus*) as affected by moisture content. *Journal of Food Science* 62: 289–294.

Dunford, N.T., M. Goto, and F. Temelli. 1998. Modeling of oil extraction with supercritical CO_2 from Atlantic Mackerel (*Scomber scombrus*) at different moisture contents. *The Journal of Supercritical Fluids* 13: 303–309.

Esquível, M.M., N.M. Bandarra, I. Fontan, M.G. Bernardo-Gil, I. Batista, M.L. Nunes, et al. 1997. Supercritical carbon dioxide extraction of sardine *Sardina pilchardus* oil. *LWT-Food Science and Technology* 30: 715–720.

Létisse, M., M. Rozières, A. Hiol, M. Sergent, and L. Comeaua. 2006. Enrichment of EPA and DHA from sardine by supercritical fluid extraction without organic modifier I. Optimization of extraction conditions. *Journal of Supercritical Fluids* 38: 27–36.

Rubio-Rodríguez, N., S.M. de Diego, S. Beltrán, I. Jaime, M.T. Sanz, and J. Rovira. 2008. Supercritical fluid extraction of the omega-3 rich oil contained in hake (*Merluccius capensis–Merluccius paradoxus*) by-products: study of the influence of process parameters on the extraction yield and oil quality. *Journal of Supercritical Fluids* 47: 215–226.

Sahena, F., I.S.M. Zaidul, S. Jinap, M.H.A. Jahurul, A. Khatib, and N.A.N. Norulaini. 2010. Extraction of fish oil from the skin of Indian mackerel using supercritical fluids. *Journal of Food Engineering* 99: 63–69.

Kang, K.-Y., D.-H. Ahn, S.-M. Jung, D.-H. Kim, and B.-S. Chun. 2005b. Separation of protein and fatty acids from tuna viscera using supercritical carbon dioxide. *Biotechnology and Bioprocess Engineering* 10: 315–321.

Létisse, M., and L. Comeau. 2008. Enrichment of eicosapentaenoic acid and docosahexaenoic acid from sardine by-products by supercritical fluid fractionation. *Journal of Separation Science* 31: 1374–1380.

Chang, L.-H., C.-T. Shen, S.-J. Hsieh, S.-L. Hsu, H.-C. Chang, J. Chieh-Ming, et al. 2008. Recovery and enhancement of unsaturated fatty acids in soft-shelled turtle fish oil using supercritical carbon dioxide and associated catalase release activity. *Separation and Purification Technology* 64: 213–220.

Fleck, U., C. Tiegs, and G. Brunner. 1998. Fractionation of fatty acid ethyl esters by supercritical CO_2: High separation efficiency using an automated countercurrent column. *Journal of Supercritical Fluids* 14: 67–74.

Perretti, G., A. Motori, E. Bravi, F. Favati, L. Montanari, and P. Fantozzi. 2007. Supercritical carbon dioxide fractionation of fish oil fatty acid ethyl esters. *Journal of Supercritical Fluids* 40: 349–353.

Davarnejad, R., K.M. Kassim, A. Zainal, and S.A. Sata. 2008. Extraction of fish oil by fractionation through supercritical carbon dioxide. *Journal of Chemical & Engineering Data* 53: 2128–2132.

Riha, V., and G. Brunner. 1999. Phase equilibrium of fish oil ethyl esters with supercritical carbon dioxide. *Journal of Supercritical Fluids* 15: 33–50.

Riha, V., and G. Brunner. 2000. Separation of fish oil ethyl esters with supercritical carbon dioxide. *Journal of Supercritical Fluids* 17: 55–64.

Brunner, G. 2000. Fractionation of fats with supercritical carbon dioxide. *European Journal of Lipid Science and Technology* 102: 240–244.

Espinosa, S., S. Diaz, and E.A. Brignole. 2002. Thermodynamic modeling and process optimization of supercritical fluid fractionation of fish oil fatty acid ethyl esters. *Industrial and Engineering Chemistry Research* 41: 1516–1527.

Espinosa, S., M.S. Diaz, and E.A. Brignole. 2008. Food additives obtained by supercritical extraction from natural sources. *Journal of Supercritical Fluids* 45: 213–219.

Gironi, F., and M. Maschietti. 2006. Separation of fish oils ethyl esters by means of supercritical carbon dioxide: thermodynamic analysis and process modelling. *Chemical Engineering Science* 61: 5114–5126.

Martín, A., and M.J. Cocero. 2007. Mathematical modeling of the fractionation of liquids with supercritical CO_2 in a countercurrent packed column. *Journal of Supercritical Fluids* 39: 304–314.

Catchpole, O.J., J.B. Grey, and K.A. Noermark. 2000. Fractionation of fish oils using supercritical CO_2 and CO_2_ethanol mixtures. *Journal of Supercritical Fluids* 19: 25–37.

Sarrade, S.J., G.M. Rios, and M. Carlés. 1998. Supercritical CO2 extraction coupled with nanofiltration separation. Applications to natural products. *Separation and Purification Technology* 14: 19–25.

Antunes Corrêa, A.P., C. Arantes Peixoto, L.A. Guaraldo Gonçalves, and F.A. Cabral. 2008. Fractionation of fish oil with supercritical carbon dioxide. *Journal of Food Engineering* 88: 381–387.

Alkio, M., C. González, M. Jäntti, and O. Aaltonen. 2000. Purification of polyunsaturated fatty acid esters from tuna oil with supercritical fluid chromatography. *Journal of the American Oil Chemists' Society* 77: 315–321.

Petinello, G., A. Bertucco, P. Pallado, and A. Stassi. 2000. Production of EPA enriched mixtures by supercritical fluid chromatography: From the laboratory scale to the pilot plant. *Journal of Supercritical Fluids* 19: 51–60.

Nieto, A., F. Borrull, E. Pocurull, and R.M. Marcé. 2010. Pressurized liquid extraction: a useful technique to extract pharmaceuticals and personal-care products from sewage sludge. *TrAC Trends in Analytical Chemistry* 29: 752–764.

Richter, B.E., B.A. Jones, J.L. Ezzell, N.L. Porter, N. Avdalovic, and C. Pohl. 1996. Accelerated solvent extraction: a technique for sample preparation. *Analytical Chemistry* 68: 1033–1039.

Breithaupt, D.E. 2004. Simultaneous HPLC determination of carotenoids used as food coloring additives: applicability of accelerated solvent extraction. *Food Chemistry* 86(3): 449–456.

Herrero, M., L. Jaime, P.J. Martín-Álvarez, A. Cifuentes, and E. Ibáñez. 2006b. Optimization of the extraction of antioxidants from *Dunaliella salina* microalga by pressurized liquids. *Journal of Agricultural and Food Chemistry* 54: 5597–5603.

Plaza, M., S. Santoyo, L. Jaime, G. Garcia-Blairsy, M. Herrero, F.J. Señorans, et al. 2010a. Screening for bioactive compounds from algae. *Journal of Pharmaceutical and Biomedical Analysis* 51: 450–455.

Santoyo, S., I. Rodriguez-Meizoso, A. Cifuentes, L. Jaime, G. García-Blairsy Reina, F.J. Señoráns, et al. 2009. Green processes based on the extraction with pressurized fluids to obtain potent antimicrobials from *Haematococcus pluvialis* microalgae. *LWT—Food Science and Technology* 42: 1213–1218.

Herrero, M., P.J. Martín-Álvarez, F.J. Señoráns, A. Cifuentes, and E. Ibáñez. 2005a. Optimization of accelerated solvent extraction of antioxidants from *Spirulina platensis* microalga. *Food Chemistry* 93: 417–423.

Herrero, M., M.J. Vicente, A. Cifuentes, and E. Ibáñez. 2007. Characterization by high-performance liquid chromatography/electrospray ionization quadrupole time-of-flight mass spectrometry of the lipid fraction of *Spirulina platensis* pressurized ethanol extract. *Rapid Communications in Mass Spectrometry* 21: 1729–1738.

Jaime, L., J.A. Mendiola, M. Herrero, C. Soler, S. Santoyo, F.J. Señoráns, et al. 2005. Separation and characterizationof antioxidants from *Spirulina platensis* microalga combining pressurized liquid extraction, TLC and HPLC-DAD. *Journal of Separation Science* 28: 2111–2119.

Shang, Y.F., S.M. Kim, W.J. Lee, and B.-H. Um. 2011. Pressurized liquid method for fucoxanthin extraction from *Eisenia bicyclis* (Kjellman) Setchell. *Journal of Bioscience and Bioengineering* 111(2): 237–241.

Plaza, M. 2010. Búsqueda de nuevos ingredientes funcionales naturales procedentes de algas (PhD Thesis), Universidad Autónoma de Madrid, http://hdl.handle.net/10486/5984.

López, A., M. Rico, and A. Rivero. 2011. The effects of solvents on the phenolic contents and antioxidant activity of *Stypocaulon scoparium* algae extracts. *Food Chemistry* 125(3): 1104–1109.

Onofrejová, L., J. Vašíčková, B. Klejdus, P. Stratil, L. Mišurcová, S. Kráčmar, et al. 2010. Bioactive phenols in algae: the application of pressurized-liquid and solid-phase extraction techniques. *Journal of Pharmaceutical and Biomedical Analysis* 51(2): 464–470.

Quan, C., and C. Turner. 2009. Extraction of astaxanthin from shrimp waste using pressurized hot ethanol. *Chromatographia* 70: 247–251.

Rubio, B.K., R.W.M. van Soest, and P. Crews. 2007. Extending the record of meroditerpenes from cacospongia marine sponges. *Journal of Natural Products* 70: 628–631.

Johnson, T.A., T. Amagata, K.V. Sashidhrar, A.G. Oliver, K. Tenney, T. Matainaho, et al. 2009. The aignopsanes, a new class of sesquiterpenes from selectes chemotypes of the sponge *cascospongia mycofijiensis*. *Organic Letters* 11: 1975–1978.

Johnson, T.A., M.V.C. Morgan, N.A. Aratow, S.A. Estee, K.V. Sashidhara, S.T. Loveridge, et al. 2010. Assessing pressurized liquid extraction for the high throughput extraction of marine-sponge-derived natural products. *Journal of Natural Products* 73: 359–364.

Dodds, E., M.R. McCoy, A. Geldenhuys, L.D. Rea, and J.M. Kennish. 2004. Microscale recovery of total lipids from fish tissues by accelerated solvent extraction. *Journal of the American Oil Chemists' Society* 81: 835–840.

Isaac, G., M. Waldebäck, U. Eriksson, G. Odham, and K.E. Markides. 2005. Total lipid extraction of homogenized and intact lean fish muscles using pressurized fluid extraction and batch extraction techniques. *Journal of Agricultural and Food Chemistry* 53: 5506–5512.

Spiric, A., D. Trbovic, D. Vrabic, J. Djinovic, R. Petronijevic, and V. Matekalo-Sverak. 2010. Statistical evaluation of fatty acid profile and cholesterol content in fish (common carp) lipids obtained by different sample preparation procedures. *Analytica Chimica Acta* 672: 66–71.

Turner, C., and E. Ibañez. 2011. Pressurized hot water extraction. In *Enhancing Extraction Processes in the Food Industry*, ed. N. Lebovka, E. Vorobiev, and F. Chemat. Boca Ratón FL: Taylor & Francis Group, LLC. expected September 2011.

Teo, C.C., S.N. Tan, J.W.H. Yong, C.S. Hew, and E.S. Ong. 2010. Pressurized hot water extraction (PHWE). *Journal of Chromatography. A* 1217: 2484–2494.

Golmohamad, F., M.H. Eikani, and S. Shokrollahzadeh. 2008. Review on extraction of medicinal plants constituents by superheated water. *Journal of Medicinal Plants* 7: 1–24.

Ong, E.S., J.S.H. Cheong, and D. Goh. 2006. Pressurized hot water extraction of bioactive or marker compounds in botanicals and medicinal plant materials. *Journal of Chromatography. A* 1112: 92–102.

Wang, L.J., and C.L. Weller. 2006. Recent advances in extraction of nutraceuticals from plants. *Trends in Food Science & Technology* 17: 300–312.

King, J.W. 2006. Pressurized water extraction: resources and techniques for optimizing analytical applications. *ACS Symposium Series* 926: 79–95.

Rodriguez-Meizoso, I., L. Jaime, S. Santoyo, F.J. Señoráns, A. Cifuentes, and E. Ibáñez. 2010. Subcritical water extraction and characterization of bioactive compounds from Haematococcus pluvialis microalga. *Journal of Pharmaceutical and Biomedical Analysis* 51(2): 456–463.

Plaza, M., M. Amigo-Benavent, M.D. del Castillo, E. Ibáñez, and M. Herrero. 2010b. Facts about the formation of new antioxidants in natural samples after subcritical water extraction. *Food Research International* 43(10): 2341–2348.

Herrero, M., C. Simo, E. Ibáñez, and A. Cifuentes. 2005b. Capillary electrophoresis-mass spectrometry of *Spirulina platensis* proteins obtained by pressurized liquid extraction. *Electrophoresis* 26: 4215–4224.

Tavakoli, O., and H. Yoshida. 2006a. Conversion of scallop viscera wastes to valuable compounds using sub-critical water. *Green Chemistry* 8: 100–106.

Yoshida, H., and O. Tavakoli. 2004. Sub-critical water hydrolysis treatment for waste squid entrails and production of amino acids, organic acids, and fatty acids. *Journal of Chemical Engineering of Japan* 37: 253–260.

Tavakoli, O., and H. Yoshida. 2006b. Squid oil and fat production from squid wasted using sub-critical water hydrolysis: free fatty acids and transesterification. *Industrial and Engineering Chemistry Research* 45: 5675–5680.

Uddin, Md.S., H.-Y. Ahn, H. Kishimura, and B.-S. Chun. 2010. Production of valued materials from squid viscera by supercritical water hydrolysis. *Journal of Environmental Biology* 31: 675–679.

MD Uddin, S., H.-M. Ahn, H. Kishimura, and B.-S. Chun. 2009. Comparative study of digestive enzymes of squid (*Todarodes pacifius*) viscera after supercritical carbon dioxide and organic solvent extraction. *Biotechnology and Bioprocess Engineering* 14: 338–344.

Yoshida, H., M. Terashima, and Y. Takahashi. 1999. Production of organic acids and amino acids from fish meat by subcritical water hydrolysis. *Biotechnology Progress* 15: 1090–1094.

Yoshida, H., Y. Takahashi, and M. Terashima. 2003. A simplified reaction model for production of oil, amino acids, and organic acids from fish meat by hidrolysis under sub-critical and super-critical conditions. *Journal of Chemical Engineering of Japan* 36: 441–448.

Kang, K., A.T. Quitain, H. Daimon, R. Noda, N. Goto, H.-Y. Hu, et al. 2001. Optimization of amino acids production from waste fish entrails by hydrolysis in sub- and supercritical water. *The Canadian Journal of Chemical Engineering* 79: 65–70.

Ying, Z., X. Han, and J. Li. 2011. Ultrasound-assisted extraction of polysaccharides from mulberry leaves. *Food Chemistry* 127: 1273–1279.

Ötles, S. 2009. *Handbook of Food Analysis Instruments*, 1st ed. Boca Raton: CRC Press Taylor & Francis Group.

Hu, A.J., S. Zhao, H. Liang, T.Q. Qiu, and G. Chen. 2007. Ultrasound assisted supercritical fluid extraction of oil and coixenolide from adlay seed. *Ultrasonics Sonochemistry* 14: 219–224.

Ganzler, K., A. Salgó, and K. Valkó. 1986. Microwave extraction: a novel sample preparation method for chromatography. *Journal of Chromatography. A* 371: 299–306.

Worsfold, P., A. Townshend, and C. Poole. 2005. *Encyclopedia of Analytical Science*, 2nd ed. Boston: Elsevier.

Garcia-Ayuso, L.E., M. Sanchez, A.A. de Fernandez, and C.M.D. Luque De. 1998. Focused microwave-assisted soxhlet: an advantageous tool for sample extraction. *Analytical Chemistry* 70: 2426–2431.

Pasquet, V., J.R. Chérouvrier, F. Farhat, V. Thiéry, J.M. Piot, J.B. Bérard, et al. 2011. Study on the microalgal pigments extraction process: performance of microwave assisted extraction. *Process Biochemistry* 46(1): 59–67.

Macías-Sánchez, M.D., C. Mantell, M. Rodríguez, O.E. de la Martínez, L.M. Lubián, and O. Montero. 2009. Comparison of supercritical fluid and ultrasound-assisted extraction of carotenoids and chlorophyll a from *Dunaliella salina*. *Talanta* 77(3): 948–952.

Araujo, G.S., L.J.B.L. Matos, L.R.B. Gonçalves, F.A.N. Fernandes, and W.R.L. Farias. 2011. Bioprospecting for oil producing microalgal strains: evaluation of oil and biomass production for ten microalgal strains. *Bioresource Technology* 102(8): 5248–5250.

Mendiola, J.A., C.F. Torres, A. Toré, P.J. Martín-Álvarez, S. Santoyo, B.O. Arredondo, et al. 2007b. Use of supercritical CO_2 to obtain extracts with antimicrobial activity from *Chaetoceros muelleri* microalga. A correlation with their lipidic content. *European Food Research and Technology* 224(4): 505–510.

Cravotto, G., L. Boffa, S. Mantegna, P. Perego, M. Avogadro, and P. Cintas. 2008. Improved extraction of vegetable oils under high-intensity ultrasound and/or microwaves. *Ultrasonics Sonochemistry* 15(5): 898–902.

Batista, A., W. Vetter, and B. Luckas. 2001. Use of focused open vessel microwave-assisted extraction as prelude for the determination of the fatty acid profile of fish – a comparison with results obtained after liquid-liquid extraction according to Bligh and Dyer. *European Food Research and Technology* 212: 377–384.

Pronyk, C., and G. Mazza. 2009. Design and scale-up of pressurized fluid extractors for food and bioproducts. Review. *Journal of Food Engineering* 95: 215–226.

del Valle, J.M., and J.C. de la Fuente. 2006. Supercritical CO_2 extraction of oilseeds: Review of kinetic and equilibrium models. *Critical Reviews in Food Science and Nutrition* 46: 131–160.

Meireles, M.A.A. 2003. Supercritical extraction from solid: Process design data (2001–2003). *Current Opinion in Solid State and Materials Science* 7: 321–330.

Berna, A., A. Tarrega, M. Blasco, and S. Subirats. 2002. Supercritical CO_2 extraction of essential oil from orange peel; effect of the height of the bed. *Journal of Supercritical Fluids* 18: 227–237.

Lagadec, A.J.M., D.J. Miller, A.L. Lilke, and S.B. Hawthorne. 2000. Pilot-scale subcritical water remediation of polycyclic aromatic hydrocarbon – and pesticide contaminated soil. *Environmental Science & Technology* 34: 1542–1548.

Terigar, B.G., S. Balasubramanian, C.M. Sabliov, M. Lima, and D. Boldor. 2011. Soybean and rice bran oil extraction in a continuous microwave system: from laboratory- to pilot-scale. *Journal of Food Engineering* 104: 208–217.

Boonkird, S., C. Phisalaphong, and M. Phisalaphong. 2008. Ultrasound-assisted extraction of capsaicinoids from *Capsicum frutescens* on a lab- and pilot-plant scale. *Ultrasonics Sonochemistry* 15: 1075–1079.

King, J.W., and K. Srinivas. 2009. Multiple unit processing using sub- and supercritical fluids. *Journal of Supercritical Fluids* 47: 598–610.

Liau, B.H., C.T. Shen, F.P. Liang, S.E. Hong, S.L. Hsu, T.T. Jong, et al. 2010. Supercritical fluids extraction and anti-solvent purification of carotenoids from microalgae and associated bioactivity. *Journal of Supercritical Fluids* 55: 169–175.

Siriwardhana, N., K.N. Kim, K.W. Lee, S.H. Kim, J.H. Ha, C.B. Song, et al. 2008. Optimisation of hydrophilic antioxidant extraction from *Hizikia fusiformis* by integrating treatments of enzymes, heat and pH control. *International Journal of Food Science and Technology* 43: 587–596.

Athukorala, Y., K.N. Kim, and Y.J. Jeon. 2006. Antiproliferative and antioxidant properties of an enzymatic hydrolysate from brown alga, *Ecklonia cava*. *Food and Chemical Toxicology* 44: 1065–1074.

Moreda-Piñeiro, A., A. Bermejo-Barrera, P. Bermejo-Barrera, J. Moreda-Piñeiro, E. Alonso-Rodriguez, S. Muniategui-Lorenzo, et al. 2007. Feasibility of pressurization to speed up enzymatic hydrolysis of biological materials for multielement determinations. *Analytical Chemistry* 79: 1797–1805.

Turner, C., P. Turner, G. Jacobson, K. Almgren, M. Waldebäck, P. Sjöberg, et al. 2006. Subcritical water extraction and b-glucosidase-catalyzed hydrolysis of quercetin glycosides in onion waste. *Green Chemistry* 8: 949–959.

Lindahl, S., A. Ekman, S. Khan, C. Wennerberg, P. Borjesson, P.J.R. Sjoberg, et al. 2010. Exploring the possibility of using a thermostable mutant of beta-glucosidase for rapid hydrolysis of quercetin glucosides in hot water. *Green Chemistry* 12: 159–168.

Heo, S.J., E.J. Park, K.W. Lee, and Y.J. Jeon. 2005. Antioxidant activities of enzymatic extracts from brown seaweeds. *Bioresource Technology* 96: 1613–1623.

Chapter 3
Marine Bioactive Peptides and Protein Hydrolysates: Generation, Isolation Procedures, and Biological and Chemical Characterizations

Protein and Peptides from Marine Byproducts

Turid Rustad and Maria Hayes

3.1 Introduction

Fish resources are limited, and there is therefore a need to optimize utilization of the catch. In Norway, *by-products* are defined as products that are not regarded as ordinary saleable products (fillet, round, eviscerated, or beheaded fish), but which can be recycled after treatment. *Waste* includes products that cannot be used for feed or value-added products, but which have to be composted, burned, or destroyed (Bekkevold and Olafsen 2007). The E.C. regulation on animal by-products (EC Nr 1774/2002), adopted on 3 October 2002, defines animal by-products as whole carcasses or parts of animals not intended for human consumption. Marine by-products intended for human consumption are not included in this definition.

There are different estimates as to how much marine by-products are available. The FAO estimates postharvest losses to be 25% of the catch. The amount of by-products in fish varies depending on species, size, season, and fishing ground (Falch et al. 2006). However, up to 50% of any fish is commonly discarded when preparing seafood industrially (Guérard et al. 2005). Others claim that seafood processing discards and by-products make up around 75% of the total weight of the catch (Shahidi 1994; Pastoriza et al. 2004; Torres et al. 2007). When viscera are included, by-products represent up to 2/3 of the weight of round cod (Mackie 1974; Slizyte et al. 2005a; Falch et al. 2006). Solid wastes generated from seafood factories range from 30% to 85% of the weight of landed fish.

T. Rustad (✉)
Department of Biotechnology, Norwegian University of Science and Technology, 7491, Trondheim, Norway
e-mail: turid.rustad@biotech.ntnu.no

M. Hayes
Food BioSciences Department, Teagasc Food Research Centre, Ashtown, Dublin 15, Ireland

M. Hayes (ed.), *Marine Bioactive Compounds: Sources, Characterization and Applications*, DOI 10.1007/978-1-4614-1247-2_3,

Fish by-products contain valuable lipid and protein fractions as well as other interesting and valuable compounds. Marine oil contains omega-3 fatty acids such as EPA and DHA, phospholipids, squalen, fat-soluble vitamins, and cholesterols. The protein-rich by-product fractions include cut-offs, backbone, heads, skins, roe, milt, stomach, viscera, and blood. The protein fraction is easily digestible and can be used for production of hydrolysates, surimi, thermostable protein dispersions, different peptides and amino acids, gelatin, and collagen as well as protamine. In addition to marine proteins and oil, other valuable components include nucleic acids, calcium, phosporous, and other bioactive compounds such as asthaxanthin as discussed in Chapter two. By-products are a very complex group of fractions and their composition as well as their stability may therefore vary. The composition of marine by-products was reviewed by Falch et al. (2006), Kerry and Murphy (2007), and Rustad (2007) previously.

The protein content in marine seaweed varies and is generally lower in brown seaweed (3–15% of dry weight) than in red seaweed (10–47% of dry weight; Fleurence 1999). The use of marine seaweed to extract biologically active proteins and peptides has been reviewed previously (Aneiros and Garateix 2004).

3.2 Processing of Protein Fractions from Marine Rest Raw Materials

Different types of protein fractions can be produced from fish rest raw material. Early work on chemical recovery of proteins from by-products used chemical hydrolysis to produce fish protein concentrates (FPC) to increase biological availability. This was intended to increase the protein intake in developing countries (Kim and Park 2007). Recently, chemical hydrolysis was exchanged for enzymatic hydrolysis, which is a processing method for recovering and modifying fish oil and protein from underutilized fish biomass and fish by-products, resulting in a number of products with a wide range of applications in the food and pharmaceutical industries (Kristinsson and Rasco 2000; Gildberg et al. 2002; Dauksas et al. 2005; Slizyte et al. 2005, 2009). Accelerated hydrolysis using commercial proteases offers more possibilities than autolysis for controlling the properties of the product: choice of enzyme, reaction conditions, and time of hydrolysis allows for efficient control of the hydrolysis process and production of products with defined and desirable properties.

Hydrolysates have good nutritional properties (Shahidi et al. 1995a, b; Slizyte et al. 2005) and exhibit bioactive properties such as antioxidative, antihypertensive, antithrombotic, and immunomodulatory properties (Kim and Mendis 2006). The chain length of the peptides is of special interest (in relation to the organoleptic and functional characteristics) because properties such as solubility, emulsifying capacity, and bitterness depend, at least in part, on molecular size. The type of raw material, as well as the properties (enzyme activity, state of degradation) of the raw material, choice of enzyme, and process conditions will determine both the yield and the properties of the resulting fish protein hydrolysates (Guérard et al. 2005; Slizyte et al. 2005).

In order to run industrial hydrolysis and have control over the process, it is important to have an overview of the raw material quality and composition including the degree of mincing/cutting, the activity and variety of endogenous enzymes, inactivation of endogenous enzymes, the type and specificity of commercial enzymes used for hydrolysis, and the conditions for hydrolysis (amount of added water, time, temperature, pH). All these factors influence the hydrolysis process and are important for the yield and quality of the end products. It is highly important to produce products with reproducible properties and this is one of the most important challenges.

Controlled enzymatic hydrolysis allows one to choose between several processing tools to obtain products with the desired bioactive and technological properties. However, the hydrolysis process often creates a bitter taste in the final product. This undesirable taste is one of the biggest shortcomings on the way to successful implementation of fish protein hydrolysate (FPH) in the marketplace. The possible sources of the bitter taste in the final product may include the composition of the starting material, or the hydrolysis process itself. Bitterness is related to the average hydrophobicity of the peptides contained in the hydrolysate (Mohr 1979) as well as the degree of hydrolysis (Shahidi 1994). Restricting the degree of hydrolysis (DH) to values of 3–5% will usually result in low bitterness in the end hydrolysate product. Alternatively, a high DH indicating a complete hydrolysis to free amino acids decreases the bitterness, because hydrophobic peptides are considerably more bitter than the corresponding mixture of free amino acids (Belitz and Wieser 1976). The highest risk of bitterness is when the DH is between 4 and 40. Debittering a hydrolysate can be achieved through the use of specific peptidases (Sugiyama et al. 1991), by iso-electric precipitation of bitter peptides (Adler-Nissen 1984), or by use of active carbon (Suh et al. 2000). Performing hydrolysis at slightly alkaline pH conditions can help minimize sensory problems (Thorkelsson and Kristinsson 2009).

Powders of FPH produced from fish viscera with gall bladder had a more bitter taste than powders produced without gall bladder (Dauksas et al. 2004), indicating that sorting of by-products could be used to obtain FPH with better sensory profiles. The bitter-tasting components in bile are cholic and taurocholic acid. The most common way to remove bile compounds is to complex the bile acids to compounds that can be easily removed from the mixture and these include soy protein, pinto beans, black beans, or wheat gluten (Kahlon and Woodruff 2002). Cholestyramine resin is also known to bind bile acids due to its cationic properties.

FPH can also be used as growth media for micro-organisms in fermentation processes. Several studies have shown that fish peptones are as good as commercially available peptones from meat, casein, or vegetable proteins (Aspmo et al. 2005). However, to date, few studies on the use of fish peptones exist (Thormodsen 2009).

3.3 Food Functionalities

FPHs have good functional properties and can contribute to water holding, texture, gelling, foaming, and emulsification properties in different food systems (Kristinsson 2007a). One, often-used definition of *functional properties* is: those physical and chemical properties that influence the behavior of proteins in food systems during processing, storage, cooking, and consumption (Kinsella 1976a). A description of the properties of the proteins important for functional properties was given by Damodaran (1997): The physicochemical properties that influence functional behavior of proteins in food include their size, shape, amino acid composition and sequence, net charge, distribution, hydrophobicity, hydrophilicity, structures (secondary, tertiary, and quaternary), molecular flexibility/rigidity in response to external environment (pH, temperature, salt concentration), or interaction with other food constituents. In some cases nutritional, sensory, and biological values are included in the functional properties.

Functional properties can be divided in several groups. It is usual to classify them into three main groups based on the mechanism of action: (1) properties related to hydration (absorption of water/oil, solubility, thickening, wettability); (2) properties related to the protein structure and rheological characteristics (viscosity, elasticity, adhesiveness, aggregation, and gelification); and (3) properties related to the protein surface (emulsifying and foaming activities, formation of protein–lipid films, whippability).

The degree of hydrolysis is a measure of the percentage of peptide bonds cleaved during hydrolysis and it is one of the most important factors determining the functional and bioactive properties of hydrolysates (Slizyte et al. 2005c, 2009bc). Hydrolysis of proteins increases their solubility and hydrolysates are usually able to tolerate heating without precipitation and are soluble over a wide pH range. This makes it possible to add FPH to a wide variety of food systems (Kristinsson 2007a). Water-holding capacity describes the ability of a protein to absorb water and retain it against outer forces. Several studies have shown that fish protein hydrolysates have good water-holding properties (Slizyte et al. 2005). It was previously shown that adding FPH from salmon to minced salmon muscle led to less drip loss on thawing (Kristinsson and Rasco 2000c) and a higher cook yield was demonstrated when FPH was added to hamburgers and minced pork (Vareltzis et al. 1990). Fish fillets can be tumbled or injected with solutions containing FPH to increase product yield significantly on cooking, equal to or better than injection/tumbling fillets in phosphates (Kristinsson 2007b).

Proteins also have interfacial properties due to the content of both hydrophobic and hydrophilic groups and are therefore often used as surfactants in foods. It has been shown that the emulsifying capacity of proteins could be improved significantly by gentle hydrolysis (Adler-Nissen and Olsen 1992). However, extensive hydrolysis will reduce the emulsifying properties, as small peptides are less efficient in forming stable protein films at the oil–water interface. Mahmoud (1994) showed that the emulsifying activity of hydrolysates decreased linearly with increasing DH in the range of 25–67%. Also Adler-Nissen and Olsen (1992) claimed that there is an optimal peptide length for good emulsifying properties. However, as many different methods are used to determine the DH – and these methods lead to varying results – it is difficult to decide on the optimal DH for functional properties of FPH (Adler-Nissen and Olsen 1992).

FPHs have good foaming and emulsifying properties and thus may be used as emulsifying and emulsion-stabilizing ingredients in a variety of products (e.g., dressing, margarine, and meat batters) as well as aids in the formation and stability of foam-based products (e.g., whipped cream, meringues, and mousse). The level of hydrolysis plays an important role because the size of the peptides is very important for interfacial/surface activity of the FPH (Jeon et al. 2000). Several reports suggest that there is an optimal molecular size or chain length for peptides to provide good foaming and emulsifying properties. Limited hydrolysis with larger peptides generally leads to improved emulsification and foaming properties of fish proteins, whereas extensive hydrolysis with small peptides reduces these properties (Quaglia and Orban 1990; Jeon et al. 2000; Kristinsson and Rasco 2000; Adler-Nissen and Olsen 1979; Lee et al. 1987). The ability of FPH to absorb and retain or hold oil influences the taste of the product and is also an important functional characteristic especially for the meat industry (Arihara et al. 2001; Kristinsson and Rasco 2000). The mechanism of fat absorption is attributed mostly to physical entrapment of the oil (Kinsella 1976b) and correlates with surface hydrophobicity (Kristinsson and Rasco 2000c). On the other hand, lipid residues retained in dried FPH after hydrolysis must be lower than 0.5% in order to prevent negative lipid changes and to reduce development of rancid taste during storage (Spinelli et al. 1972).

FPH may possess cryoprotective properties. Hydrolyzed bovine gelatin was found to have cryoprotective properties (Wang et al. 2009) and a recent study (Cheung et al. 2009) showed that hydrolysates from Pacific hake worked as cryoprotectants in frozen fish mince. The cryoprotective activity was ascribed to the content of oligopeptides and the high content of free amino acids including Asp, Glu, Arg, and Lys. Compared to surimi with no FPH added, surimi with FPH exhibited better gel-forming ability and also demonstrated higher residual ATPase activity, which is a sign that it had a protective effect on myosin during freezing and thawing (Khan et al. 2003). FPH may also protect proteins on drying (Zhang et al. 2002; Khan et al. 2003).

3.4 Physiological Activities In Vitro and In Vivo

Proteins/peptides from fish sources are valuable nutritional components that could be used as ingredients in food and feed industries. However, recently many studies have identified a number of bioactive compounds from fish by-products and shellfish and crustacean shells (Kim and Mendis 2006; Kim and Wijesekara 2010). These compounds can be extracted and purified with technologies of varying complexity. Development of new technologies to extract new bioactive compounds from marine processing by-products may bring more value out of what is today considered a waste.

Several studies have indicated that peptides derived from fish proteins have antioxidative properties in different oxidative systems (Jeon et al. 2000; Jung et al. 2003; Rajapakse et al. 2005; Klompong et al. 2007; Kristinsson 2007; Klompong et al. 2008; Samaranayaka and Li-Chan 2008; Yang et al. 2008; Slizyte et al. 2009). The antioxidant activity of proteins and peptides may result from specific scavenging of radicals formed during peroxidation, scavenging of oxygen containing compounds, or metal-chelating ability (Gutierrez et al. 2003; Kristinsson 2007). The antioxidative activity displayed by the FPH is not necessarily due to only peptides formed during the hydrolysis process (Amarowicz and Shahidi 1997; Jeon et al. 2000; Kristinsson 2007). The "press juice" of fish muscle also has a strong antioxidative activity on hemoglobin-mediated oxidation in a washed fish muscle matrix (Undeland et al. 2003). This indicates that there are a number of compounds in fish muscle that have good antioxidative properties and which can appear in the FPH.

FPHs also have other bioactive properties, but they have not been as extensively studied as peptides from other sources such as milk (Undeland et al. 2009a). Beneficial health effects are linked to fish consumption in general, but it is suggested by some authors that these effects are improved by intake of FPH due to the high content of easily digestible bioactive peptides. The bioactivity is closely related to the amino acid composition and sequence. Undeland et al. reviewed the health effect of different seafood products, including both fish proteins and FPH (Undeland et al. 2009b). The bioactive effects of marine by-products are discussed in a review by Kim and Mendis (2006).

Peptides obtained from various FPHs have shown antihypertensive (blood pressure lowering, ACE-I inhibiting effects on the renin–angiotensin system (RAS); (Vercruysse et al. 2002; Kawasaki et al. 2002; Je et al. 2009), antithrombotic, immunomodulatory, antiobesity (Docmar 2007; Liaset and Espe 2008), growth inhibitory on cancer cell lines (Picot et al. 2006), and antioxidative activities (Kim and Mendis 2006). Hydrolyzed fish proteins also showed anticoagulant (Fig. 3.1) and antiplatelet properties in vitro (Rajapakse et al. 2005). It is reported that fish peptides are capable of accelerating calcium absorption and possess hormone-like activities and growth factors (Fouchereau-Peron et al. 1999). All these are highly desirable in functional food applications. Recently, Kim and Wijesekara (2010) have reviewed the development and biological activities of marine-derived bioactive peptides. Naqash and Nazeer (2011) demonstrated antimicrobial activity against several bacteria – both Gram-negative and Gram-positive – by purified peptide fractions generated from the backbone of flying fish (*Exocoetus volitans*). Loach (*Misgurnus anguillicaudatus*) peptides were shown to have both in vitro antioxidant activity and in vivo antifatigue effects in mice (You et al. 2011). Marine oligopeptides from Chum salmon were found to have immunostimulatory effects in mice. The oligopeptides were rich in the amino acids Glu, Asp, Lys, Leu, Arg, and Gly with molecular weights of between 100 and 860 Da. The mechanism behind the observed immunomodulation is believed to be due to enhanced lymphocyte proliferation and antibody synthesis.

Cholecystokinin (CCK) is the most studied satiety signal (Cudennec et al. 2008). Peptides from fish (blue whiting) and crustacean (brown shrimp) were able to

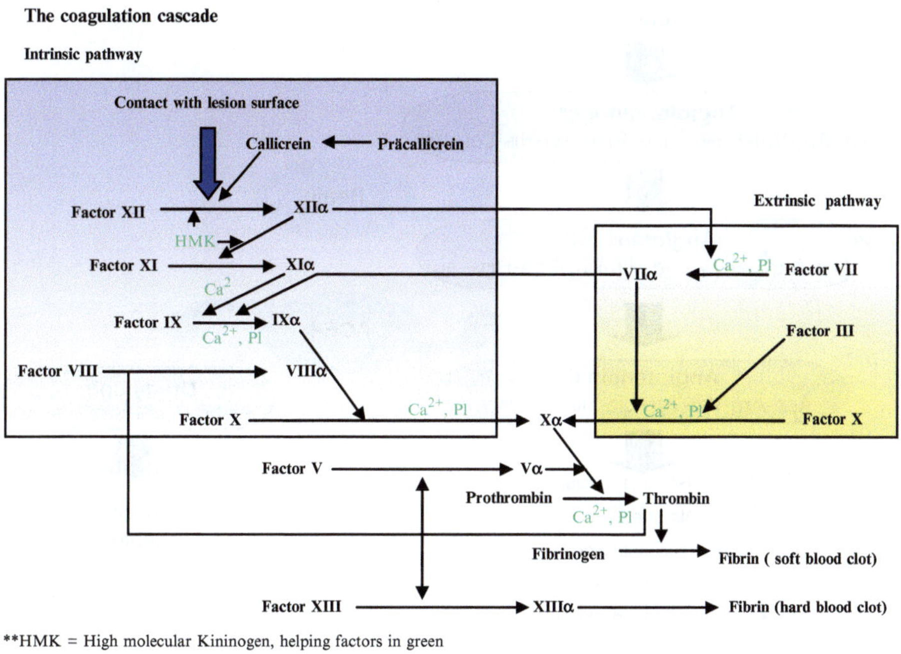

Fig. 3.1 *The coagulation cascade.* Due to the dominant role of platelets in thrombosis, current strategies to inhibit thrombogenesis focus mainly on drugs that block platelet function, but also include anticoagulants for prevention of cardioembolic events. Figure 3.1 shows the coagulation cascade and the three pathways that are responsible for the eventual formation of a thrombus (clot)

stimulate CCK secretion in intestinal STC-1 cells to a high degree and the stimulating activity was found to be related to peptides with molecular weights between 1,000 and 1,500 Da. This demonstrated that FPH could be used as a potential appetite-suppressing product. The effect should be tested in rats and humans to clarify the bioavailability of the peptides in vivo. (Cudennec et al. 2008).

A peptide fraction with neuroprotective effects (determined in rats) was isolated from Chum salmon collagen (Pei et al. 2010). The peptide fraction consisted of proline- and hydroxyproline-rich oligopeptides with molecular weights of between 100 and 860 Da. The neuroprotective effect was due to reduced oxidative damage in the brain and increased expression levels of brain-derived neurotrophic factor (BNDF) and postsynaptic density protein 95 (PSD95).

Several studies have been carried out on the antihypertensive effect of peptides from marine rest raw material. High blood pressure is one of the most important risk factors for cardiovascular diseases (Kim and Wijesekara 2010). Different marine peptides have been found to exhibit antihypertensive properties. This blood pressure reducing effect is primarily ascribed to the inhibitory effect the peptides have on angiotensin-I converting enzyme (ACE-I). ACE-I is an important regulator of blood pressure as it converts the decapeptide angiotensin-I to the potent vasoconstrictor

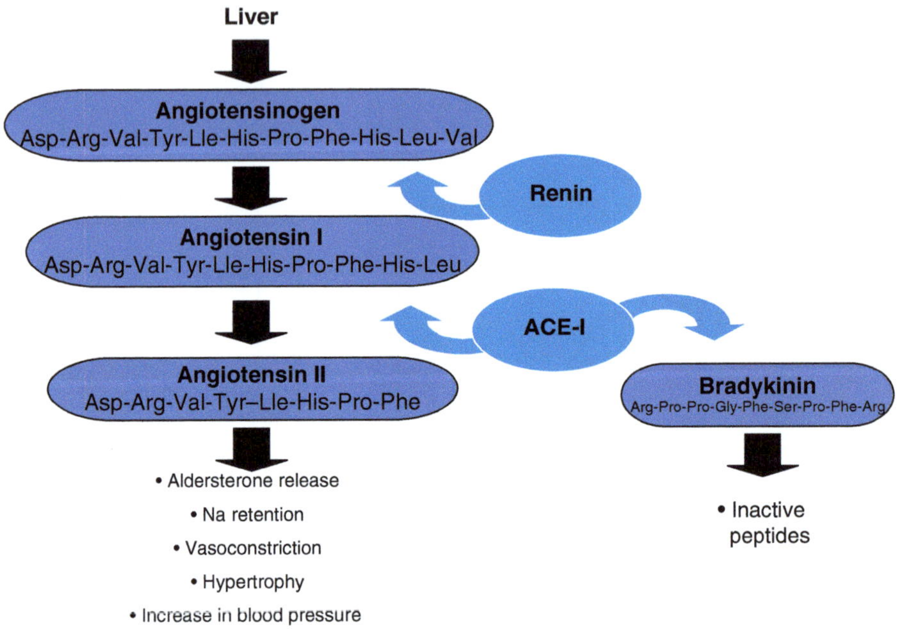

Fig. 3.2 *The Renin–angiotensin system.* Renin hydrolyzes angiotensinogen, its only known substrate forming angiotensin-I. The enzyme, ACE-I, in turn hydrolyzes the decapeptide, angiotensin-I, to form angiotensin-II, a potent vasoconstrictive molecule. ACE-I also acts on the vasodilatory peptide bradykinin further activating hypertension

angiotensin-II, thereby increasing blood pressure (Fig. 3.2). In addition, ACE-I inactivates bradykinin, which has vasodilating properties. The evidence for ACE-I inhibitory activity is largely derived from in vitro studies, and it has also been challenging to evaluate the antihypertensive efficiency of the various ACE-I-inhibitory peptides, because many different assays have been applied to assess the effects in vitro. Nevertheless, mounting evidence suggests that hydrolyzed fish peptides offer substantial blood pressure reducing effects. Many in vitro studies suggest antihypertensive effects of fish peptides, usually measured as inhibition of ACE-I (Kawasaki et al. 2000). Apart from the skeletal muscle proteins myosin and troponin mentioned above, hydrolysis of tropomyosin, actin, and collagen has resulted in bioactive peptides, and these peptides have potential as compounds for the treatment of hypertension. Some in vivo animal studies have been performed in order to assess the antihypertensive effects of fish and meat peptides. (Arihara et al 2001; Jang and Lee 2005). Spontaneously hypertensive (SH) rats that were fed a diet consisting of 20% isolated fish proteins or 20% casein milk proteins, the feeds being identical in all other ways, showed significantly reduced blood pressure in the rats fed fish proteins (Ait-Yahia et al. 2003).

A number of human studies have also been performed on meat and fish peptides. An oligopeptide derived from bonito was found to have antihypertensive effects in a cross-over study with 61 borderline and mildly hypertensive subjects at doses of 1.5 g/day, and approximately 62% of test subjects were judged to have significant to moderate decreases in blood pressure (Fujita et al. 2001). In a 4-week, double-blind placebo-controlled trial on 29 hypertensive subjects, systolic and diastolic blood pressure was reduced to 9.3 and 5.2 mmHg, respectively, after 4 weeks administration of the valyl-tyrosine peptide, which was derived from enzymatic hydrolysis of sardines (Kawasaki et al. 2000).

Some studies have been carried out using gastrointestinal model systems to check if bioactive peptides are resistant to hydrolysis and are able to reach the target organs. Khantaphant et al. (2011) found that the antioxidant activity of red snapper hydrolysates was preserved after digestion in a gastrointestinal model system. Hwang (2010) found that food processing (heat, pH, and high pressure) had no significant effect on the ACE-I inhibitory activity of peptides from tuna cooking juice. This study also concluded that the peptides were resistant to digestion in the gastrointestinal tract.

When producing fish protein hydrolysates or soluble protein fractions, it is important not to create large waste streams. The insoluble fraction (sludge) constitutes a significant part after hydrolysis and contains a high percentage of protein from the raw material. It was shown to have a higher protein efficiency ratio (PER) than FPH (soluble part), contained a high amount of lipids (up to 50%) including a high concentration of phospholipids (up to 60% of lipids) and might be a bulk product for further processing and purification of phospholipids. However, it is also a useful product in itself. To achieve better utilization of all fish by-products and to find better applications for all fractions after hydrolysis it is necessary to pay more attention to the insoluble fraction (Slizyte et al. 2005).

Interesting proteins and bioactive peptides have also been isolated from seaweeds (Aneiros and Garateix 2004). These include lectins, isoagglutinin, glycoproteins, and phycobiliproteins. Phycobiliproteins have displayed potent antioxidant effects and therefore may be applied in the food industry and in cosmetics.

3.5 Challenges and Possible Solutions in Downstream Processing of Marine Peptides

In order to be able to use fish protein and peptides in functional foods or as health food supplements there is first and foremost a need for scientific evidence with regard to the effect of these components. There is also a need to produce fractions without taste and smell (Thorkelsson et al. 2009). Fish rest raw material is (as already mentioned) a highly perishable product and varies regarding composition. The composition varies with species, fishing ground, and season (Falch et al. 2006). To be able to produce protein fractions with stable reproducible quality it is important to use fresh by-products with a low degree of degradation (Thorkelsson et al. 2009). There is therefore a need to find methods to preserve the raw material either on board or at the fish processing plant, and/or develop processing methods to make bulk fractions that can be processed further into refined products. The state of the wild fish stocks, as well as the seasonality of the catches are factors that need to be considered and that can be problematic in planning large-scale production of proteins and peptides.

Several research papers have purified fish peptide fractions in order to characterize peptides with biological activity, but also to find fractions with the highest concentration of bioactive peptides. Most of these studies use different laboratory scale chromatographic techniques. However, one study describes the use of ultrafiltration technology to purify peptide fractions (Kim and Mendis 2006). For production of ACE-I inhibitors from soy protein hydrolysates a continous membrane reactor has been used (Thorkelsson et al. 2008).

For phycobiliproteins from marine seaweed, purification, including density gradient centrifugation, chromatography, and electrophoresis is described in the literature, Also preparative level purification procedures based on hydroxyapatite precipitation was described previously (Aneiros and Garateix 2004).

Many authors refer to DH and amino acid composition as important parameters in the production of FPH. However, as different methods have been used to determine DH, and these methods lead to varying results, it is difficult to decide on the optimal DH for bioactive properties. Both chain length and amino acid composition have effects on the antioxidant activity (Khantaphant et al. 2011). For example, short-chain peptides may lose their ability to chelate metals.

Enzymes with different specificities will lead to formation of different peptides. Alcalase is a commercial protease that is widely used in the production of hydrolysates. It is an endopeptidase with a broad specificity and prefers uncharged residues. Another commonly used enzyme, Flavourzyme, has both endo- and expopeptidase activities.

3.6 Commercial Products

A number of commercially available marine protein/peptide products use fish fillet as a starting material: SEACURE® (Intestive® is SEACURE®), NutriPeptin™, Sardine Peptide SP100N, and Danish fish protein are just a few of the commercially available products. Many producers do not define the part of the fish they use for the production of marine peptides, but claim only the species used. This leads to the assumption that whole fish or fillets were utilized. In addition, several producers note that their products are made from fish and consumers could notice a slight fish odor. This odor is completely normal and in no way indicates spoilage.

The market potential for marine proteins and derivatives has not been fully realized. Today's producers of fish meal and fish protein products mainly supply products for aquaculture and pet food applications. However, several commercially available fish protein hydrolysates for use in food are found today, especially in

American and Asian markets. Researchers have identified that FPH may also possess hormonal-like peptides to accelerate calcium absorption and provide satiety such as calcitonin/CGRP-like peptides and/or cholecystokinin-like peptides (Cudennec et al. 2008; Guerard et al. 2010).

There is a market for marine water retaining proteins used to reduce drip loss, increase shelf life, and improve juiciness. The protein isolate, NutriLean, produced by Proteus industries (www.proteusindustries.com) reduces fat uptake in fried foods and increases juiciness. In Iceland, a protein isolate called Iceprotein is being developed (www.marifunc.org).

Compared to milk and soy proteins, the market for marine proteins and peptides is not large (Thorkelsson et al. 2009) and they have so far mainly been used as seafood flavors. The sale of marine proteins for more advanced applications is still low and development has been slow. Altavida AS has done a survey of the American market for protein ingredients (Skjævestad 2010). The potential market value has been estimated to be $45–60 million. Marine protein ingredients may be used as an alternative to milk- and soy-based products. Several commercially available products with bioactive peptides (Thorkelsson et al. 2009) including products from bonito (www.nippon-sapuri.com/english, www.metagenics.com), sardines, collagen, and hydrolyzed whitefish (www.propernutrition.com, www.copalis.fr, www.nutrimarine.com) are available. A new protein ingredient needs to have a competitive price, good flavor (little or no flavor), documentation of health benefits, and an adequate shelf life. Others products include Stabilium 200, an Atlantic fish autolysate (www.yalacta.com) and PROTIZEN®, a white fish hydrolysate said to have relaxing effects, which would most likely be due to opioid peptides. SEACURE®, is a fish fillet hydrolysate obtained by fermentation using marine micro-organisms and which is mainly composed of di- and tripeptides (www.propernutrition.com, USA; Guerard et al. 2010).

3.7 Future Work

To date, the primary source of bioactive peptides has been dairy products. However, the marine environment represents a myriad of protein resources including algae, and fisheries by-products, and microalgae, and the environments in which they are found makes these resources a new and relatively untapped source for new bioactive compound generation. Greater efforts are needed to fully exploit their potential for use and delivery to consumers in food products. One market that could be targeted is sports beverages. The market for sports nutrition products grows globally, on average 5–7% annually. The main ingredients are amino acids and proteins. Sports nutrition is therefore an area that is well suited for marine proteins (Skjævestad 2010) in so far as they have excellent amino acid composition and high digestibility. However, the biggest challenge in the utilization of marine protein sources is to ensure good quality raw materials and consistent composition in all batches following enzymatic treatment of marine substrates.

References

Adler-Nissen, J. 1984. Control of the proteolytic reaction and of the level of bitterness in protein hydrolysis processes. *Journal of Chemical Technology and Biotechnology* 34B: 215–222.

Adler-Nissen, J., and H.S. Olsen. 1979. *The influence of peptide chain length on taste and functional properties of enzymatically modified soy protein. Functionality and protein structure. A. Pour-El.* Washington, DC: American Chemical Society Symposium Series 92.

Adler-Nissen, J., and H.W. Olsen. 1992. *Functionality and protein structure. A. Pour-El.* Washington, DC: American Chemical Society. 92.

Ait-Yahia, D., Madani, S., Sawelli, J. L., Prost, J., Bouchenak, M., Belleville, J., (2003). Dietary fish proteins lowers blood pressure and alter tissue polyunsaturate fatty acid composition in spontaneously hypertensive rats. nutrition, 19, 4, 324–346.

Amarowicz, R., and F. Shahidi. 1997. Antioxidant activity of peptide fractions of capelin protein hydrolysates. *Food Chemistry* 58: 355–359.

Aneiros, A., and A. Garateix. 2004. Bioactive peptides from marine sources: pharmacological properties and isolation procedures. *Journal of Chromatography B* 803(1): 41–53.

Arihara, K., Y. Nakashima, et al. 2001. Peptide inhibitors for angiotensin I-converting enzyme from enzymatic hydrolysates of porcine skeletal muscle proteins. *Meat Science* 57(3): 319–324.

Aspmo, S.I., S.J. Horn, et al. 2005. Hydrolysates from Atlantic cod (Gadus morhua L.) viscera as components of microbial growth media. *Process Biochemistry* 40(12): 3714–3722.

Bekkevold, S., and T. Olafsen. 2007. *Råvarer med muligheter*. Trondheim: RUBIN AS.

Belitz, H.D., and H. Wieser. 1976. Steric arrangement of sweet and bitter taste of amino acids and peptides. *Zeitschrift für Lebensmittel-Untersuchung und -Forschung* 160(3): 251–253.

Cheung, I.W.Y., A.M. Liceaga, et al. 2009. Pacific hake (merluccius productus) hydrolysates as cryoprotective agents in frozen pacific cod fillet mince. *Journal of Food Science* 74(8): C588–C594.

Cudennec, B., R. Ravallec-Plé, et al. 2008. Peptides from fish and crustacean by-products hydrolysates stimulate cholecystokinin release in STC-1 cells. *Food Chemistry* 111(4): 970–975.

Damodaran, S. 1997. Food proteins: an overview. In *Food proteins and their applications*, ed. S. Damodaran and A. Paraf, 1–24. New York: Marcel Dekker.

Dauksas, E., R. Slizyte, et al. 2004. Bitterness in fish protein hydrolysates and methods for removal. *Journal of Aquatic Food Product Technology* 13(2): 101–114.

Dauksas, E., E. Falch, et al. 2005. Composition of fatty acids and lipid classes in bulk products generated during enzymic hydrolysis of cod (Gadus morhua) by-products. *Process Biochemistry* 40(8): 2659–2670.

Docmar (2007) "Delprosjekt: Peptide."

Falch, E., T. Rustad, et al. 2006. Geographical and seasonal differences in lipid composition and relative weight of by-products from gadiform species. *Journal of Food Composition and Analysis* 19(6–7): 727–736.

Fleurence, J. 1999. Seaweed proteins: biochemical, nutritional aspects and potential uses. *Trends in Food Science & Technology* 10(1): 25–28.

Fouchereau-Peron, M., L. Duvail, et al. 1999. Isolation of an acid fraction from a fish protein hydrolysate with a calcitonin-gene-related-peptide-like biological activity. *Biotechnology and Applied Biochemistry* 29: 87–92.

Fujita, H., Yamagami, Tomohide, B.S., Kanzunori, O., Effect of an ace-inhibitory agent, katsuobushi oligopeptide, in spontanously hypertensive rat and in borderline and mildly hypertensive subjects (2001). nutrition Research, 21, 8, 1149–1158.

Gildberg, A., J.A. Arnesen, et al. 2002. Utilisation of cod backbone by biochemical fractionation. *Process Biochemistry* 38(4): 475–480.

Guerard, F.G., Fabienne, D. Sellos, et al. Enzymatic methods for marine by-products recovery, Fish and Shellfish Upgrading, Traceability. Maximising the value of marine by-products, Marine Biotechnology I. F. Shahidi, Roland and Y. Le Gal: 107-143127-143163.

Guerard, F., N. Decourcelle, C. Sabourin, C. Floch-Laizet, L. Le Grel, P. Le Floch, F. Gourlay, R. Le Delezir, P. Jaquen, and P. Bourseau. 2010. Recent developments of marine ingredients for food and nutraceutical applications: a review. *J Sci Hal Aquat* 2: 21–27.

Guérard, F., D. Sellos, et al. 2005. Fish and shellfish upgrading, traceability. *Advances in Biochemical Engineering/Biotechnology* 96: 127–163.
Gutierrez, M.E., A.F. Garcia, et al. 2003. Interaction of tocopherols and phenolic compounds with membrane lipid components: evaluation of their antioxidant activity in a liposomal model system. *Life Sciences* 72(21): 2337–2360.
Hwang, J.-S. 2010. Impact of processing on stability of angiotensin I-converting enzyme (ACE) inhibitory peptides obtained from tuna cooking juice. *Food Research International* 43(3): 902–906.
Jang, A., and M. Lee. 2005. Purification and identification of angiotensin converting enzyme inhibitory peptides from beef hydrolysates. *Meat Science* 69(4): 653–661.
Je, J.Y., K.H. Lee, et al. 2009. Antioxidant and antihypertensive protein hydrolysates produced from tuna liver by enzymatic hydrolysis. *Food Research International* 42(9): 1266–1272.
Jeon, Y.J., H.G. Byun, et al. 2000. Improvement of functional properties of cod frame protein hydrolysates using ultrafiltration membranes. *Process Biochemistry* 35(5): 471–478.
Jung, W.K., P.J. Park, et al. 2003. Purification and characterization of a new lectin from the hard roe of skipjack tuna, Katsuwonus pelamis. *The International Journal of Biochemistry & Cell Biology* 35(2): 255–265.
Kahlon, T.S., and C.L. Woodruff. 2002. In vitro binding of bile acids by soy protein, pinto beans, black beans and wheat gluten. *Food Chemistry* 79(4): 425–429.
Kawasaki, T., E. Seki, et al. 2000. Antihypertensive effect of Valyl-Tyrosine, a short chain peptide derived from sardine muscle hydrolyzate, on mild hypertensive subjects. *Journal of Human Hypertension* 14(8): 519–523.
Kawasaki, T.C.J., Y. Jun, et al. 2002. Antihypertensive effect and safety evaluation of vegetable drink with peptides derived from sardine protein hydrolysates on mild hypertensive, high-normal blood pressure subjects. *Fukuoka Igaku Zasshi* 93(10): 208–218.
Kerry, J.P., and S.C. Murphy. 2007. Physical and chemical properties of lipid by-products from seafood waste. In *Maximising the value of marine by-products*, ed. F. Shahidi, 22–46. Cambridge: Woodhead.
Khan, M.A., M.A. Hossain, et al. 2003. Effect of enzymatic fish protein hydrolysate from fish scrap on the state of water and denaturation of lizard fish (Saurida wanieso) myofibrils during dehydration. *Food Science and Technology Research* 9(3): 257–263.
Khantaphant, S., S. Benjakul, et al. 2011. Antioxidative and ACE inhibitory activities of protein hydrolysates from the muscle of brownstripe red snapper prepared using pyloric caeca and commercial proteases. *Process Biochemistry* 46(1): 318–327.
Kim, S.-K., and E. Mendis. 2006. Bioactive compounds from marine processing byproducts – a review. *Food Research International* 39(4): 383–393.
Kim, J.-S., and J.W. Park. 2007. Mince from seafood processing by-product and surimi as food ingredient. In *Maximising the value of marine by-products*, ed. F. Shahidi, 196–228. Cambridge: Woodhead.
Kim, S.-K., and I. Wijesekara. 2010. Development and biological activities of marine-derived bioactive peptides: a review. *Journal of Functional Foods* 2(1): 1–9.
Kinsella, J.E. 1976a. Functional properties of food proteins: a survey. *CRC Critical Reviews in Food Science and Nutrition* 7: 219–280.
Kinsella, J.E. 1976b. Functional properties of proteins in food: a survey. *Critical Reviews in Food Science and Nutrition* 8: 219–280.
Klompong, V., S. Benjakul, et al. 2007. Antioxidative activity and functional properties of protein hydrolysate of yellow stripe trevally (Selaroides leptolepis) as influenced by the degree of hydrolysis and enzyme type. *Food Chemistry* 102(4): 1317–1327.
Klompong, V., S. Benjakul, et al. 2008. Comparative study on antioxidative activity of yellow stripe trevally protein hydrolysate produced from alcalase and flavourzyme. *International Journal of Food Science and Technology* 43(6): 1019–1026.
Kristinsson, H.G. 2007a. Aquatic food protein hydrolysates. In *Maximising the value of marine by-products*, ed. F. Shahidi. Cambridge: Woodhead.
Kristinsson, H.G. 2007b. Aquatic food protein hydrolysates. In *Maximising the value of marine by-products*, ed. F. Shahidi. Cambridge: Woodhead.

Kristinsson, H.G., and B.A. Rasco. 2000a. Biochemical and functional properties of Atlantic salmon (Salmo salar) muscle proteins hydrolyzed with various alkaline proteases. *Journal of Agricultural and Food Chemistry* 48(3): 657–666.

Kristinsson, H.G., and B.A. Rasco. 2000b. Biochemical and functional properties of Atlantic salmon (Salmo salar) muscle proteins hydrolyzed with various alkaline proteases. *Journal of Agricultural and Food Chemistry* 48(3): 657–666.

Kristinsson, H.G., and B.A. Rasco. 2000c. Fish protein hydrolysates: production, biochemical, and functional properties. *Critical Reviews in Food Science and Nutrition* 40(1): 43–81.

Lee, S.W., M. Shimizu, et al. 1987. Emulsifying properties of peptides obtained from the hydrolysates of b-casein. *Agricultural and Biological Chemistry* 51: 1661–1666.

Liaset, B., and M. Espe. 2008. Nutritional composition of soluble and insoluble fractions obtained by enzymatic hydrolysis of fish-raw materials. *Process Biochemistry* 43(1): 42–48.

Mackie, I.M. (1974). Proteolytic enzymes in recovery of proteins from fish waste. Process biochemistry, 9, 12–14.

Mahmoud, M.I. 1994. Physicochemical and functional properties of protein hydrolysates in nutritional products. *Food Technology* 48: 89–94.

Mohr, V. 1979. *Enzymes technology in the meat and fisheries industries*. Paris: The International Microbiology and Food industry Congress.

Naqash, S.Y., and R.A. Nazeer. 2011. Evaluation of bioactive properties of peptide isolated from Exocoetus volitans backbone. *International Journal of Food Science & Technology* 46(1): 37–43.

Pastoriza, L., Sampedro, S., Cabo, M. L., Herrera, J.J.R. & Bernardez, M. (2004). Solubilisation of proteins from rayfish residues by endogenous and commercial enzymes. Journal of the Science of Food and Agric 84, 83–8.

Pei, X., R. Yang, et al. 2010. Marine collagen peptide isolated from Chum Salmon (Oncorhynchus keta) skin facilitates learning and memory in aged C57BL/6 J mice. *Food Chemistry* 118(2): 333–340.

Picot, L., S. Bordenave, et al. 2006. Antiproliferative activity of fish protein hydrolysates on human breast cancer cell lines. *Process Biochemistry* 41(5): 1217–1222.

Quaglia, G.B., and E. Orban. 1990. Influence of enzymatic-hydrolysis on structure and emulsifying properties of sardine (Sardina-Pilchardus) protein hydrolysates. *Journal of Food Science* 55(6): 1571–1573.

Rajapakse, N., W.K. Jung, et al. 2005. A novel anticoagulant purified from fish protein hydrolysate inhibits factor XIIa and platelet aggregation. *Life Sciences* 76(22): 2607–2619.

Rustad, T. 2007. Physical and chemical properties of protein seafood by-products. In *Maximising the value of marine by-products*, ed. F. Shahidi, 3–21. Cambridge: Woodhead.

Samaranayaka, A.G.P., and E.C.Y. Li-Chan. 2008. Autolysis-assisted production of fish protein hydrolysates with antioxidant properties from Pacific hake (Merluccius productus). *Food Chemistry* 107(2): 768–776.

Shahidi, F. 1994. Seafood processing by-products. In *Seafoods chemistry*, ed. F. Shahidi and J.R. Botta. London: Blackie Academic & Professional.

Shahidi, F., X.Q. Han, et al. 1995a. Production and characteristics of protein hydrolysates from capelin (Mallotus-Villosus). *Food Chemistry* 53(3): 285–293.

Shahidi, F., X.Q. Han, et al. 1995b. Production and characteristics of protein hydrolysates from capelin (Mallotus villosus). *Food Chemistry* 53(3): 285–293.

Skjævestad, B. 2010. Muligheter for marine proteiningredienser i det amerikanske helse- og ernæringsmarkedet, RUBIN. 186.

Slizyte, R., E. Dauksas, et al. 2005a. Characteristics of protein fractions generated from hydrolysed cod (Gadus morhua) by-products. *Process Biochemistry* 40(6): 2021–2033.

Slizyte, R., E. Dauksas, et al. 2005b. Characteristics of protein fractions generated from hydrolysed cod (Gadus morhua) by-products. *Process Biochemistry* 40(6): 2021–2033.

Slizyte, R., T. Rustad, et al. 2005c. Enzymatic hydrolysis of cod (Gadus morhua) by-products: optimization of yield and properties of lipid and protein fractions. *Process Biochemistry* 40(12): 3680–3692.

Slizyte, R., R. Mozuraityte, et al. 2009a. Functional, bioactive and antioxidative properties of hydrolysates obtained from cod (Gadus morhua) backbones. *Process Biochemistry* 44(6): 668–677.

Slizyte, R., R. Mozuraityte, et al. 2009b. Functional, bioactive and antioxidative properties of hydrolysates obtained from cod (Gadus morhua) backbones. *Process Biochemistry* 44: 668–677.

Spinelli, J., B. Koury, et al. 1972. Approaches to the utilisation of fish for the preparation of protein isolates; enzymatic modification of myofibrillal fish proteins. *Journal of Food Science* 37: 604–608.

Sugiyama, K., M. Egawa, et al. 1991. Characteristics of sardine muscle hydrolysates prepared by various enzymatic treatments. *Nippon Suisan Gakkaishi* 57(3): 475–479.

Suh, H.J., S.H. Bae, et al. 2000. Debittering of corn gluten hydrolysate with active carbon. *Journal of the Science of Food and Agriculture* 80(5): 614–618.

Thorkelsson, G., and H.G. Kristinsson. 2009. Bioactive peptides from marine sources. State of art. Report to the NORA fund. Skýrsla Matís14-09: 19.

Thorkelsson, G., S. Sigurgisladottir, et al. 2008. Mild processing techniques and development of functional marine protein and peptide ingredients. In *Improving seafood products for the consumer*, ed. T. Børresen, 363–398. Cambridge: Woodhead.

Thorkelsson, G., R. Slizyte, et al. 2009. Fish proteins and peptide products:processing methods, quality and functional properties. In *Marine functional food*, ed. J.B. Luten, 115–139. Wageningen: Wageningen Academic Publishers.

Thormodsen, T. 2009. *Industriell utvikling av peptoner fra biråstoff av laks*. Will find: RUBIN.

Torres, J.A., Y.C. Chen, et al. 2007. Recovery of by-products from seafood processing streams. In *Maximising the value of marine by-products*, ed. F. Shahidi, 65–90. Boca Raton: CRC Press.

Undeland, I., H.O. Hultin, et al. 2003. Aqueous extracts from some muscles inhibit hemoglobin-mediated oxidation of cod muscle membrane lipids. *Journal of Agricultural and Food Chemistry* 51(10): 3111–3119.

Undeland, I., H. Lindqvist, et al. 2009a. Seafood and health: what is the full story? In *Marine functional food*, ed. J.B. Luten. Wageningen: Wageningen Academic Publishers.

Undeland, I., H. Linquist, et al. 2009b. Seafood and health: what is the full story? In *Marine functional food*, ed. J.B. Luten, 17–87. Wageningen: Wageningen Academic Publishers.

Vareltzis, K., N. Soultos, et al. 1990. Proximate composition and quality of a hamburger type product made from minced beef and fish-protein concentrate. *Lebensmittel-Wissenschaft & Technologie* 23(2): 112–116.

Vercruysse, L., J. Van Camp, et al. 2005. ACE inhibitory peptides derived from enzymatic hydrolysates of animal muscle protein: a review. *Journal of Agricultural and Food Chemistry* 53(21): 8106–8115.

Wang, S., K. Agyare, et al. 2009. Optimisation of hydrolysis conditions and fractionation of peptide cryoprotectants from gelatin hydrolysate. *Food Chemistry* 115(2): 620–630.

Yang, J.L., H.Y. Ho, et al. 2008. Characteristic and antioxidant activity of retorted gelatin hydrolysates from cobia (Rachycentron canadum) skin. *Food Chemistry* 110(1): 128–136.

You, L., M. Zhao, et al. 2011. In vitro antioxidant activity and in vivo anti-fatigue effect of loach (Misgurnus anguillicaudatus) peptides prepared by papain digestion. *Food Chemistry* 124(1): 188–194.

Zhang, N., Y. Yamashita, et al. 2002. Effect of protein hydrolysate from antarctic krill on the state of water and denaturation of lizard fish myofibrils during frozen storage. *Food Science and Technology Research* 8(3): 200–206.

Chapter 4
Chitin, Chitosan and their Derivatives from Marine Rest Raw Materials: Potential Food and Pharmaceutical Applications

Chitin, Chitosan and Chitooligosaccharides

Maria Hayes

4.1 Introduction

Chitin and chitosan are valuable and versatile natural biopolymers derived from a number of different marine and terrestrial sources. The word *chitin* comes from Greek etymology, meaning "a coat of mail." The product was first used in 1823. The importance of chitin was discovered in the 1970s, when fishing companies were prevented from dumping shells of crabs and lobsters at sea. Research studies on these shells indicated that shell chitin has properties for a wide variety of industrial applications. Commercial interest in these biopolymers is due to the high percentage of nitrogen (6.89%) found in chitin and chitosan compared to synthetically substituted cellulose (1.25%). The first patent on chitosan production was introduced in the 1920s, and today there are several hundred patents on production of chitin and other applications. Large-scale production of the biopolymer began approximately two decades ago. Until recently, the chitin and chitosan industry suffered from issues such as the unavailability of a large and reliable supply of raw material, inconsistent quality, the presence of pollutants such as heavy metals, ash, and other foreign materials, along with high price for production and poor economic returns.

Chitin (β-1, 4-poly-*N*-acetyl-*D*-glucosamine) is second only to cellulose as the most plentiful natural polymer found on Earth and it is obtained from a number of sources including marine crustacean shell-waste material, insects, and the exoskeleton of invertebrates (Muzzarelli 1977; Shahidi et al. 1999). It is also found as a structural component in the cell walls of the Basidomycete fungi (Knorr 1984). Chitin and chitosan are nontoxic and biodegradable. Chitin consists of a linear chain of poly ((β-1, 4)-*N*-acetyl *D*-glucosamine) and was first identified by

M. Hayes (✉)
Food BioSciences Department, Teagasc Food Research Centre, Ashtown, Dublin 15, Ireland
e-mail: maria.hayes@teagasc.ie

M. Hayes (ed.), *Marine Bioactive Compounds: Sources, Characterization and Applications*, DOI 10.1007/978-1-4614-1247-2_4,

Bradconnot in 1811 (Skaugrud et al. 1990). Chitosan refers to low acetyl-substituted forms of chitin and is composed primarily of glucosamine (2-amino-2-deoxy-β-*D*-glucose) known as (1–4)-2-amino-2-deoxy-β-*D*-glucose. Chitin has three different types of reactive functional groups, an amino group and a primary and secondary hydroxyl group at the C-2, C-3, and C-6 positions, respectively (Roberts 2008; Furusaki et al. 1996). However, naturally occurring chitin has <30% of the functional groups on the C-2 deacetylated, making it a natural linear copolymer of β-(1–4)-2-amino-2-deoxy-D-glucan and β-(1–4)-2-acetomido-2-deoxy-D-glucan.

Chitin, chitosan, and chito-oligosaccharides are found in a variety of end-use applications, the major ones including use in wastewater treatment facilities, food and beverages, cosmetics, agrochemicals, cell culture, and others. Applications of chitinous products in foods and pharmaceuticals as well as processing aids, have received considerable attention in recent years as exotic synthetic compounds but are losing their appeal. The potential for innovation with chitin and chitin-containing compounds may come from separation and process technologies. There is a particular need for low-cost extraction methodologies that can be scaled up to a commercial level.

Shrimp and prawn shell waste are the principal sources of chitin and chitosan due to aquaculture, which allows for a reliable and continuous supply of marine shell material, and secondly due to the increase in consumption of shellfish, particularly in Asia and the Middle East (Roberts 2008). Other raw material sources of chitin and chitosan include crab, lobster, crayfish, squid pen, and krill as well as chitin derived from fungal mycelium waste sources (Roberts 2008). Krill has undergone considerable study as a source of chitosan in Chile but also in Poland where a pilot plant scale production line was established in Gdynia at the Sea Fisheries Institute (Roberts 2008). However, commercially available chitin and chitosan from krill sources are not currently available. Crab, prawn, shrimp, and squid pen chitin sources are commercially available. Chitin, chitosan, and their derivatives from crustacean shell are a huge resource and have many potential uses. There are opportunities to include the recovery of chitin and chitosan in environmentally friendly process operations using biotechnology (Hall 1996). This chapter highlights the production methods of chitin, chitosan, and chitosanoligosaccharides, the various methods that have been developed to determine the degree of N-acetylation (DA) for chitin and chitosan, and the primary world markets and applications of chitin, chitosan, and chito-oligosaccharides.

4.2 Sources and Structures

Shellfisheries' waste streams are the predominant source of chitin; crab, shrimp, and prawn derived chitin are commercially available today. Crab, lobster, crayfish, and krill have also been the subject of considerable study for the generation of chitin and chitosan. Approximately 1 t of chitin can be produced from 7 t of crab shells and 2 t can be processed from 7 t of shrimp shells. In addition, fungal chitin and chitosan from mycelium waste streams have received attention in recent times and a Belgian company called Kitozyme produces fungal-derived chitosan. Indeed, in February 2011, this company obtained authorization from the European commission and European Food Safety Authority (EFSA) to place a chitin–glucan product derived from *Aspergillus niger* on the market as a novel food ingredient.

The degree of N-acetylation (DA) is used to distinguish between chitin and chitosan. It is called chitin when the DA is greater than a certain value (DA > 50%) and the sample is not soluble in weak acid and water; otherwise it is referred to as chitosan (Kasaai 2009). Chitin is a high molecular weight linear polymer, white in color and hard and inelastic. It has a hydrophobic nature and therefore is insoluble in water and organic solvents. Alpha-chitin (α-chitin) is the most abundant of the chitin polymers in nature and has a structure of antiparallel chains. It is found in crab, lobster, and shrimp shells. Beta-chitin (β-chitin) is found in squid and has intra-sheet hydrogen bonding by parallel chains and γ-chitin consists of a mixture of parallel and antiparallel chains, which is a combination of α- and β-chitin (Jang et al. 2004).

Chitosan is the N-deacetylated form of chitin and is soluble in dilute organic acids such as acetic and lactic acids at low pH (Peniston and Johnson 1980). Chitosan does not refer to a single unique substance but refers to a number of copolymers varying in the ratio of anhydro-*N*-acetyl-D-glucosamine to anhydro-D-glucosamine residues (Roberts 2008). It is cationic in nature, and this is the reason for many of its observed bioactivities. The solubility of the polymer in aqueous acidic solutions is used to distinguish between chitin and chitosan. When chitosan has a DA = 0% the sample is insoluble and highly crystalline. Solubility depends on the molecular weight and the degree of neutralization of amine groups, ionic strength of solvent, pH of the chitosan solution, and the distribution of *N*-acetyl glucosamine along the backbone of the molecule (*Official Journal of the European Union*). Roberts proposed a nomenclature system for chitosan in his excellent paper "Thirty Years of Progress in Chitin and Chitosan" published in 2008 (Roberts 2008). He proposed that:

- All β-(1–4) linked copolymers of anhydro-2-acetamido-2-deoxy-D-glucopyranose and anhydro-2-amino-2-deoxy-D-glucopyranose units should be termed chitin or chitosan based on their solubility in 0.1 M acetic acid.
- The literature should make reference to the mole fraction (F_A) of anhydro-2-acetamido-2-deoxy-D-glucopyranose units. Roberts also suggests that if the F_A is known, it should be placed in brackets after the name of the compound (either chitin or chitosan).
- A distinction should be made between chitosan prepared directly from chitin using deacetylation and material prepared by re-N-acetylation of chitosan (N-acetylchitosans). He also suggested that if deacetylation was performed under homogeneous conditions that it should be indicated by the use of an italicized "*h*". Figure 4.1 shows the chemical structure of chitin and chitosan and their properties at a glance.

Chitosan oligosaccharides (COS), like chitosan, have positive charges due to the removal of acetyl units from D-glucosamine units. COS have shorter chain lengths than chitosan and free amino groups in *D*-glucosamine units (Percot et al. 2003). COS have better solubility than low molecular weight chitosans (LMWC). No

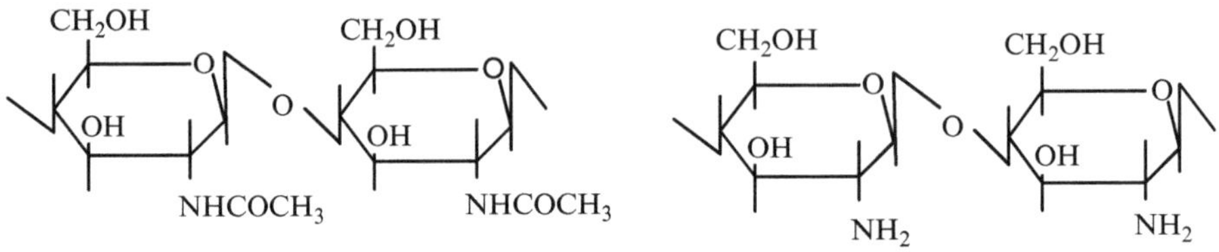

Chitin

- Natural biopolymer
- Contains the N-acetyl group
- Biodegradable
- Antibacterial
- Antifungal
- Antiviral
- Anti-allergenic
- Non-toxic
- Insoluble in water and weak acids

Chitosan

- Natural biopolymer
- Deacetylated chitin
- Biodegradable
- Antibacterial
- Antifungal
- Antiviral
- Anti-allergenic
- Non-toxic
- Soluble in weak acids
- Binds fat

Fig. 4.1 Chitin and chitosan structures and features at a glance

specific degree of polymerization (DP) exists to distinguish LMWC from COS. However, the molecular weight of COS can be considered to be up to and equal to 10 kDa and viscosity is normally used as the parameter to determine the molecular weight of the substance.

4.3 Preparation of Chitin

Crustacean shell consists primarily of protein (30–40%), mineral salts (30–50%), and chitin (13–42%). The main industrial techniques used to extract chitin from different shell-waste sources rely on chemical processes for the hydrolysis of the protein and removal of inorganic matter. Chitin isolation consists of three steps: demineralization, deproteinization, and bleaching where demineralization and deproteinization are interchangeable in terms of order depending on the proposed use of chitin (Hayes et al. 2008a). Demineralization can be achieved using diluted HCL (1–8%) at room temperature for 1–3 h. Demineralization using other acids such as acetic acid and sulfuric acid was also reported previously. One study has shown that demineralization of shrimp waste can be completed in 15 min by using 0.25 M HCL at room temperature and a ratio of solvent to waste of 40:1 (Percot et al. 2003; Kjartansson et al. 2006). Industrial methods usually utilize aqueous base solutions such as NaOH or KOH for

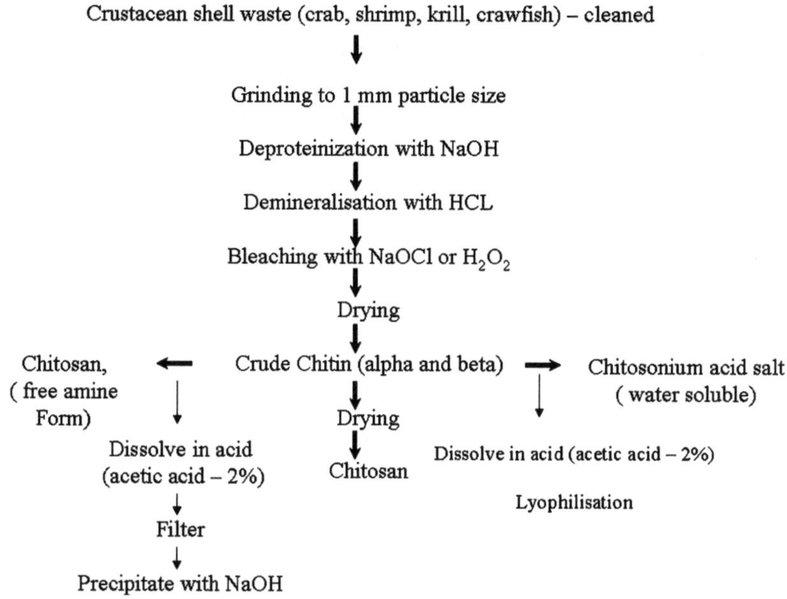

Fig. 4.2 Preparation of chitin and chitosan using traditional acid-based methods

the deproteinization step and the effectiveness depends on the shell-to-solution ratio, temperature, concentration of base used, and the duration of the reaction. Prolonged reaction times of more than 6 h have been reported to cause depolymerization and deacetylation of chitin. Industrial methods may also include a decolorization step that improves chitin color and these methods employ bleaching agents including NaOCl or H_2O_2 solutions. Enzymes can also be used in the deproteinization step, and methods used often employ lactic acid bacteria (LAB) during a fermentation process (Guerrero et al. 1996; Healy and Healy 2003). Usually a carbohydrate source is required for growth of LAB and in turn lactic acid is produced which causes a drop in the pH and subsequent dissolution of $CaCO_3$. It has been suggested that when energy and waste disposal costs are taken into account, the use of lactic acid bacteria to generate chitin from shellfisheries' waste streams is more economical and yields several potential products (Healy and Healy 2003). Figure 4.2 shows a simple schematic demonstrating the preparation of chitin and chitosan from shellfisheries' waste.

4.4 Preparation of Chitosan

Chitosan is manufactured industrially by hydrolyzing amino acetyl groups of chitin. It is available commercially in several physical states including solution, flake, fine powder, bead, and fiber. Deacetylation of chitin to produce the more soluble polymer chitosan usually involves the use of 50% NaOH and relatively high temperatures of between 60°C and 120°C. There is a requirement for lowering temperatures and reducing the amounts of NaOH used and much research has been carried out looking at enzymatic deacetylation of chitin using chitin deacetylases (Tsigos et al. 1996). Advances in relation to lower temperature deacetylation, reduced alkali deacetylation, and enzymatic deacetylation have occurred (Yoshiichi et al. 1987). For example, one detailed study examined the deacetylation of chitin at 30°C using NaOH solutions having concentrations ranging from 23 wt.% to 39 wt.% and varying liquor ratios. At the lowest solvent to substrate ratio (14:1), acid-soluble products were only obtained with NaOH solutions of 33 wt.% or higher, with the time required to obtain them decreasing with increase in NaOH concentration: 6 days for 33 wt.%, 5 days for 35 wt.%, and 4 days for 37.5 or 39.5 wt.% (Roberts 2008). The traditional deacetylation method for converting chitin to chitosan uses an alkali solvent: chitin ratio of between 10:1 to 15:1. Industrial processes use slightly lower ratios and this leads to problems regarding disposal of effluents, chemical costs, and the corrosive effects of the alkali. Several attempts have been made to reduce the amount of alkali used in the production of chitosan. One approach used one part chitin with two to five parts NaOH solution and subsequently the solution was heated (Peniston and Johnson 1980). Other methods involve steeping of the chitin in excess NaOH prior to heating (Kim and Rajapakse 2005). Enzymatic deacetylation of chitin using chitin deacetylases has also been studied. One of the first attempts involved the use of chitin deacetylases extracted from the bacteria *Colletotrichum lindemuthianum* and *Mucor rouxii* which were used to successfully deacetylate water-soluble glycol chitin (Tsigos et al. 1996). However, when applied to insoluble chitin, results were not promising. Enzymatic deacetylation is an attractive method due to:

- Its perception as being an environmentally friendly, "green" processing method
- Lower chemical costs
- The nondegradative effect on chitosan resulting in higher molecular weight chitosan with more potential bioactivities
- Ambient temperature use for processing

Drawbacks of using enzymatic deacetylation in the production of chitosan from chitin include the limited accessibility of chitin deacetylases to crystalline chitin.

4.5 Preparation of Chito-Oligosaccharides (COS)

Chitosan can be cleaved by hydrolyzing agents due to the presence of unstable glycosidic bonds. Degradation of *O*-glycosidic linkages leads to production of COS varying in the degree of polymerization and the number and sequence of glucosamine (GlcN) and GlcNAc units. Chemical and enzymatic methods are used in COS production with chemical hydrolysis used more commonly in industrial-scale COS manufacture. However, chemical hydrolysis may result in the formation of toxic compounds and more environmental pollution problems. Lower production yields are also a negative outcome of this process. Enzymatic processes carried out in batch reactors are preferable to chemical hydrolysis methods. Chitosanolytic enzymes are employed in the enzymatic conversion of chitosan to COS and have been reported from different micro-organisms including fungi and bacteria. Furthermore, carbohydrases and proteases are also capable of hydrolyzing chitosan to produce COS with varying molecular weights.

A clear distinction between chitosanase and chitinase for the hydrolysis of differentially deacetylated chitosans cannot be made (Kim and Rajapakse 2005). It has been generally observed that chitosanases obtained from microbes produce a higher yield of COS compared to chitosanases from other sources, such as fungi. Drawbacks regarding the implementation of enzymatic deacetylation in the production of COS include the high cost of chitosan deacetylases and often other commercial enzymes are employed to produce COS. Kim et al. (2005) suggested the use of a dual reactor system combining a column reactor and an ultrafiltration (UF) membrane reactor for continuous production of COS where chitosan is partially hydrolyzed by an immobilized enzyme prepacked in the column reactor and this product is supplied to the UF membrane for production of COS.

4.6 Methods of Analysis Used to Differentiate Chitin, Chitosan, and COS

Several different methods are used to determine the degree of *N*-acetylation for chitin and chitosan (Heux et al. 2000). These methods include spectroscopy (IR, ^{1}H NMR, ^{13}C NMR, ^{15}N NMR, and UV); destructive methods including elemental analysis, acid and enzymatic hydrolysis of chitin and chitosan; DA measurement by colorimetry or HPLC; and pyrolysis using gas chromatography and scanning calorimetry. Conventional methods used include titration, conductometry, potentiometry, and ninhydrin assay. The various methods are discussed in a review by Kasaai (2009) and Heux et al. (2000).

The DA of chitin or chitosan is the parameter that influences the various bioactivities of the molecules the most and also the physicochemical and mechanical properties of the polymers. Therefore, DA determination is essential in order to properly study the chemical and structure–properties relationships of the compounds and to maximize the potential applications of the compounds.

In order to determine the DA of chitin/chitosan various spectroscopy methods and NMR methods may be used and these compare the values of A_m/A_R for unknown samples with similar ratios of reference samples whose DA is known. A calibration curve is made by plotting the absorbance ratio of chitin/chitosan samples having known DA versus their corresponding DA values (Heux et al. 2000). When ^{1}H NMR is used, dilute solutions of chitosan samples in aqueous acid are prepared and their ^{1}H NMR spectra recorded from 0 to 10 ppm using a proton NMR spectrometer. When ^{15}N NMR is used, the spectra of chitin/chitosan samples in the solid state are recorded between 0 and 200 with the spectrometer operating at 30–200 MHz (Heux et al. 2000; Yu et al. 1999).

Table 4.1 Different methods for determination of DA of chitin and chitosan

Method	DA range	Advantages	Disadvantages
Conventional	Applicable for soluble chitin/ chitosan	Availablity of instrumentation	Only useful for soluble chitin/chitosan
		Easy to use	Impurities such as proteins and organic acids may interfere
		Humidity does not affect analysis	Obtain unreliable results in the case of the presence of impurities
Spectroscopy (13 C NMR, 15 N NMR)	0–100%	Useful for both soluble and insoluble chitin/chitosan	Impurities may create interference peaks
			Methods are not sensitive enough to detect low values of the DA
			Instrument availability and cost are limitations in using this method
Spectroscopy (1H NMR, IR, near-IR, UV)	Applicable for soluble samples only	1HNMR and UV techniques are precise	Careful chitin/chitosan reference sample selection is required
		Accurate data obtained	
Destructive methods	0–100%	Entire DA range can be detected	Two-step process required for analysis
			Long time needed for DA analysis

The biggest problems faced when determining the DA of chitin/chitosan samples are the associated impurities including proteins, organic pigments, and minerals (No et al. 1995), and the DA of chitin and chitosan may change depending on the nature and level of impurities in each sample. The water content and impurity level of each sample is therefore important. The solubility of chitin and chitosan also makes it difficult to obtain accurate DA values. Chitin with β-(1–4)-2-acetamido-2-deoxy-D-glucan (GlcNAc) as a structural unit, and B-(1–4)-D-glucoside linkages, forms a linear chain through the many inter- and intramolecular hydrogen bonds (Heux et al. 2000). The linearity of chitin makes it easy for the molecules to produce strong intermolecular forces which results in a high degree of crystallinity (Heux et al. 2000; Kumar et al. 2004). The DA and the degree of crystallinity are different for chitin from different sources (β, α). A large DA and greater degree of crystallinity lead to stronger intermolecular interactions and reduced solubility of chitin/chitosan in water and acidic solvents. The solubility of chitosan in a weak acid depends on the molecular weight, degree of neutralization of amine groups, the ionic strength, pH, and concentration of the polymer (Heux et al. 2000). In order to determine the DA of chitosan effectively it is advised to use a small concentration of the polymer and smaller macromolecules, which enables dissolution of chitosan and determination of a wider range of the DA in solution techniques. Table 4.1 shows the advantages and disadvantages of different methods used to determine the DA of chitin.

4.7 Applications of Chitin, Chitosan, and Chitosan Oligosaccharides (COS)

Functional foods first evolved in Japan, where the concept of food designed for specific medical benefit to the consumer evolved in the 1980s. Functional foods involve fortifying foods with added ingredients that can confer health benefits beyond basic nutrition (Stanton et al. 2005). The FDA has approved the use of chitosan for certain food applications, such as an edible film to protect foods. Chitosan produced by Primex® of Norway has Generally Recognized as Safe (GRAS) status (Preuss et al. 2006) and is considered a functional food. Functional foods may encompass a wide range of products and uses, including dietary supplements, medical foods, and food additives. Careful consideration of the potential market for a product must be weighed against the expense and time required in order to know the exact status of a product under the laws and regulations of the country where the product will be sold (see Chapter 8). As a dietary supplement, chitosan doesn't need FDA approval but chitosan is not approved as a food additive in many countries. However, despite this, in 2010, the fiber-boosting weight management ingredient chitin–glucan (branded as Artinia) won E.U. Novel Foods approval after EFSA approved its safety in 2010. This chitin product is extracted from the fungal source, *Aspergillus niger*.

4.8 Food Applications of Chitin and Its Derivatives

Chitin, chitosan, and chito-oligosaccharides have a myriad of food applications and conversion of shellfisheries' by-products into valuable, value-added ingredients has been identified as a key area of research and development. The uses of chitinous biopolymers range from preservative agents against food microbial contamination (Chen et al. 1998) to the recovery of wastewater materials from food processing plants (Pinott et al. 1997). Furthermore, chitin and chitosan find applications as food additives, anticholesterol agents, and as dietary supplements. They also find use as flavor ingredients and as substitutes for fat, and they are used additionally in edible films for preserving food quality and texture.

4.9 Fat Absorbance and Anticholesterol Agent

Chitosan prevents fat and bile acids from being absorbed into the bloodstream and allows them to be passed out of the body before they are absorbed. In addition, chitosan lowers the unfavorable LDL cholesterol and triglycerides and boosts the more favorable HDL cholesterol. It does not absorb vitamins and minerals (Chitin & Chitosan, April 2008).

4.10 Low-Density Lipoprotein–Cholesterol Reduction

The positively charged nature of chitosan and its polyelectrolyte properties, gel-forming abilities, and the presence of reactive functional groups govern the biological activities of chitosan. It has been studied for its effectiveness in reducing low-density lipoprotein (LDL) cholesterol levels and its fat binding capacity (Einbu et al. 2007). A study carried out by Jameela et al. (1994) demonstrated that chitosan decreased blood cholesterol levels by more than 50%, demonstrating its potential as a hypocholesterolemic agent. In addition, a preliminary human trial where 3–6 g of chitosan were consumed each day for 2 weeks resulted in decreased LDL cholesterol by approximately 6% and a boost in high-density lipoprotein (HDL) cholesterol by 10%. In the liver, chitosan and COS are able to reduce LDL cholesterol levels where they can prevent the action of hepatotrope poisons that are responsible in part for fatty liver (Maezake et al. 1993). It is thought that chitosan and COS act as fat scavengers in the digestive tract and remove cholesterol via ionic binding mechanisms where COS or chitosan bind to bile salts and acids, preventing the formation of lipid micelles. COS is also thought to decrease LDL cholesterol by increasing neutral sterol excretion. However, certain low molecular weight COS, which are usually less than 2,000 Da in size, fail to absorb dietary fats (Kim and Rajapakse 2005).

4.11 Hypertension and ACE-I-Inhibitors

Hypertension is one of the main risk factors for cardiovascular diseases including stroke and also renal failure. Angiotensin-I-converting enzyme (ACE-I) is a nonspecific but highly selective ectoenzyme involved in the regulation of blood pressure (Skeggs et al. 1957). ACE-I-inhibitors reduce peripheral blood pressure and exert an antihypertensive effect in vivo. CPS is a nonpeptidic group of competitive ACE-I inhibitors. Hong et al. (1998) identified chitosan trimer as the most effective oligomer in lowering blood pressure. Park et al. (2003) prepared deacetylated chitosans (90%, 75%, and 50%) from crab chitin and tested these for ACE-I-inhibition. In COS hydrolyzed from 50% deacetylated chitosan, the lowest degree of deacetylation tested showed the highest ACE-I-inhibitory activity (IC_{50} 1.22 ± 0.13 mg/ml). The food and general applications of chitin, chitosan, and COS are reviewed further in several papers and include their use in fire extinguishers, as flocculants in the recovery of milk proteins, in wastewater treatments, and in medicinal applications (Hayes et al. 2008b).

4.12 Markets

There is a significant and growing global market demand for chitin and its derivatives across many applications (Table 4.2). In 2008, the world market for chitin was estimated to be 11.4 thousand tons with a projected compound annual growth rate

Table 4.2 World chitosan consumption by application (t), 2010, 2015

Application	2010	2015 (Projected)	Projected CAGR (%)
Water treatment	6,670	11,436	11.55
Cosmetics and toiletries	2,031	3,776	13.39
Food and beverages	1,641	3,154	14.12
Healthcare/medical applications	1,474	3,063	15.93
Agrochemicals	1,181	2,604	17.36
Biotechnology	508	925	12.81
Pulp and paper	252	456	12.77
Textile finishing	172	336	14.48
Photography products	116	222	14.04
Miscellaneous	225	407	12.75

of approximately 20% to 2010. The market for chitin derivatives in the same period was estimated to be 33.4 thousand tons with projected growth rates of similar orders of magnitude. Strong growth rates are expected to continue, however, the rate of growth is expected to decline (Chitin & Chitosan, April 2008).

4.13 Markets by Application

A number of characteristics of chitin and its derivatives (e.g., natural, biopolymer, biodegradable, antibacterial, antifungal, antiviral, antiallergenic, and nontoxic) lend the product to many end uses in the industrial and biomedical fields. It acts as a flocculating/filtrating agent to purify drinking water, treat industrial wastewater including nuclear wastewater, and clean swimming pools.

Cosmetics and toiletries, food and beverages, healthcare/medical applications, and agrochemicals are also significant user segments. The natural and biodegradable characteristics of chitin and chitosan make it attractive in the cosmetics and toiletries segment where it is used, for example, as a moisturizing agent. The fat-absorption and cholesterol-lowering properties of chitin and chitosan make it attractive to the food (and petfood) area. Some of their derivatives are also used as dietary supplements to address joint pain as outlined earlier in this chapter. Developments in end-use applications in medical textiles, such as chitosan–alginate fiber for dressing wounds and barbed bidirectional surgical sutures, are expanding the healthcare/medical end-use segment. In agriculture, they are used as soil conditioners, in chemicals, and in pesticides.

Smaller, but nonetheless, rapidly growing segments include biotechnology, pulp and paper, textile furnishings, and photography products. In the pulp and paper arena, chitin can be used to produce biodegradable environmentally friendly packaging.

Lack of awareness of potential applications along with lack of commercialization mean that the end-use applications of chitin and its derivatives still have significant potential.

4.14 Geographical Markets

Japan is the largest market internationally for chitin derivatives, with water treatments being significant applications. The United States is the second largest market for chitin with projected growth rates larger than some other regions of the world including Japan, Europe, and Canada. Water purification is the single largest application of chitin in the United States whereas the major end-use segment for chitosan is cosmetics and toiletries applications (projected to be 918.7 t in 2010). Most production within Europe is destined for the U.S. market, however, demand within Europe is projected to increase significantly with Germany expected to be the biggest country market within Europe due to pressure from environmental groups to use natural/biodegradable products. Growth rates in Asia–Pacific are considerable to the extent that the market will surpass Europe in size. The major end-use segment for chitosan in this region is for agrochemical applications. (China is the largest producer of chitin in the world.), (Chitin & Chitosan, April 2008).

The price of chitin or its derivatives in these markets depends on its end application and its grade. Grades include biomedical, medical, or nonmedical with the biomedical grade chitin being expensive relative to nonmedical grades.

4.15 Market Prospects

Although double-digit growth is projected across all end-use segments and most geographic regions, there are a number of factors constraining growth in the market including:

- Insufficient supply to meet demand
- High costs of production
- Environmental impact of processing
- Quality control regarding molecular weight and deacetylation difficulty
- Limited availability of high-quality product

There are also some issues around specific applications, for example, conflicting media reports regarding the effectiveness of glucosamine and chondroitin as supplements for osteoarthritis. Other issues affecting the market include global trade issues. For example, in 2002/2003 there was a decline in the supply of chitin in China due to a European Union ban on seafood imported into China. This affected China's shrimp-meat sales and hence chitin production and ultimately price. Energy prices also affect the market as increased energy prices affect fishing as well as production costs.

Nonetheless, there continues to be significant improvements in technology to produce and extract chitin and its derivatives and also in its application domain indicating strong continued market demand. Key global players in the manufacture of chitin and chitosan include Primex (Norway), Kitozyme (Brussels), and United Chittechnologies, Inc. (United States) but there are several others (Global Industry Analysts, Inc 2008).

References

Chen, C., W. Liau, and G. Tsai. 1998. Antibacterial effects of N-sulfonated and N-sulfobenzoyl chitosan and application to oyster preservation. *Journal of Food Protection* 61: 1124–1128.

Einbu, A. 2007. Characterisation of chitin and a study of its acid-catalyzed hydrolysis. Ph.D. thesis, Norwegian University of Science and Technology, p. 21.

Furusaki, E., Y. Ueno, N. Sakairi, N. Nishi, and S. Tokura. 1996. Facile preparation and inclusion ability of a chitosan derivative bearing carboxymethyl-β-cyclodextrin. *Carbohydrate Polymers* 9: 29–34.

Global Industry Analysts. April 2008, Chitin and chitosan, A Global Strategic Business Report.

Hall, G.M. 1996. Biotechnology approaches to chitin recovery. In *Chitin and chitosan environmentally friendly and versatile biomaterials, proceedings of the second Asia Pacific symposium*, eds. Stevens, W.J., Rao, M.S., and Chandrkrachang, S. Asian Institute of Technology, Bangkok, pp. 26–33, 21–23 Nov 1996.

Hayes, M., B. Carney, J. Slater, and W. Bruck. 2008a. Mining marine shellfish waste for bioactive molecules: chitin and chitosan – Part A: Extraction methods. *Biotechnology Journal* 3: 871–877.

Hayes, M., B. Carney, J. Slater, and W. Bruck. 2008b. Mining marine shellfish waste for bioactive molecules: chitin and chitosan – Part B: Applications. *Biotechnology Journal* 3: 878–889.

Healy, M., A. Green, and A. Healy. 2003. Bioprocessing of marine crustacean shell waste. *Acta Biotechnologica* 23(2–3): 151–160.

Heux, L., J. Brugnerotto, J. Desbieres, M.-F. Versali, and M. Rinaudo. 2000. Solid state NMR for determination of degree of acetylation of chitin and chitosan. *Biomacromolecules* 1: 746–751.

Hong, S.P., M.H. Kim, S.W. Oh, C.H. Han, and C.H. Kim. 1998. ACE inhibitory and antihypertensive effect of chitosan oligosaccharides in SHR. *Korean Journal of Food Science* 30: 1471–1479.

Jameela, S.R., A. Misra, and A. Jayakrishnan. 1994. Cross-linked chitosan microspheres as carriers for prolonged delivery of macromolecular drugs. *Journal of Biomaterials Science, Polymer Edition* 6: 621–632.

Jang, M.K., B.G. Kong, Y.I. Jeong, C.H. Lee, and J.W. Nah. 2004. Physicochemical characterization of alpha – chitin, beta-chitin and gama – chitin separated from natural resources. *Journal of Polymer Science Part A: Polymer Chemistry* 42: 3423–3432.

Kasaai, M.R. 2009. Various methods for determination of the degree of N-Acetylation of chitin and chitosan: a review. *Journal of Agricultural and Food Chemistry* 2009(57): 1667–1676.

Kim, S.-K., and N. Rajapakse. 2005. Enzymatic production and biological activities of chitosan oligosaccharides (COS): a review. *Carbohydrate Polymers* 62: 357–368.

Kjartansson, G.T., S. Zivanovic, K. Kristbergsson, and J. Weiss. 2006. Sonication assisted extraction of chitin from shells of fresh water prawns (Macrobrachium rosenbergii). *Journal of Agricultural and Food Chemistry* 54: 3317–3323.

Knorr, D. 1984. Use of chitinous polymers in food. *Food Technology* 38: 85–97.

Kumar, M.N.V.R., R.A.A. Muzzarelli, C. Muzzarelli, H. Sashiwa, and A.J. Domb. 2004. Chitosan chemistry and pharmaceutical perspectives. *Chemical Reviews* 104: 6017–6084.

Legarreta, G.I., Z. Zakaria, and G.M. Hall. 1996 Lactic fermentation of prawn waste: comparison of commercial and isolated starter cultures. In *Advances in chitin science* Vol. I eds. A. Domard, C. Jeuniaux, R. Muzzarelli and G. Roberts; Jacques Andre publishers, 1996, Lyon ISBN 2-907922-40-8. pp. 399–406.

Maezake, Y., K. Tsuji, and Y. Nakagawa. 1993. Hypocholesterolemic effect of chitosan in adult males. *Bioscience, Biotechnology, and Biochemistry* 57: 1439–1444.

Muzzarelli, R.A.A. 1977. *Chitin*. Oxford: Pergamon.

No, H.K. 1995. Preparation and characterisation of chitin and chitosan. A review. *Journal of Aquatic Food Product Technology* 4(2): 27–52.

Park, P.J., J.Y. Je, and S.K. Kim. 2003. Angiotensin I converting (ACE) inhibitory activity of hetero-chitooligosaccharides prepared from partially different deacetylated chitosans. *Journal of Agricultural and Food Chemistry* 51: 4930–4934.

Peniston, Q. P., and E. Johnson. 1980 Process of the manufacture of chitosan. US Patent No. 4,195,175, pp. 5–15.
Percot, A., C. Viton, and A. Domard. 2003. Optimisation of chitin extraction from shrimp shells. *Biomacromolecules* 4: 12–18.
Pinott, A., A. Bevilacqua, and N. Zaritzky. 1997. Optimisation of the flocculation stage in a model system of a food emulsion waste using chitosan as polyelectrolyte. *Journal of Food Engineering* 32: 69–81.
Preuss, H.G., and G.R. Kaats. 2006. Chitosan as a dietary supplement for weight loss: a review. *Current Nutrition & Food Science* 2: 297–311.
Roberts, G.A.F. 2008. Thirty years of progress in chitin and chitosan. *Progress on Chemistry and Application of Chitin and Its Derivatives* XIII: 7–15.
Shahidi, F., J.K.V. Arachchi, and Y.-J. Jeon. 1999. Food applications of chitin and chitosans. *Trends in Food Science & Technology* 10: 37–51.
Skaugrud, O., and G. Sargent. 1990 'Chitin and chitosan: Crustacean biopolymers with potential' international by-products conference, Anchorage, Alaska, pp. 61–72.
Skeggs, L.T., J.E. Kahn, and N.P. Shumway. 1957. The preparation and function of the angiotensin-converting enzyme. *Journal of Experimental Medicine* 103: 295–299.
Stanton, C., R.P. Ross, G.F. Fitzgerald, and D. van Sinderen. 2005. Fermented functional foods based on probiotics and their biogenic metabolites. *Current Opinion in Biotechnology* 16: 198–203.
Tsigos, I., A. Martinou, K.M. Varum, and V. Bouriotis. 1996 Enzymatic deacetylation of chitinous substrates employing chitin deacetylases. In *Advances in chitin science* Vol. I eds.A. Domard, C Jeuniaux, R. Muzzarelli and G. Roberts; Jacques Andre publishers, 1996, Lyon ISBN 2-907922-40-8. 59–69.
Yoshiichi, A., Tomoya, T. and Akira, A. 1987. Japanese Patent No. 87179503.
Yu, G., F.G. Morin, G.A.R. Nobes, and R.H. Marchessault. 1999. Degree of acetylation of chitin and extent of grafting PHB on chitosan determined by solid state 15 N NMR. *Macromolecules* 32: 518–520.

Chapter 5
Industry Potential of Marine Bioactive Components: Downstream Processing and Vehicles for Efficient Delivery In Situ

Wolfram M. Brück, Steven Reisse, Daniel Garbe, and Thomas B. Brück

5.1 Introduction

The marine environment is a diverse source of compounds which, through the use of biotechnology, yield a great variety of new products for industrial development. To date, thousands of unique chemical compounds for use in pharmaceuticals, cosmetics, nutritional supplements, molecular probes, enzymes, fine chemicals, and agrichemicals have been identified, each with potential multimillion-dollar market value (BioScience 1996). Due to this variety, the challenge that is faced by the marine biotechnology industry is to identify new sources of marine bioproducts, develop new screening techniques, and provide a sustainable source of supply (Pomponi et al. 2007). However, the biggest challenge lies in the broad spectrum of separation techniques that have to be applied for the downstream processing of marine compounds, inasmuch as it has been identified as the most ineffective and expensive part of the overall bioprocess. In general, each downstream process consists of the following steps.

1. Removal of insoluble particles
2. Isolation of the product
3. Purification
4. Polishing (Betler et al. 1988)

W.M. Brück (✉)
Centre of Applied Marine Biotechnology, Letterkenny Institute of Technology, Donegal, Ireland

Bioanalytical Science, Food and Health Microbiology, Nestlé Research Center, Lausanne, Switzerland
e-mail: wolfram.brueck@rdls.nestle.com

S. Reisse • D. Garbe • T.B. Brück
Department of Chemistry, Division of Industrial Biocatalysis, Technische Universität München (TUM), Garching bei München, Germany

M. Hayes (ed.), *Marine Bioactive Compounds: Sources, Characterization and Applications*, DOI 10.1007/978-1-4614-1247-2_5,

The aim is to purify a certain compound. This typically means that compounds are separated in a dilute suspension with the aim of creating a highly purified dry product. Due to the type and concentration of most compounds in their natural source, however, this highly purified dry product may only be available in micro- or nanogram amounts. On a laboratory scale, one is only limited by available equipment and product yield, however, industrial production of marine bioactives faces several other problems, including the following.

1. Efficient and cost-effective production
2. Consistent product quality
3. Impurities
4. Location of the product in the process stream (Muffler and Ulber 2005)

In addition, it is also important to consider that the product's behavior may change in large-scale production compared to laboratory-based purification as the materials and extraction protocols for production may be modified. Furthermore, several upstream parameters that may influence product recovery and product properties during downstream processing (Muffler and Ulber 2005) should be taken into consideration. These include:

1. Characteristic properties of the producing organism
2. Product location in the producing organism
3. Product stability
4. By-products and impurities
5. Product concentration in recovery medium

On the other side of the process, finding efficient in situ delivery vehicles for the application of purified marine compounds presents another neglected bottleneck. Although certain marine compounds may utilize the same delivery vehicles as traditional products they aim to replace (i.e., biofuels, bioplastics, bioemulsifiers, and biosurfactants), other products may require more novel and innovative ways for efficient release. In this chapter a variety of available and potential downstream processes for marine bioactive compounds and novel vehicles for their efficient and effective in situ delivery are discussed based on their various applications.

5.2 Oligo- and Polysaccharides

The marine environment is an extraordinary source of polysaccharides that find a multitude of uses in the pharmaceutical, chemical, and food industries. Carrageenans and agar that are used for gel formation are largely found in red algae such as *Gelidium*, *Grateloupia*, and others (BeMiller and Whistler 1996). Fucans and fucanoids with their application as neutraceuticals are found in brown algae, sea urchin eggs, and sea cucumbers (Berteau and Mulloy 2003). Chitin, chitosan, and their derivatives may also be used as a gelling agent. However, due to their unique structure, they represent a unique molecule for which novel applications abound.

5.3 Chitin, Chitosan, and COS

As discussed earlier in Chapter 4, chitin is a cationic amino polysaccharide composed of N-acetyl-D-glucosamine (GlcNAc, 2-acetamido-2-deoxy-D-glucose; ~50–100%) with β (1→ 4) glycosidic bonds between each monomer (Beaney et al. 2005). It occurs in three polymorphic solid-state forms designated as α, β, and γ chitin which differ in their degree of hydration, size of unit cell, and number of chitin chains per unit cell (Einbu 2007). Worldwide production of chitin has been estimated to be approximately 10^{11} tons per annum making it the most abundant nitrogen-bearing biopolymer in nature (Kurita 2006). Crustacean and mollusk waste streams constitute a rich source of chitin and the full exploitation/bioconversion of this easily accessible resource has attracted much interest. Proportions of chitin from these sources may vary with season and species but in general, exoskeletons contain 15–40% chitin, 20–40% protein, and 20–50% calcium carbonate, with other components such as pigments, lipids, and other metal salts present as minor components (Kurita 2006; Singer and Wooten 2003; No and Meyers 1995). Chitosan is a copolymer of GlcNAc (approx. 20%) and glucosamine (GlcN, approx. 80%) and is obtained through the deacetylation of chitin using hot alkali (Tharanathan and Kittur 2003). When the degree of deacetylation (DD) is over 60% the term chitosan is preferred (Peter et al. 1986). Like chitin, chitosan is a β (1→ 4) glycan and has been described as nature's most versatile biomaterial as it is biodegradable, renewable, and almost nontoxic (Sandford 1989). Although at a molecular level, chitin and chitosan appear similar, both possessing reactive hydroxyl and amino groups, chitosan is more accessible to reagents which may be due to its less crystalline structure. This makes chitosan readily dissolvable in aqueous acids, with formic acid and acetic acid being most commonly used (Rudrapatnam et al. 2002; Tsugita 1990; Muzzarelli 1985). Further acid hydrolysis cleaves the glycosidic linkages of the chitin and chitosan main chains, producing chitin/chitosan oligomers (chito-oligosaccharides, COS) that are β (1→4) linked homo- or hetero-oligomers of GlcNAc and/or GlcN with a chain length of up to $n = 15$ and a molecular weight of up to 10 kDa (Kurita 2006; Kim and Rajapakse 2005; Jeon and Kim 2000; Domard and Cartier 1989). Unlike chitin or chitosan, COS have a lower viscosity with greater solubility in water at neutral pH due to their shorter chain length and free amino groups in GlcN units (Kim and Rajapakse 2005).

5.3.1 Industrial Processing of Chitinous Raw Material

In general, chitin is isolated from shellfish waste in three steps: deproteinization (DP), demineralization (DM), and decolorization (DC) whereby the order of the first two steps is generally considered irrelevant if protein or pigment recovery is not an objective (Fig. 5.1; Shahidi and Synowiecki 1991). Industrial techniques for chitin and chitosan extraction from different shell waste streams normally rely on harsh chemical processes due to covalent associations with other shell constituents.

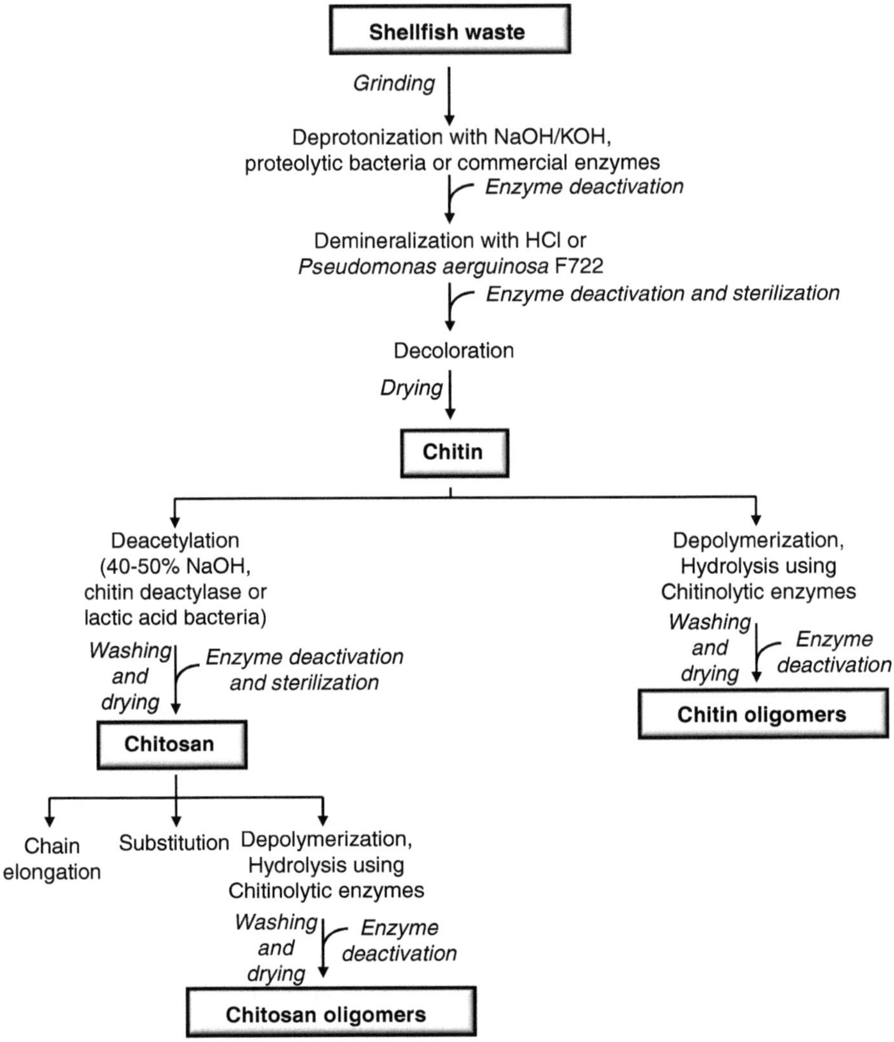

Fig. 5.1 Flow diagram for the production of Chitin, Chitosan, and COS from shellfish waste using chemical and enzymatic digestion (Modified from Brück et al. 2011)

Initially, proteins are removed from ground shells by treating with mild sodium hydroxide or potassium hydroxide solution at elevated temperature. Alkali (i.e., NaOH, KOH) concentrations between 1% and 10% with temperatures ranging from 30°C to 100°C, independent of starting material, are most common with a reaction time between 30 min and 12 h (Shahidi and Synowiecki 1991; Whistler and BeMiller 1962). The removal of calcium carbonate, calcium phosphate and other mineral salts found in shell waste is accomplished by extraction with dilute acids. It is important that the amount of acid is stoichiometrically equal to or greater than all the minerals present in the shell to ensure complete DM (Shahidi and Synowiecki 1991). Hydrochloric acid at room temperature and a reaction time of 2–3 h was most commonly used, which produces a brown to brownish-white product that may be bleached using a variety of reagents (No and Meyers 1995). Commonly used bleaching reagents include ethanol, ether, acetone, sodium hypochlorite, hydrogen peroxide, or a combination, chiefly used at ambient temperature (No and Meyers 1995). White chitin and chitosan without the need for DC were obtained by using 30% HCl for DM and 1.5 N NaOH for DP from Hariana Shrimp (Naznin 2005). Effective DA of chitin was achieved by washing the intermediate product in two or more changes of water during DP. This resulted in nondegraded, 100% deacetylated chitosan (Mima et al. 1983).

Another method that required considerably less severe conditions and resulted in better DA with less degradation was described by Domard and Rinaudo (1983) who added thiophenol as an oxygen trap and catalyst during DA. The technique was later refined by Suryanarayana Rao et al. (1987) who impregnated chitin with a four times excess NaOH (w/v) by mixing and heating to 60°C for 2 h. For the industrial-scale production of COS, chitin and chitosan may be hydrolyzed with concentrated HCl at elevated temperatures to form monomeric glucosamine, whereas less severe methods yield a series of GlcNAc oligomers (chitin) or GlcN oligomers (chitosan; Kurita 2006; Domard and Cartier 1989). This, however, often results in low yields of COS with desired molecular weights and a large amount of monomers (Uchida et al. 1989). Commercially available proteolytic enzymes such as Alcalase (EC 3.4.21.62), chymotrypsin (EC 3.4.21.2), and papain (EC 4.3.22.2) have been used to extract protein and chitin from shell waste (He et al. 2006; Wang and Chio 1998). Bacterial release of chitin, chitosan, and COS was achieved using a bacterial protease from *Pseudomonas maltophilia* for DP or a protease from *Bacillus* sp. TKU004 (Wang et al. 2006; Tsai et al. 2002). Chitinolytic enzymes such as chitinase, chitonsanase, papain, and lysozyme are widely distributed in all kingdoms and enzymatic hydrolysis for COS production is preferable because these methods obtain greater yields of oligomers with a higher degree of polymerization (Kim and Mendis 2006; Chen et al. 2003; Jeon and Kim 2000; Muzzarelli et al. 1994; Uchida et al. 1989). In particular, chitin deacetylase from *Colletotrichum lindemuthianum*, *Mucor rouxii*, *Abisidia butleri*, or *Aspergillus nidulans* may convert chitin from shell waste to chitosan and COS (Tsai et al. 2002; Arcidiacono et al. 1989; Kauss and Bauch 1988). However, different preparation methods result in different DD, distribution of acetyl groups, chain length, and conformational structures (Synowiecki 2007; Kurita 2006; Rudrapatnam et al. 2002).

5.4 Commercial Applications of Chitin and Derivatives

Chitosan can be used for the clarification of waste and effluent water and together with its derivatives carboxymethyl chitosan and cross-linked chitosan have been used for the removal of Pb^{2+}, Cu^{2+}, and Cd^{2+} and organic pollutants from drinking water and aquatic environments (Aksu 2005; Deans and Dixon 1992; Onsoyen and Skaugred 1990; Muzzarelli et al. 1989). Barriada et al. (2007) further demonstrated that chitin was as good as other waste materials as other low-cost sorbents for the removal of Cd and Pb but was poorer compared with raw crab shells from spider crab, *Maia squinado*. It has also proved more efficient than charcoal in the removal of polychlorinated biphenyls from contaminated waters (Tharanathan and Kittur 2003).

In recent years, chitosan has attracted the interest of the pharmaceutical industry as it possesses the ability of controlled drug release, wound healing, use as a scaffold for tissue regeneration, and blood cholesterol control (Khoushab and Yamabhai 2010; Muzzarelli et al. 1999; Baba et al. 1989). It has been shown that finely divided chitin, partially deacetylated chitin, and chitin derivatives possess the ability to promote the healing of wounds and would be suitable for use in bandages and surgical sutures, or may be blown directly into the wound using an atomizer to aid blood coagulation (Balassa 1971). Okamoto et al. (2003) further reported that chitin is an effective agent for hemostasis maintenance, helping blood coagulation through aggregating platelets. To use chitin as a functional wound dressing, a hydrogel sheet composed of a blended powder of alginate, chitin/chitosan, and fucoidan (ACF-HS; 60:20:2:4 w/w) has been developed. In a rat model comparing wounds treated with commercially available calcium alginate fiber dressings, ACF-HS, and untreated wounds, examination demonstrated significantly advanced granulation tissue and capillary formation in healing-impaired wounds treated with ACF-HS after 7 days (Murakami et al. 2010). Chitosan and its derivatives have also been shown to reduce scar tissue formation by inhibiting fibrin formation and affecting macrophage activity (Muzzarelli et al. 1999). Chitin and its derivatives have also been used as scaffolds for bone and tissue regeneration as they are able to form a temporary porous matrix that allows tissue to grow while being biodegradable and nontoxic after digestion (Tsioptsias et al. 2009; Brandl et al. 2007). In addition, chitosan acetate bandages have been shown to have antibacterial activity when applied to burned skin contaminated with *Pseudomonas aeruginosa* (Dai et al. 2009). Antifungal and other antibacterial activity against *Fusarium oxysporum*, *Colletotrichum gloesporioides*, *E. coli*, and *Xanthomonas* sp. were also investigated (Li et al. 2008a; Bautista-Baños et al. 2003; San-Lang et al. 2002; Tsai and Su 1999).

5.5 Enzymes

Marine enzymes exhibit some unique properties such as high salt tolerance, hyperthermostability, barophilicity, cold adaptivity, and easy large-scale cultivation useful for various industrial applications (Debashish et al. 2005). The sources of marine enzymes have been widely studied and include marine micro-organisms (extremophiles, symbionts, and sediment and water assemblages), marine animals, and marine plants.

5.6 Sources of Marine Enzymes and Their Application

Extremophiles isolated near thermal vents on the ocean floor produce the second generation of thermostable polymerase chain reaction enzymes, marketed as Vent^{Tm} and Deep Vent^{Tm} polymerases (Grace 1997). Similarly, enzymes, such as proteases and carbohydrases from psychrophilic micro-organisms, could be used in commercial laundry detergents capable of washing clothes in cold water, thus dramatically reducing electricity consumption (Deming 2002). Other interesting enzymes with industrial potential have been isolated from micro-organisms associated with marine animals. *Pseudoalteromonas* sp. isolated from the Antarctic krill *Thyssanoessa macrura* produces a cold-adapted β-galactosidase, and a *Sphingomonas paucimobilis* isolated from the krill *Euphausia superba* produced an extracellular protease (Turkiewicz et al. 2003; Turkiewicz et al. 1999). A *Mucor* sp. from the marine sponge *Spirastrella* sp. produces a novel amylase (Mohapatra et al. 1998). Free-living bacteria from near-shore sediments have produced numerous proteases, carbohydrases, and peroxidases, with commercial applications such as biofuel production, detergent and baking additives, or as diagnostic reporter enzymes. Among those, *Vibrio* sp. were found to produce a host of different proteases, including a detergent-resistant alkaline serine exoprotease, produced by *Vibrio alginolyticus* (Debashish et al. 2005). An endonuclease from *Alteromonas espejiana* BAL 31 produces BAL 31, a nuclease with endonuclease activity specific to single-stranded DNA. In addition BAL 31 also has exonuclease activity and degrades double-stranded nucleic acids from both the 3′- and 5′-ends when single-stranded nucleic acids are not present (Gray et al. 1981). The enzyme is currently manufactured by Takara Bio Inc., Japan.

Marine animals and plants also represent a large group of organisms from which commercial enzymes have been isolated. From marine animals, enzymes such as proteases (Arreguín et al. 1993), amylases (Munilla-Morán and Saborido-Rey 1996), cholinesterases (Mora et al. 1999), and protein kinases (MacDonald and Sptorey 1999) among others have been isolated. Two of the most utilized and well-known enzymes originate from the bioluminescent jellyfish, *Aequora victoria*, namely the luminescent protein aequorin and the fluorescent molecule GFP (green fluorescent protein; Hannick et al. 1993; Prasher et al. 1992). Marine plants and especially marine algae have also been good candidates for the isolation of novel enzymes with commercial importance (Debashish et al. 2005). Haloperoxidases such as vanadium bromoperoxidase with a high degree of stability have been identified in red, brown, and green algae, with *Rhydophyta* being the richest in bromoperoxidase activity, with the genus *Corallina* possessing the highest activity (Winter and Moore 2009). Halogenated compounds with pharmaceutical and agrochemical applications are of commercial significance. Examples are the chlorinated and fluorinated antibiotics vancomycin and ciprofloxacin, respectively. Although chemical synthesis of organohalogens requires harsh conditions, biosynthesis of these compounds occurs under relatively mild conditions and often with a significant stereospecificity. Although the mechanism of haloperoxidases is not well understood, these enzymes have growing applications in sustainable production processes for agrochemicals and pharmaceuticals (Murphy 2003).

Cyanobacteria, red algae, and cryptomonads are also an important source of Phycoerythrin, a red protein from the light-harvesting phycobiliprotein family which is an accessory pigment to the main chlorophyll pigments responsible for photosynthesis (MacColl et al. 1996; French and Young 1952). R-Phycoerythrin and B-Phycoerythrin are among the brightest fluorescent dyes ever identified and are currently marketed by various life science companies as fluorescent conjugates to other molecules to make fluorescent probes or antibodies. Luciferase from the marine dinoflagellates *Lingulodinium polyedrum* and *Pyrocystis lunula* is responsible for the enzymatic oxidation of luciferin (Morishita et al. 2002). Luciferase is commonly used as a reporter molecule to assess the transcriptional activity of transfected cells or to assess the level of cellular ATP in cell viability assays (Fan and Wood 2007). Other enzymes isolated from marine plants are ATPase (Muramatsu et al. 2002), urease (Fan et al. 2003), ligase (Maurin and le Gal 1997), and isomerase (Zheng et al. 2002) among others.

5.7 Isolation and Formulation of Industrially Relevant Proteins

The isolation and formulation of industrially relevant proteins depends on whether the entities are produced using a recombinant production system or are isolated from the natural source (Fig. 5.2). Certainly, if the protein of interest originates from a marine mammal, plant, or an unculturable microbe, it would be advantageous to utilize modern genomic tools to transfer the enzyme encoding genes into industrial production systems such as *E. coli*, *S. cerevisiae*, or *Aspergillus niger* (Jiang et al. 2010; Kovar et al. 2010; Tang and Zhao 2009). Protein production in recombinant systems often results in intracellular expression. The isolation of recombinant proteins follows standard protocols, including cell lysis by osmotic shock in the absence or presence of detergents and removal of cell debris using centrifugation. The protein of interest is then commonly found in the soluble fraction remaining after centrifugation (Wang et al. 2010). The superior advantage of using recombinant production methods for marine proteins is the possibility for merging the gene sequence of the target protein with a DNA sequence encoding a conventional purification tag, such as an His_6, CBP, or Strep^{Tm}-tag prior to protein expression. On a protein level, the presence of these tags allows for single-step purification of the protein of interest, even from dilute solutions, via specific affinity chromatography (Walls and Loughran 2011; Lichty et al. 2005).

Production of marine enzymes in "standard" recombinant systems, however, does not guarantee high production titers or correct protein folding inasmuch as endogenous promoters and chaperones are mostly not available. Therefore, generation of industrially relevant production titers of functional marine enzymes may be more successful by "classical" (i.e., chemical mutagenesis) optimization of the natural hosts (Jiang et al. 2011). Methods for processing and isolation of natural marine enzymes highly depend on whether the target enzyme is localized intracellularly or

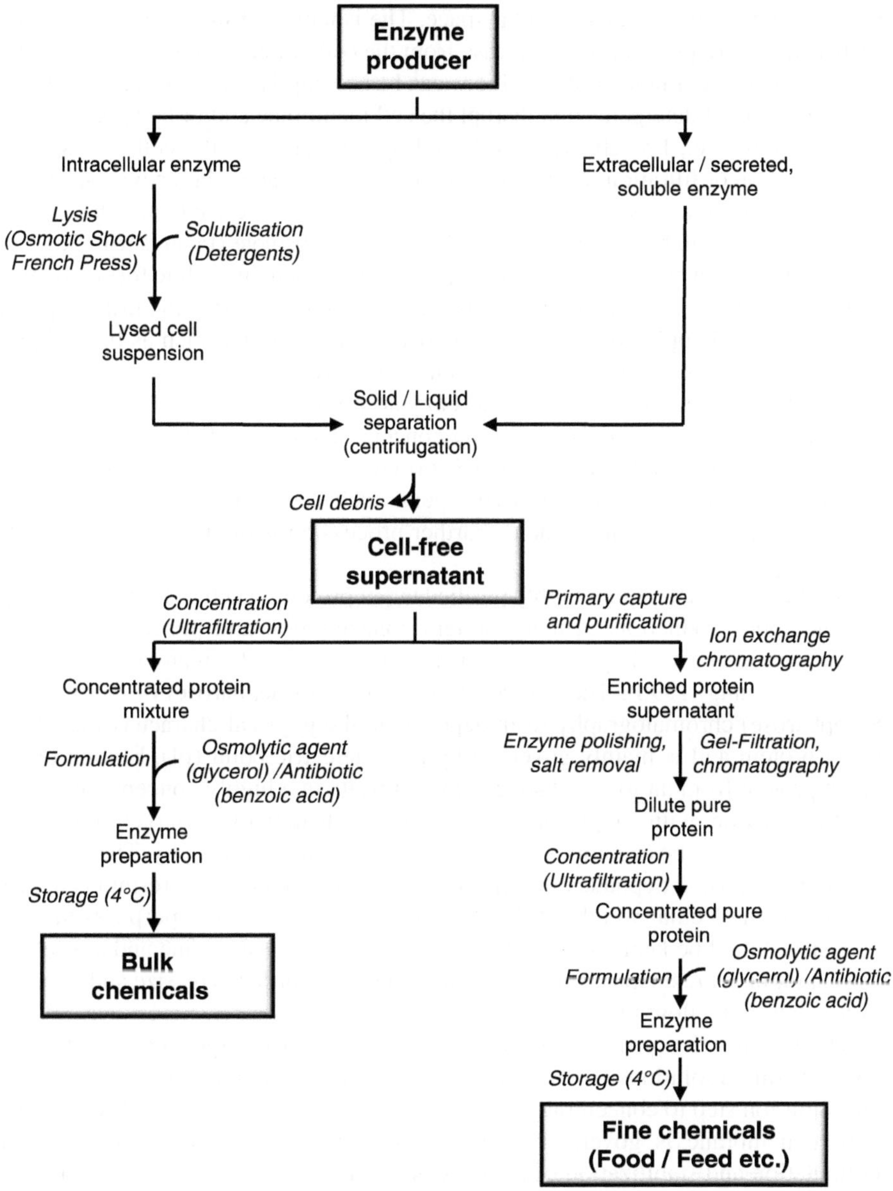

Fig. 5.2 Flow diagram for the industrial production of enzymes and proteins from marine sources

if it is secreted into the extracellular space. The isolation of intracellular enzymes, such as ligase, requires efficient release from the cellular environment and solubilization, which under industrial conditions can be accomplished using osmotic shock, by application of detergents that disrupt the cell membrane and/or by application of high pressure (i.e., French Press) followed by separation of the cell debris using centrifugation or filtration techniques. In the case of extracellular enzymes, such as proteases and other hydrolases, which are commonly secreted into the growth medium and are highly soluble, the cell lysis step is obsolete (Fu and Kim 2010).

Once the target enzyme is solubilized, it is often present in a dilute form (Harakas and Builder, 1987). Purification options then depend on the protein purity required for the intended application. For bulk industrial applications, such as biofuel production that utilize lipases, cellulose, and other hydrolase activities to perform chemical transformations, only crude enzyme preparations are required. Hence, the primary protein isolate is often only concentrated using filtration technologies such as ultrafiltration. Stabilization and formulation of the enzyme system can be accomplished by addition of osmolytic agents such as glycerol (2–25% v/v) and/or antibacterials such as benzoic acid. Further processing is often not desired due to processing costs.

Applications that require highly purified target proteins, such as for pharmaceutical and food production, primary protein concentration and purification can be accomplished in a single process step using ion exchange chromatography (Christi and Moo-Young, 1990). The choice of anion- (i.e., Q-Sepharose) or cation- (i.e., S-Sepharose) chromatography resin depends on the physical characteristics of the target protein and is mainly governed by its isolelectric point (pI). Elution of the target protein from an ion-exchange column results in a highly concentrated semi-purified fraction, with a high salt concentration. Polishing of the target protein and desalting can then be accomplished by gel-filtration chromatography. The choice of filtration resin (i.e., Sephadex or Sephacryl S-75, S-100, or S-200 resins) is dependent on the molecular weight of the target protein. For industrial purposes the column size should be kept to a minimum to decrease processing time and cost of the chromatography resin material. Gel filtration separates proteins by size and removes excess salt in the flow-through fraction.

The gel filtration process can result in highly purified target proteins but also results in dilute solutions. Therefore, gel filtration will commonly be followed by an ultrafiltration step to concentrate the target protein. Again the choice of the ultrafiltration membrane is governed by the molecular weight of the target protein. Formulation and stabilization of the concentration enzyme fraction can be accomplished by the use of osmolytic additives (i.e., glycerol or sucrose). An alternative to filtration procedures is lyophilization of freeze-drying of the target protein preparation. The lyophilized protein preparation can be further stabilized by addition of inert inorganic carriers, such as bentonites and zirconia (Salazar et al. 2008). Storage conditions, particularly storage temperature are highly dependent on the protein, its stability in the storage buffer system, and the presence of any additives. For the short term, protein preparations can be stored at 4°C, whereas storage in excess of 3 months is commonly carried out at −20°C.

5.8 Lipids

Microalgae contain a host of secondary metabolites, some of which are produced in high concentrations, making them attractive organisms for further exploration (Oh et al. 2003; Richmond 2003; Chen and Jiang 2000).

5.9 Market Opportunities

With markets for microalgae products already existing, their growth is currently limited by the production technologies employed (Pulz and Gross 2004). To develop microalgal bioproducts on an industrial scale, large amounts of biomass must be produced in an efficient and reasonably compact installation, resulting in biomass that is of sufficient concentration of the desired product (Wijffels 2008). Current commercial applications of microalgae concentrate on low-volume, high-value lipids and lipophilic pigment products such as carotenoids and ω-3 fatty acids and polyunsaturated fatty acids. To be a feasible alternative for the production of biofuel the current production costs need to be reduced by at least two orders of magnitude. In addition, the scale of production for microalgal lipids to produce biodiesel would have to be increased by three orders of magnitude (Chisti 2007). The theoretical production maximum in areas of high irradiation would yield an annual productivity of approximately 280 t of dry biomass. With an assumed lipid content of 40% in the microalgae, the total amount of oil that could thus be produced for the biofuel industry would be 150,000 L/ha/year. However, with today's state-of-the-art technology, a maximum of 20,000 L/ha/year could only be obtained, which is still significantly more than the productivity of other high-energy crops such as palm oil with a productivity of 6,000 L/ha/year (Molina Grima et al. 2003). Once production of biofuel from microalgae has been optimized, it would further offer the following advantages over land-based biofuel crops.

1. High growth rate of microalgae makes it possible to satisfy the current and future biofuel demand while using a minimum of resources.
2. Microalgal cultivation consumes less water than land crops.
3. Microalgal cultivation allows for high-efficiency CO_2 mitigation.
4. Nitrous oxide release could be minimized.
5. Microalgal farming could be more cost-effective than conventional farming of crop biofuels (Li et al. 2008b).

5.10 Industrial Production and Application of Microalgal Lipids

Current bioreactors for the commercial production of microalgae consist of open systems such as raceway ponds which, due to their simplicity and low investment costs are assumed to be the best system for biofuel and food production.

However, due to their construction, the concentration of biomass is very low and requires extensive labor costs for harvesting. Open culture systems require large surface areas and shallow depth (ca. 12–15 cm) to improve light penetration. Furthermore, agitation of the culture prevents the cells from sinking to the bottom and facilitates efficient cell growth. In raceway ponds, agitation often is accomplished by using a paddle wheel (Chaumont 1993; Richmond 1992). Contamination by different algal species is the biggest problem in open culture systems. As a result, *Chlorella*, *Dunaliella*, and *Spirulina* are especially attractive strains for open culture due to their tolerance to extreme environments such as high nutrient concentrations, high salinity, and high pH (Matsunaga et al. 2005). In addition, open culture systems are also easily affected by weather conditions that cause contamination due to a reduction in salinity from rain. As a result, chemicals and metabolites that require more restricted conditions for efficient growth are grown in closed culture systems. Closed photobioreactors, such as a flat panel of tubular bioreactors, can result in higher productivity but require a larger primary investment. Because their produced biomass is in high density, costs for harvesting are negligible whereas costs for efficient mixing of the biomass could yield an overall negative energy balance (Lee 2001). However, considering the investment and production costs required and the productivity of the systems, both open and closed systems are currently similar in cost (Wijffels 2008). In the future, closed systems could, however, be the most economical and energy-efficient process for large-scale microalgal food, lipid, and biodiesel production inasmuch as these photobioreactors offer several advantages:

1. Facilitates maintenance of monoalgal cultures by protecting them from contamination
2. Reduces water loss and the subsequent increase of salinity
3. Results in higher productivity with greater cell densities
4. Are applicable for various microalgal species
5. Provides an optimized environment

Prior to extraction of cellular components, algal biomass has to be separated from the liquid growth medium. Industrial techniques for solid/liquid separation applied to algal culture include centrifugation, filtration, and flocculation (Harun et al. 2010). Under process conditions, algal biomass is commonly dewatered using a continuous centrifugation. An alternative means for dewatering algae biomass is the application of filtration. Filtration is energetically more advantageous than centrifugation but requires more process time. Filtration methods applied to the harvesting of algal biomass include dead-end filtration, microfiltration, ultrafiltration, pressure filtration, vacuum filtration, and tangential flow filtration (TFF). Harvesting algal cells by filtration involves running the cultivation medium through filters on which the algae will settle, whereas the medium will pass through the filter system. The culture medium will be continually cycled through the microfilters until a thick algae paste is obtained. Also the use of filter presses under pressure or a vacuum are adequate methods to concentrate microalgae and are considered to isolate algal cells as large as *Spirulina* sp. efficiently. Small algae cells such as those of *Dunaliella* and *Chlorella* species cannot be recovered by the pressure or vacuum filtration methods.

Recent literature studies indicate that tangential flow filtration and pressure filtration are energy-efficient dewatering methods, as they consume adequate amounts of energy when considering the output and initial concentration of the feedstock. In contrast to TFF, simple filtration methods such as dead-end filtration are not adequate as dewatering techniques for industrial processes due to issues of back mixing. The most cost-and energy-effective method for separating algal biomass is flocculation. Certain algal strains form aggregates naturally at high cell densities and when CO_2 supply is interrupted, as commonly practiced during batch cultivation at the end of the cultivation phase. This process of autoflocculation will result in algal cells separating out of the medium by gravity once medium mixing has stopped. When algal cells do not form aggregates, the addition of flocculation agents, such as the inorganic salts aluminum sulfate and ferric chloride, organo-polymers (i.e., polyethylene amine, polystyrene sulfanate, polyacrylamide, chitosan), or functional clays (i.e., bentonites) will induce gravitational settling of biomass. Multivalent salts, such as aluminum sulfate are used with varying effectiveness, which is mainly due to the ionic charge of the flocculant. However, there is literature precedence that Fe^{3+} flocs with induced pH to harvest various kinds of algae can achieve around 80% efficiencies.

Cationic organo-polymers are polyelectrolytes, which have the advantage of physically linking cells. However, the efficiency of flocculation will largely depend on the physical characteristics of the polymer. Key polymer properties include ionic charge, molecular weight, and concentration. It has been demonstrated that increasing the molecular weight and charge on the polymers results in increased binding properties. The choice of organo-polymer used for flocculation will depend on the properties of the algal culture, including biomass concentration, charge, and pH of the culture medium. Organo-polymers have the desired properties that they are less sensitive to medium pH and require a lower dosage to induce flocculation compared to their inorganic counterparts.

The most cost-efficient method to induce flocculation is the use of functional clays, such as bentonites, which have large surface areas and will bind algal cells though ionic interactions (Shelef et al. 1984). An overview of industrially relevant flocculants and their dosage for inducing effective flocculation are shown in Table 5.1. Once algal biomass has been dewatered the resulting algal cake, which now comprises 15–30% w/w dry solids will be processed to extract and fractionate value-adding chemical components of the algal cell, such as lipids, carbohydrates, and secondary metabolites (i.e., carotenoids, terpenes, etc.). As certain algal strains may accumulate up to 70% w/w lipids in form of neutral oils and fats, polyunsaturated fatty acids, or glycolipids, this compound has found interest in the biofuel, food, and pharmaceutical industries, respectively. Extraction processes for these components from the cellular environment largely depend on the chemical nature of the compound to be extracted and the purity required in downstream processing steps. For biofuel applications, neutral lipids such as oils are extracted using an expellor/oil pressing, organic solvent extraction, supercritical fluid extraction (SFE), and ultrasound techniques. The use of expellor presses for oil extraction, which are used for processing oil seeds, is an industrially well-established method for oil extraction. The main issue for extracting algal oil using this methodology, however,

Table 5.1 Industrial flocculation reagent pH and dosing regimens for effective algal biomass separation (Modified from Harun et al. 2010)

Flocculant	Type	Optimal dose mg/l	Optimal pH	Testing scale
Alum, $Al_2(SO_4)_3 \cdot 18H_2O$	Polyvalent metal ion	80–250	5.3–5.6	Pilot scale
Ferric sulfate	Polyvalent metal ion	50–90	3–9	Batch/Pilot scale
Zetay 51 (cationic polymer)	Polyethylene amine	10	>9	Batch
Dow 21 M (cationic polymer)	Polyethylene amine	10	4–7	Batch
Dow C31 (cationic polymer)	Polyamine	1–5	2–4	Batch
Chitosan	Diacetylated polymer of natural chitin	100	8.4	Batch

Table 5.2 Comparison of extraction methods for chemical constituents of algal cells (Modified from Harun et al. 2010)

Extraction method	Advantages	Limitations
Oil press	Easy operation, no solvents	Slow process, processing of dried biomass only, large sample mass required
Solvent extraction	Reproduceable, inexpensive solvent, solvent recycling possible	Solvent highly flammable/toxic, solvent recovery energy/cost intensive
Supercritical fluid extraction with CO_2 as extracting agent	No-toxicity, "green" solvent, nonflammable, easy operation	No quantitative extraction of polar compounds, insufficient contact of CO_2 with sample
Ultrasound in combination with organic solvent extraction	Reduced operation time, reduced solvent use,	High power consumption, scale-up to industrial scale not possible

is that it requires completely dried biomass, which is only accomplished at the expense of processing energy in the form of heat. The advantage of this technology is that it is based on robust mechanics, does not require specifically skilled labor, and can achieve extraction yields of up to 75% w/w. The main drawback of this process is the long processing time. The industrial relevance of each method is shown in Table 5.2.

Purification of polyunsaturated fatty acids (PUFAs) must take into account the oxygen sensitivity of nonconjugated methylene groups (Muffler and Ulber 2005). Usually, cultivated microalgae are separated from the culture broth by filtration of centrifugation, followed by ultrasonication, freezing, and grinding or disruption by bead mills (Garrido et al. 1994; Chisti and Moo-Young 1986). Lipids are commonly solvent extracted in the wet state as solvents also induce breakage of cells. The organic solvents applied should be inexpensive, volatile, free of impurities, and able to form a two-phase system with water to facilitate the removal of nonlipid compounds. Here, the solvent is also strongly dependent on the type of lipids to be extracted and how they link to other cell components. Lipids that are bound by van der Waals interaction can easily be separated by nonpolar solvents whereas polar solvents can break existing hydrogen bonds of membrane-associated polar lipids while simultaneously inactivating several lipid-degrading enzymes (Muffler and Ulber 2005). Undesirable by-products of the extraction have to be removed in subsequent purification steps. Ionically bound lipids require a harsh shift in pH value for extraction which may also alter the lipid structure. Thus, care has to be taken to select the most gentle extraction procedure that is able to release the lipid. This usually involves deaerating the solvent by bubbling with inert gases to prevent lipid oxidation (Kates 1988). A gentler and widely used technique for the extraction of ion-bound lipids from microalgae is the use of chloroform–methanol–water (1:1:0.8) which is used as a single-phase extractant and is further diluted with chloroform and water (2.2:1.8) to obtain a biphasic system (Bligh and Dyer 1959). Here the lipid is extracted in the lower nonpolar phase and unwanted components of the crude extract are solved in the methanol–water phase. Other systems based on this extraction protocol but adapted for industrial use and focused on low toxicity and high performance use hexane–isopropanol and butanol–ethanol (Ahlgren and Merino 1991). For the production of free fatty acids, the oils resulting from the extraction are saponified through the addition of alkali (Molina Grima et al. 2003).

Supercritical fluid extraction has been used for the extraction of oils and seems to be favorable for PUFA extraction as the carbon dioxide atmosphere protects from auto-oxidation. However, supercritical fluid extraction has only been carried out on a few marine algae to date, including *Scenedesmus obliquus*, *Nannochloropsis* sp., *Dilophus ligulatus*, and *Ochromonas danica* (Andrich et al. 2005; Subra et al. 1991; Polak et al. 1989; Choi et al. 1987).

Next to lipid-based constituents algal stains can also accumulate significant amounts (>50% w/w dry cell mass) of proteins and polymeric carbohydrates mainly in the form of starch as storage compounds (Subhadraa and Grinson 2011). Although proteins and starch both find application in the food and feed industries, algal-based starch is also utilized in the production of bioethanol for fuel applications. The extraction of these cellular components also requires cellular lysis prior to down stream processing of the intermediate product streams. Cell lysis is commonly accomplished by French press apparatus or by use of organic solvents, such as hexane. Proteins and starch are then released into the reaction liquid medium. Isolation of proteins and starch is accomplished by changes in the ionic strength of the reaction medium usually by addition of high salt concentrations. The water molecules in the reaction medium are attracted to the salt ions, which decreases the number of water molecules available to interact with the charged part of the protein or starch. As a result of the increased demand for water molecules, the interactions between individual solute molecules are stronger than the solvent–solute interactions. As a result the protein or starch molecules coagulate by forming hydrophobic interactions with each other. The technique of salting-out proteins and starch is commonly accomplished by addition of 1–10 M sodium chloride or ammonium sulfate, which are both cost-efficient reagents for industrial applications. High-value secondary metabolites such as carotenoids and other pigments are commonly extracted using solvent extraction techniques, which have been described above.

For pharmaceutical applications, which require high compound purity product, polishing can be accomplished by an additional ion exchange or hydrophobic interaction chromatography step.

5.11 An Integrated Biorefinery Approach for Microalgal Lipid Production

By using an integrated biorefinery approach for the commercial production of microalgae lipids, productivity could be enhanced considerably (Fig. 5.3; Subhadraa and Grinson 2011). Utilizing waste CO_2 and nutrient effluents from various waste streams to feed

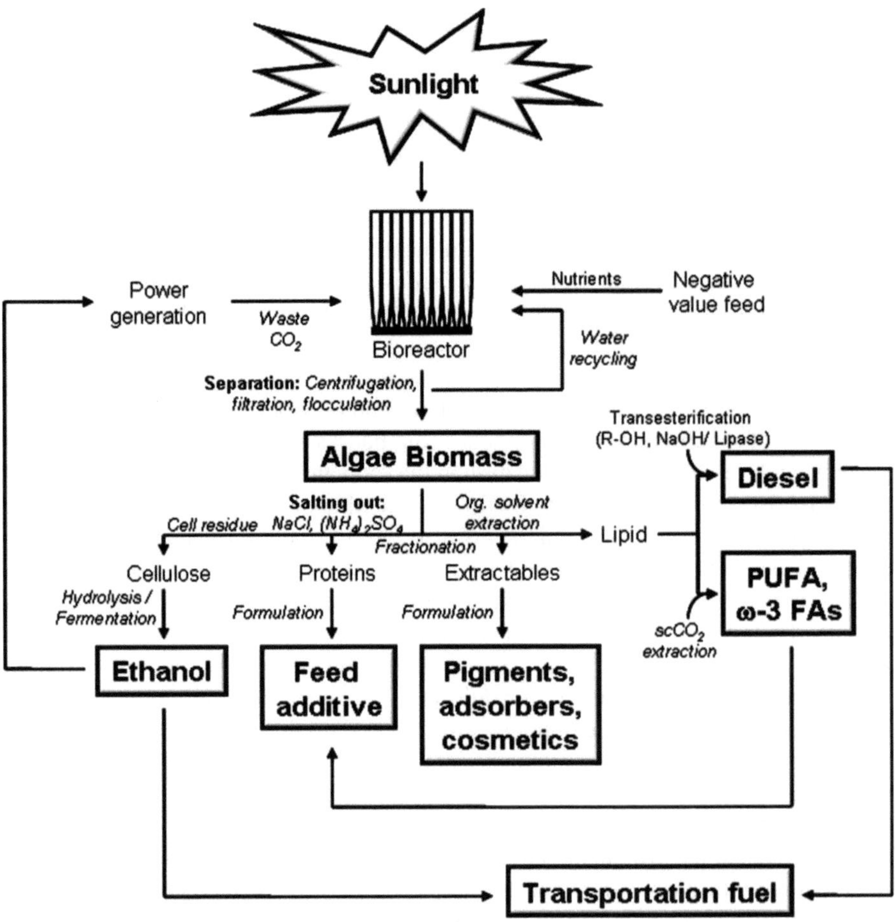

Fig. 5.3 Flow diagram for an integrated biorefinery approach for lipid and biofuel production from marine algae

closed reactors while simultaneously improving productivity through strain optimization and reducing the energy input required for mixing, large-scale microalgal production could become a realistic approach for the food, lipid, and biofuel markets (Richmond 2003; Richmond et al. 2003; Qiang et al. 1998). This would be further improved by utilizing a coproduct strategy by sequentially cultivating the microalgae for CO_2 mitigation; extracting bioactive products from harvested biomass; thermal processing (pyrolysis, liquefaction, gasification); extracting high-value chemicals (fatty acids, PUFAs, pigments, carotenoids) from the resulting liquid, vapor, and solid phases; and reforming biofuels for different applications (Li et al. 2008b).

5.12 Low-Molecular-Weight Molecules

To date, about 15,000, often low-molecular-weight, natural marine compounds have been screened for pharmaceutical and cosmetic applications (Wijffels 2008). Of these, there are about 45 products in preclinical or clinical trials (Newman and Cragg 2004). Only a few marine bioactive compounds have successfully been tested and approved for drug use, among which are Ziconotide from a marine cone snail which is marketed as Prialt (Elon Pharmaceuticals) and trabectedin/ecteinascidin from a tunicate which is marketed as Yondelis by PharmaMar (Wijffels 2008). Prialt is marketed as a drug against chronic pain, and Yondelis is used against soft tissue sarcomas and ovarian cancer.

If it only were a matter of identifying a bioactive compound that has a pharmacological profile, a multitude of the marine drugs and cosmetics currently on the market would be available. Unfortunately, the establishment of a cost-effective and sustainable source for a large-scale supply and the suitability and safety profile of a compound for use in or on the human body are paramount (Sipkema et al. 2005). Many bioactive compounds that are targeted by the pharmaceutical industry originate predominately from sponges, corals, gorgonians, sea hares, nudibranchs, bryozoans, tunicates, and other marine invertebrates as well as marine micro-organisms (Faulkner 2000). These organisms often show a whole range of new chemical structures, which are presently not found in terrestrial organisms and may serve as lead structures for novel synthetic pharmaceuticals (Muffler and Ulber. 2005).

5.13 The Origin of Marine Bioactive Compounds

There has been significant discussion about the true origin of bioactive compounds from these organisms in recent years. it seems plausible that sessile macro-organisms such as sponges and coral will have developed successful defense mechanisms to protect themselves from attack, however, many compounds isolated from these organisms point towards a microbiological origin. Indeed a variety of compounds may now be attributed to associate or symbiotic micro-organisms that reside inside these invertebrates even though conclusive evidence of a compound's origin is still limited (Taylor et al. 2007; Newman and Hill 2006; Piel 2006). An example of a bioactive compound shown conclusively through cell cultures to be produced by a sponge is stevensine from *Axinella corrugata* (Pomponi and Willoughby 1994). On the other hand, a bioactive compound shown conclusively to be produced by an associated micro-organism is manzamine from *Micromonospora* sp. isolated from the sponge *Acanthostrongylophora* and theopalauamide from *Entotheonella palauensis*, an associated bacterium from the lithistid sponge *Theonella swinhoei* from Palau and Mozambique (Hill et al. 2005; Schmidt et al. 2000). Interestingly, there is also preliminary evidence that *Entotheonella* sp. is responsible for the production of discodermolide in the Caribbean lithistid sponge *Discodermia dissoluta* (Brück et al. 2008; Schirmer et al. 2005).

Most compounds derived from either the marine invertebrate or their associated micro-organsims are exceedingly complex, therefore synthetic production, although possible, may require a number of reaction steps that may make production on an industrial scale infeasible in most cases (Sipkema et al. 2005). Nonetheless, several promising marine drugs such as discodermolide, halichondrin B, and ecteinascidin have been synthesized successfully and used on an industrially relevant scale (Harried et al. 2003; Cuevas et al. 2000; Aicher et al. 1992). Most often synthesis of a compound is the only option for production of marine bioactive compounds inasmuch as most compounds in vivo are produced on a nano- to picogram scale, making commercial harvest far too costly to be commercially viable. This is in addition to the environmental impact that would undoubtedly occur from the large-scale harvesting that would be required. One noted exception for this is Yondelis (ecteinascidin) from PharmaMar which had been produced by open-sea mariculture and sustainable harvest of *Ecteinascidia turbinata*, the tunicate in which the compound was originally discovered, until a total synthesis was developed (Ali and Gab-Alla 2008; Cuevas et al. 2000). Regardless of this example, it is vital for the commercial production of bioactive compounds to seek alternative ways of production (Wijffels 2008).

Knowing what genes are involved in the production of a compound of interest would enable their expression in a suitable host organism, while optimizing conditions to achieve a more efficient production and a higher yield. Using a metagenomic approach to analyze the genes present in a complex mixture of genes composed of invertebrates, associated micro-organisms, and the environment (i.e. free-living organisms from water and sediment) would be the first step of identifying the genes responsible for the synthesis of a bioactive compound. Although this technique has been shown to be effective and powerful in a marine environment, identifying the genes of interest may not be straightforward (Venter et al. 2004). The metabolic pathways of a bioactive compound may be very complex, involving a large number of genes, all of which would have to be identified before transferring them to a suitable host. This has been achieved to some degree with compounds that are produced in the highly modularized polyketide synthase pathways, but other genes may not be identified as easily (Schirmer et al. 2005).

If a marine bioactive is produced by an invertebrate-associated or free-living micro-organism, the most obvious way of industrial-scale production would be to cultivate the organism on an appropriate growth medium. To date, the highest number of structures has been isolated from *Streptomyces*, *Alteromonas*, *Vibrio*, *Agrobacteria*, *Bacillus*, *Pseudomonas*, *Actinomyces*, and "unidentified bacteria" as well as other orders in some isolated cases (Wagner-Döbler et al. 2002). However, as the majority of micro-organisms are unculturable in what has been termed the "great plate anomaly," novel ways for culture would have to be devised that not only allow the organisms to thrive but also continue to produce the compound of interest (Sfanos et al. 2005). Domestication of wild marine microbial strains has been achieved in some cases by using low nutrient media, the addition of various by-products found in the host, or the addition of signal molecules, all with varying success (Nichols et al. 2008; Bruns et al. 2003; Bruns et al. 2002; Connon and Giovannoni 2002; Janssen et al. 2002). Placing organisms in a diffusible reactor and returning them to their natural habitat allowed for a several 100-fold increase in microbial recovery, however, this would not be a realistic approach towards industrial scale culture of micro-organisms for bioactive production (Kaeberlein et al. 2002). The most promising cultivation approach has been shown by Nichols et al. (2008), who demonstrated that an uncultivable micro-organism can produce a cultivable variant by the addition of small signaling peptides to a synthetic medium. However, whether this approach works with a variety of organisms is unknown.

If a marine compound is known to be produced by the invertebrate itself, the use of axenic in vitro culture of dissociated cells may be an option. Even so, attempts to develop proliferating cell lines from marine invertebrates have thus far not moved from the laboratory scale. Sponge aqua- or mariculture on the other hand has shown some success in recent years. Here, factors such as depth, current, light, and carrier material have been shown to be important for sponge mortality and bioactive production (Rinkevich 2005; Duckworth 2004; Duckworth and Battershill 2003). Mariculture in addition would largely simulate conditions in the wild, enabling a focus on factors such as stress that may induce metabolite production and identify the structure within the invertebrate responsible. Only then could a concerted effort be made to address the needs of either sponge cell or symbiont culture for large-scale production (Koopmans et al. 2009).

5.14 Compound Isolation and Fractionation

Regardless of the product's origin, after harvest the biomass is extracted by an adequate solvent system (i.e., methanol or acetone) and partitioned by a gradient of polar and nonpolar organic solvents to separate the compound of interest from the crude extract (Fig. 5.4; Muffler and Ulber 2005). The resulting fraction can then be separated based on its polarity. Generally, there is little interest in the high polar fraction containing water-soluble alkaloids, salts, amino acids, saponins, and polyhydroxysteroids due to their difficulty and high costs involved in further fractionation and purification. The low polarity fractions (i.e., CCl_4 or hexane) containing, for example, fatty acids and terpenes as well as the medium polarity fractions (i.e., CH_2Cl_2) containing peptides and depsipeptides are most investigated (Riguera 1997). After extraction and fractionation, lipophilic compounds can be further separated and

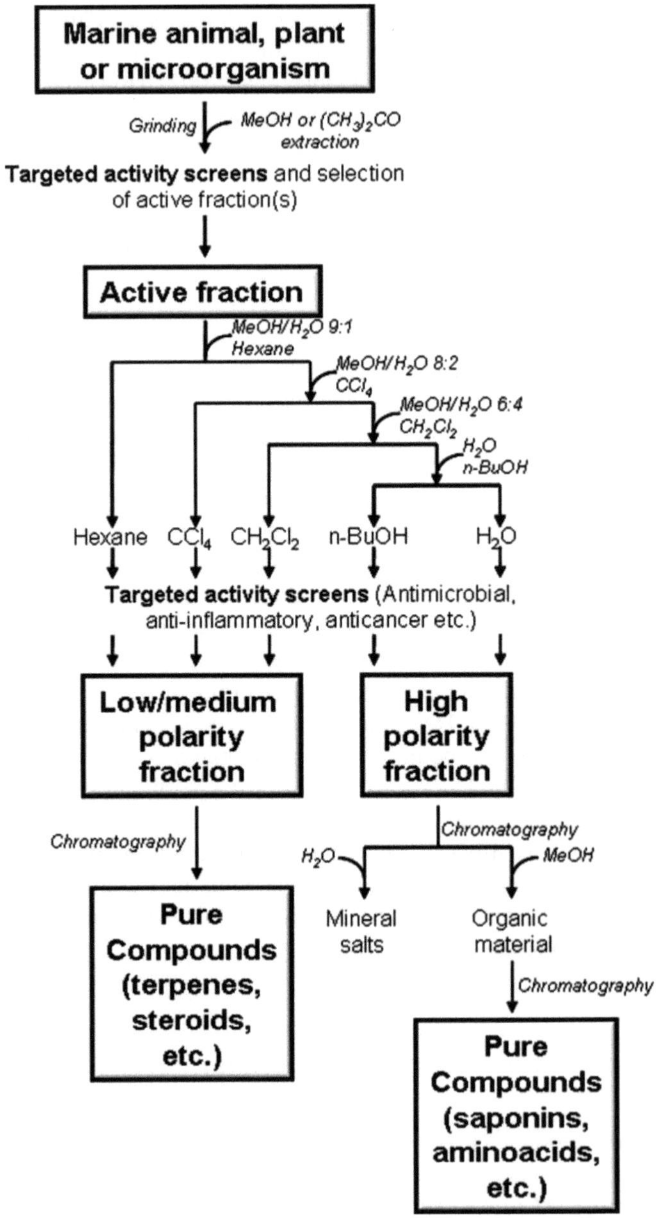

Fig. 5.4 Flow diagram for the commercial extraction of small molecules for pharmaceutical and cosmetic use (Modified from Muffler and Ulber 2005)

purified by standard or reverse-phase chromatography, followed by HPLC to obtain fractions with single molecules. In contrast, hydrophilic compounds obtained in high polarity fractions often require desalting by column chromatography before the organic material can be separated by size-exclusion chromatography and HPLC (Muffler and Ulber 2005).

5.15 Vehicles for Efficient Delivery of Small Molecules

In situ delivery vehicles for pharmaceuticals and cosmetics from marine origin may most often utilize existing methods. However, due to the often-encountered toxicity of marine pharmaceuticals, it is important that they are delivered successfully to their intended site of action. Furthermore, it is important for a drug delivery carrier to be efficiently removed after delivering the drug, meaning that it is neither toxic nor accumulates in the body (Dev et al. 2010). Marine polysaccharides and chitin derivatives such as N-succinyl-chitosan, carboxymethyl chitin, chitosan hydrogel, and hydroxyethyl chitin have been shown to be good candidates for this application (Dev et al. 2010; Khoushab and Yamabhai 2010; Kiyosawa et al. 2010; Ishihara et al. 2006; Kamiyama et al. 1999). Chitosan has been shown to form colloidal particles and entrap macromolecules through a number of mechanisms such as ionic cross-linking, whereas cationic chitosan and anionic alginate in combination with other natural polymers may enhance the drug encapsulation efficiency of liposomes via the layer-by-layer (L-b-L) self-assembly technique (Haidar et al. 2008; Janes et al. 2001). To date, the in vivo efficacy of these chitosan-based colloidal carriers has reported them to be efficient vehicles for the transport of peptides across the nasal mucosa (Janes et al. 2001). In addition, nanoparticles made of chitosan in association with polyethylene oxide have been used as a protein carrier with increased protein-loading capacity (entrapment efficiency up to 80% of the protein). This system provided a continuous release of the entrapped protein for up to 1 week, and could be useful for future drug applications such as the slow continuous release of implanted drugs at their site of action (Calvo et al. 1997). In the cosmetic industry, low-molecular-weight compounds and marine extracts may be used as an additive in skin creams. For example, the Caribbean gorgonian, *Pseudopterogorgia elisabethae*, *Octocorallia*, *Cnidaria*, produces anti-inflammatory pseudopterosins, an additive in skin creams marketed as Resilience by Estée Lauder (Look et al. 1986).

5.16 Conclusion

The marine environment with its host of environmental niches and organisms has understandably become a well-established source of novel compounds with uses that are as diverse as the environment from which they are sourced. There are successful pharmaceuticals, industrial enzymes, food ingredients, biosensors, drug delivery systems, and chemical compounds of marine origin with the number set to increase as future discoveries are made. Considering how little of the marine world has been sampled to date, the scope for discovery and breakthrough technologies can only be estimated. The next generation of marine bioactive components with industrial potential will doubtlessly require a diverse set of scientific and entrepreneurial skills to deliver not only compounds for the sake of science alone but also the technologies, delivery vehicles, and applications for their use. Here, considerable effort will have to be undertaken in order to develop downstream processes and delivery vehicles that are well understood and tailored to their specific purpose. For this, more integrated marine bioprocesses will have to be developed that lead to higher performance, higher product yields, and lower production costs.

References

Ahlgren, G., and L. Merino. 1991. Lipid analysis of freshwater microalgae: A method study. *Archives of Hydrobiology* 121: 295–306.

Aicher, T.D., K.R. Buszek, F.G. Fang, C.J. Forsyth, S.H. Jung, Y. Kishi, M.C. Matelich, P.M. Scola, D.M. Spero, and S.K. Yoon. 1992. Total synthesis of halichondrin B and norhalichondrin B. *Journal of the American Chemical Society* 114(8): 3162–3164.

Aksu, Z. 2005. Application of biosorption for the removal of organic pollutants: A review. *Process Biochemistry* 40: 997–1026.

Ali, A.-F., and A. Gab-Alla. 2008. Distribution of the sea squirt *Ecteinascidia thurstoni* Herdman, 1890 (Ascidiacea: Perophoridae) along Suez Canal and Egyptian Red Sea coasts. *Oceanologia* 50(2): 239–253.

Andrich, G., U. Nesti, F. Venturi, A. Zinnai, and R. Fiorentin. 2005. Supercritical fluid extraction of bioactive lipids from the microalga *Nannochloropsis* sp. *European Journal of Lipid Science and Technology* 107(6): 381–386.

Arcidiacono, S., S.J. Lombardi, and D.L. Kaplan. 1989. Fermentation, processing and enzyme characterization for chitosan biosynthesis by *Mucor rouxii*. In *Chitin and chitosan: Sources, chemistry, biochemistry, physical properties and applications*, ed. G. Skjak-Braek, T. Anthonsen, and P. Sandford, 319–332. London: Elsevier Applied Sci.

Arreguín, R., B. Arreguín, M. Soriano-García, A. Hernández-Arana, and A. Rodríguez-Romero. 1993. Isolation and characterization of a protease from the marine sponge *Spheciospongia vesparia*. *FEBS Letters* 320(3): 235–238.

Baba, S., Y. Uraki, Y. Miura, and S. Tokura. 1989. Controlled release and hydrolysis of prodrug using carboxymethylchitin as a drug carrier. In *Chitin and chitosan*, ed. G.S. Brack, T. Anthonsen, and P. Sandford, 703–715. New York: Elsevier.

Balassa, L.L. 1971. Chitin and chitin derivatives for promoting wound healing. US Patent No. 3903268.

Barriada, J.L., R. Herrero, D. Prada-Rodriguez, and M.E.S. de Vicente. 2007. Waste spider crab shell and derived chitin as lowcost materials for cadmium and lead removal. *Journal of Chemical Technology and Biotechnology* 82: 39–46.

Bautista-Baños, S., M. Hernández-López, E. Bosquez-Molina, and C.L. Wilson. 2003. Effects of chitosan and plant extracts on growth of *Colletotrichum gloeosporioides*, anthracnose levels and quality of papaya fruit. *Crop Protection* 22: 1087–1092.

Beaney, P., J. Lizardi-Mendoza, and M. Healy. 2005. Comparison of chitins produced by chemical and bioprocessing methods. *Journal of Chemical Technology and Biotechnology* 80: 145–150.

BeMiller, J.N., and R.L. Whistler. 1996. Carbohydrates. In *Food chemistry*, ed. O.R. Fennema, 157–223. New York: Marcel Dekker.

Berteau, O., and B. Mulloy. 2003. Sulfated fucans, fresh perspectives: Structures, functions, and biological properties of sulfated fucans and an overview of enzymes active toward this class of polysaccharide. *Glycobiology* 13: 29R–40R.

Betler, P.A., E.L. Cussler, and W.-S. Hu. 1988. *Bioseparations*. New York: Wiley.

BioScience. 1996. Marine biotechnology. Special issue, 46.

Bligh, E.G., and W.J. Dyer. 1959. A rapid methods of total lipid extraction and purification. *Canadian Journal of Biochemistry and Physiology* 37: 911–917.

Brandl, F., F. Sommer, and A. Goepferich. 2007. Rational design of hydrogels for tissue engineering: Impact of physical factors on cell behavior. *Biomaterials* 28: 134–146.

Brück, W.M., S.H. Sennett, S.A. Pomponi, P. Willenz, and P.J. McCarthy. 2008. Identification of the bacterial symbiont *Entotheonella* sp. in the mesohyl of the marine sponge *Discodermia* sp. *The ISME Journal* 2: 335–339.

Brück, W.M., J.W. Slater, and B.F. Carney. 2011. Chitin and chitosan from marine organisms. In *Chitin, chitosan, oligosaccharides and their derivatives: Biological activities and applications*, ed. S.-K. Kim, 11–23. Boca Raton: CRC Press.

Bruns, A., H. Cypionka, and J. Overmann. 2002. Cyclic AMP and acyl homoserine lactones increase the cultivation efficiency of heterotrophic bacteria from the central Baltic Sea. *Applied and Environmental Microbiology* 68: 3978–3987.

Bruns, A., U. Nubel, H. Cypionka, and J. Overmann. 2003. Effect of signal compounds and incubation conditions on the culturability of freshwater bacterioplankton. *Applied and Environmental Microbiology* 69: 1980–1989.

Calvo, P., C. Remuñán-López, J.L. Vila-Jato, and M.J. Alonso. 1997. Novel hydrophilic chitosan-polyethylene oxide nanoparticles as protein carriers. *Journal of Applied Polymer Science* 63: 125–132.

Chaumont, D. 1993. Biotechnology of algal biomass production: a review of systems for outdoor mass culture. *Journal of Applied Phycology* 5(6): 593–604.

Chen, F., and Jiang, Y. 2000. Algae and their biotechnological potential. Kluvwer Academic: Dordrecht/Boston/London.

Chen, S.-H., Y.-H. Yen, C.-L. Wang, and S.-L. Wang. 2003. Reversible immobilization of lysozyme via coupling to reversibly soluble polymer. *Enzyme and Microbial Technology* 33: 643–649.

Chisti, Y. 2007. Biodiesel from microalgae. *Biotechnology Advances* 25: 294–306.

Chisti, Y., and M. Moo-Young. 1986. Disruption of microbial cells for intracellular products. *Enzyme and Microbial Technology* 8: 194–204.

Chisti, Y., and M. Moo-Young. 1990. Large scale protein separations: Engineering aspects of Chromatography. *Biotech. Adv.* 8: 699–708.

Choi, K.J., Z. Nakhost, V.J. Krukonis, and M. Karel. 1987. Supercritical fluid extraction and characterization of lipids from algae *Scenedesmus obliquus*. *Food Biotechnology* 1(2): 263–281.

Connon, S.A., and S.J. Giovannoni. 2002. High-throughput methods for culturing microorganisms in very-low-nutrient media yield diverse new marine isolates. *Applied and Environmental Microbiology* 68: 3878–3885.

Cuevas, C., M. Pérez, M.J. Martín, J.L. Chicharro, C. Fernández-Rivas, M. Flores, A. Francesch, P. Gallego, M. Zarzuelo, F. de la Calle, J. García, C. Polanco, I. Rodríguez, and I. Manzanares. 2000. Synthesis of Ecteinascidin ET-743 and Phthalascidin Pt-650 from Cyanosafracin B. *Organic Letters* 2(16): 2545–2548.

Dai, T., G.P. Tegos, M. Burkatovskaya, A.P. Castano, and M.R. Hamblin. 2009. Chitosan acetate bandage as a topical antimicrobial dressing for infected burns. *Antimicrobial Agents and Chemotherapy* 53: 393–400.

Deans, J.R., and B.G. Dixon. 1992. Bioabsorbents for wastewater treatment. In *Advances in chitin and chitosan*, ed. C.J. Brine and P. Sandford, 648–656. New York: Elsevier.

Debashish, G., S. Malay, S. Barindra, and M. Joydeep. 2005. Marine enzymes. *Advances in Biochemical Engineering/Biotechnology* 96: 189–218.

Deming, J.W. 2002. Psychrophiles and polar regions. *Current Opinion in Microbiology* 5: 301–309.

Dev, A., J.C. Mohan, V. Sreeja, H. Tamura, G.R. Patzke, F. Hussain, S. Weyeneth, S.V. Nair, and R. Jayakumar. 2010. Novel carboxymethyl chitin nanoparticles for cancer drug delivery applications. *Carbohydrate Polymers* 79: 1073–1079.

Domard, A., and N. Cartier. 1989. Glucosamine oligomers. 1. Preparation and characterization. *International Journal of Biological Macromolecules* 11: 297–302.

Domard, A., and M. Rinaudo. 1983. Preparation and characterization of fully deacetylated chitosan. *International Journal of Biological Macromolecules* 5: 49–52.

Duckworth, A.R. 2004. Ascidian, sponge culture supplies: Bioactive metabolites. *Global Aquaculture Advocate* 7(6): 66–67.

Duckworth, A.R., and C.N. Battershill. 2003. Sponge aquaculture for the production of biologically active metabolites: The influence of farming protocols and environment. *Aquaculture* 221: 311–329.

Einbu, A. 2007. Characterisation of chitin and a study of its acid-catalysed hydrolysis. PhD diss., Norwegian University of Science and Technology, Trondheim.

Fan, F., and K.V. Wood. 2007. Bioluminescent assays for high-throughput screening. *Assay and Drug Development Technologies* 5(1): 127–136.

Fan, C., P.M. Glibert, J. Alexander, and M.W. Lomas. 2003. Characterization of urease activity in three marine phytoplankton species, *Aureococcus anophagefferens*, *Prorocentrum minimum*, and *Thalassiosira weissflogii*. *Marine Biology* 142: 949–958.

Faulkner, D.J. 2000. Marine pharmacology. *Antonie Van Leeuwenhoek* 77(2): 135–145.

French, C.S., and V.K. Young. 1952. The fluorescence spectra of red algae and the transfer of energy from phycoerythrin to phycocyanin and chlorophyll. *Journal of General Physiology* 35(6): 873–90.

Fu, X.T., and S.M. Kim. 2010. Agarase: Review of major sources, categories, purification method, enzyme characteristics and applications. *Marine Drugs* 8(1): 200–218.

Garrido, F., U.C. Banerjee, Y. Chisti, and M. Moo-Young. 1994. Disruption of a recombinant yeast for the release of beta-galactosidase. *Bioseparation* 4(5): 319–328.

Grace, E.S. 1997. Biotechnology in seas and tress, 6. In *Biotechnology unzipped: Promises and realities*, 163–190. Washington, DC: National Academic.

Gray Jr., H.B., T.P. Winston, J.L. Hodnett, R.J. Legerski, D.W. Nees, C.F. Wie, and D.L. Robberson. 1981. The extracellular nuclease from *Alteromonas espejiana*: An enzyme highly specific for nonduplex structure in nominally duplex DNAs. In *Gene amplification and analysis: Structural analysis of nucleic acids*, vol. 2, ed. J.G. Chirikjian and T.S. Papas, 169–203. New York: Elsevier Biomedical Press.

Haidar, Z.S., R.C. Hamdy, and M. Tabrizian. 2008. Protein release kinetics for core-shell hybrid nanoparticles based on the layer-by-layer assembly of alginate and chitosan on liposomes. *Biomaterials* 29: 1207–1215.

Hannick, L.I., D.C. Prasher, L.W. Schultz, J.R. Deschamps, and K.B. Ward. 1993. Preparation and initial characterization of crystals of the photoprotein aequorin from *Aequorea victoria*. *Proteins: Structure, Function, and Bioinformatics* 15(1): 103–107.

Harakas, N.K., and S.E. Builder. 1987. Industrial Scale Protein Purification: A Perspective. *Biotechnology Progress* 3(1): m2–m3.

Harried, S.S., C.P. Lee, G. Yang, T.I.H. Lee, and D.C. Myles. 2003. Total synthesis of the potent microtubule-stabilizing agent (+)-discodermolide. *Journal of Organic Chemistry* 68: 6646–6666.

Harun, R., M. Singh, G.M. Forde, and M.K. Danquah. 2010. Bioprocess engineering of microalgae to produce a variety of consumer products. *Renewable and Sustainable Energy Reviews* 14: 1037–1047.

He, H., X. Chen, C. Sun, Y. Zhang, and P. Gao. 2006. Preparation and functional evaluation of oligopeptide-enriched hydrolysate from shrimp (*Acetes chinensis*) treated with crude protease from *Bacillus* sp. SM98011. *Bioresource Technology* 97: 385–390.

Hill, R.T., O. Peraud, M.T. Hamann, and N. Kasanah. 2005. Manzamine-producing actinomycetes. US Patent No. 20050244938.

Ishihara, M., K. Obara, S. Nakamura, M. Fujita, K. Masuoka, Y. Kanatani, B. Takase, H. Hattori, Y. Morimoto, M. Ishihara, T. Maehara, and M. Kikuchi. 2006. Chitosan hydrogel as a drug delivery carrier to control angiogenesis. *Journal of Artificial Organs* 9: 8–16.

Janes, K.A., P. Calvo, and M.J. Alonso. 2001. Polysaccharide colloidal particles as delivery systems for macromolecules. *Advanced Drug Delivery Reviews* 47: 83–97.

Janssen, P., P. Yates, B. Grinton, P. Taylor, and M. Sait. 2002. Improved culturability of soil bacteria and isolation in pure culture of novel members of the divisions Acidobacteria, Actinobacteria, Proteobacteria, and Verrucomicrobia. *Applied and Environmental Microbiology* 68: 2391–2396.

Jeon, Y.J., and S.K. Kim. 2000. Continuous production of chitooligosaccharides using a dual reactor system. *Process Biochemistry* 35: 6223–6632.

Jiang, C., L.L. Wu, G.C. Zhao, P.H. Shen, K. Jin, Z.Y. Hao, S.X. Li, G.F. Ma, F.F. Luo, G.Q. Hu, W.L. Kang, X.M. Qin, Y.L. Bi, X.L. Tang, and B. Wu. 2010. Identification and characterization of a novel fumarase gene by metagenome expression cloning from marine microorganisms. *Microbial Cell Factories* 9: 91.

Jiang, X., A. Geng, N. He, and Q. Li. 2011. New isolate of *Trichoderma viride* strain for enhanced cellulolytic enzyme complex production. *Journal of Bioscience and Bioengineering* 111(2): 121–127.

Kaeberlein, T., K. Lewis, and S.S. Epstein. 2002. Isolating "uncultivable" microorganisms in pure culture in a simulated natural environment. *Science* 296: 1127–1129.

Kamiyama, K., H. Onishi, and Y. Machida. 1999. Biodisposition characteristics of N-succinyl-chitosan and glycol- chitosan in normal and tumor-bearing mice. *Biological and Pharmaceutical Bulletin* 22: 179–186.

Kates, M. 1988. Separation of lipid mixtures. In *Techniques of lipidology: Isolation, analysis and identification of lipids*, ed. R.H. Burdon and P.H. van Knippenberg, 186–278. Amsterdam: Elsevier.

Kauss, H., and B. Bauch. 1988. Chitin deacetylase from *Colletotrichum lindemuthianum*. *Methods in Enzymology* 161: 518–523.

Khoushab, F., and M. Yamabhai. 2010. Chitin research revisited. *Marine Drugs* 8(7): 1988–2012.

Kim, S.-K., and E. Mendis. 2006. Bioactive compounds from marine processing byproducts – a review. *Food Research International* 39: 383–393.

Kim, S.-K., and N. Rajapakse. 2005. Enzymatic production and biological activities of chitosan oligosaccharides (COS): A review. *Carbohydrate Polymers* 62: 357–368.

Kiyosawa, T., Y. Sato, and M. Ishihara. 2010. Hydrogel blends of chitin/chitosan, fucoidan and alginate as healing-impaired wound dressings. *Biomaterials* 31(1): 83–90.

Koopmans, M., D. Martens, and R.H. Wijffels. 2009. Towards commercial production of sponge medicines. *Marine Drugs* 7: 787–802.

Kovar, K., V. Looser, P. Hyka, T. Merseburger, and C. Meier. 2010. Recombinant yeast technology at the cutting edge: Robust tools for both designed catalysts and new biologicals. *Chimia (Aarau)* 64(11): 813–818.

Kurita, K. 2006. Chitin and chitosan: Functional biopolymers from Marine Crustaceans. *Marine Biotechnology* 8: 203–226.

Lee, Y.K. 2001. Microalgal mass culture systems and methods: Their limitations and potential. *Journal of Applied Phycology* 13(4): 307–315.

Li, B., X. Wang, R. Chen, W. Huangfu, and G. Xie. 2008a. Antibacterial activity of chitosan solution against *Xanthomonas* pathogenic bacteria isolated from *Euphorbia pulcherrima*. *Carbohydrate Polymers* 72: 287–292.

Li, Y., M. Horsman, N. Wu, C.Q. Lan, and N. Dubois-Calero. 2008b. Biofuels from microalgae. *Biotechnology Progress* 24: 815–820.

Lichty, J.J., J.L. Malecki, H.D. Agnew, D.J. Michelson-Horowitz, and S. Tan. 2005. Comparison of affinity tags for protein purification. *Protein Expression and Purification* 41(1): 98–105.

Look, S.A., W. Fenical, R.S. Jacobs, and J. Clardy. 1986. The pseudopterosins: Anti-inflammatory and analgesic natural products from the sea whip *Pseudopterogorgia elisabethae*. *Proceedings of the National Academy of Sciences of the United States of America* 83(17): 6238–6240.

MacColl, R., L.E. Eisele, E.C. Williams, and S.S. Bowser. 1996. The discovery of a novel R-phycoerythrin from an Antarctic Red Alga. *The Journal of Biological Chemistry* 271: 17157–17160.

MacDonald, J.A., and K.B. Sptorey. 1999. Cyclic AMP-dependent protein kinase: Role in anoxia and freezing tolerance of the marine periwinkle *Littorina littorea*. *Marine Biology* 133(2): 193–203.

Matsunaga, T., H. Takeyama, H. Miyashita, and H. Yokouchi. 2005. Marine microalgae. *Advances in Biochemical Engineering/Biotechnology* 96: 165–188.

Maurin, C., and Y. le Gal. 1997. Glutamine synthetase in the marine coccolithophorid *Emiliania huxleyi* (Prymnesiophyceae): Regulation of activity in relation to light and nitrogen availability. *Plant Science* 122(1): 61–69.

Mima, S., M. Miya, R. Iwamoto, and S. Yoshikawa. 1983. Highly deacetylated chitosan and its properties. *Journal of Applied Polymer Science* 28: 1909–1917.

Mohapatra, B.R., U.C. Banerjee, and M. Bapuji. 1998. Characterization of a fungal amylase from *Mucor* sp. associated with the marine sponge *Spirastrella* sp. *Journal of Biotechnology* 60(1–2): 113–117.

Molina Grima, E., E.-H. Belarbi, F.G. Acién Fernández, A. Robles Medina, and Y. Chisti. 2003. Recovery of microalgal biomass and metabolites: Process options and economics. *Biotechnology Advances* 20: 491–515.

Mora, P., D. Fournier, and J.-F. Narbonne. 1999. Cholinesterases from the marine mussels *Mytilus galloprovincialis* Lmk. and *M. edulis* L. and from the freshwater bivalve *Corbicula fluminea* Müller. *Comparative Biochemistry and Physiology Part C: Pharmacology, Toxicology and Endocrinology* 122(3): 353–361.

Morishita, H., S. Ohashi, T. Oku, Y. Nakajima, S. Kojima, M. Ryufuku, H. Nakamura, and Y. Ohmiya. 2002. Cloning and characterization of an active fragment of luciferase from a luminescent marine alga, *Pyrocystis lunula*. *Photochemistry and Photobiology* 75(3): 311–315.

Muffler, K., and R. Ulber. 2005. Downstream processing in marine biotechnology. *Advances in Biochemical Engineering/Biotechnology* 97: 63–103.

Munilla-Morán, R., and F. Saborido-Rey. 1996. Digestive enzymes in marine species. II. Amylase activities in gut from seabream (*Sparus aurata*), turbot (*Scophthalmus maximus*) and redfish (*Sebastes mentella*). *Comparative Biochemistry and Physiology Part B: Biochemistry and Molecular Biology* 113(4): 827–834.

Murakami, K., H. Aoki, S. Nakamura, S.-I. Nakamura, M. Takikawa, M. Hanzawa, S. Kishimoto, H. Hattori, Y. Tanaka, T. Kiyosawa, Y. Sato, and M. Ishihara. 2010. Hydrogel blends of chitin/chitosan, fucoidan and alginate as healing-impaired wound dressings. *Biomaterials* 31(1): 83–90.

Muramatsu, Y., A. Harada, Y. Ohwaki, Y. Kasahara, S. Takagi, and T. Fukuhara. 2002. Salt-tolerant ATPase activity in the plasma membrane of the Marine Angiosperm *Zostera marina* L. *Plant and Cell Physiology* 43(10): 1137–1145.

Murphy, C.D. 2003. New frontiers in biological halogenation. *Journal of Applied Microbiology* 94: 539–548.

Muzzarelli, R.A.A. 1985. Chitin. In *The polysaccharides*, vol. 3, ed. G.O. Aspinall. Orlando: Academic.

Muzzarelli, R.A.A., M. Weckx, O. Filippini, and F. Sigon. 1989. Removal of trace metal ions from industrial waters, nuclear effluents and drinking water, with the aid of cross-linked N-carboxymethyl chitosan. *Carbohydrate Polymers* 11(4): 293–306.

Muzzarelli, R.A.A., M. Tomasetti, and P. Ilari. 1994. Molecular parameters of chitosans depolymerized with the aid of papain. *Enzyme and Microbial Technology* 16: 110–114.

Muzzarelli, R.A., M. Mattioli-Belmonte, A. Pugnaloni, and G. Biagini. 1999. Biochemistry, histology and clinical uses of chitins and chitosans in wound healing. *EXS* 87: 251–264.

Naznin, R. 2005. Extraction of chitin and chitosan from shrimp (*Metapenaeus monoceros*) shell by chemical method. *Pakistan Journal of Biological Sciences* 8: 1051–1054.

Newman, D.J., and G.M. Cragg. 2004. Marine natural products and related compounds in clinical and advanced preclinical trials. *Journal of Natural Products* 67: 1216–1238.

Newman, D.J., and R.T. Hill. 2006. New drugs from marine microbes: The tide is turning. *Journal of Industrial Microbiology and Biotechnology* 33: 539–544.

Nichols, D., K. Lewis, J. Orjala, S. Mo, R. Ortenberg, P. O'Connor, C. Zhao, P. Vouros, T. Kaeberlein, and S.S. Epstein. 2008. Short peptide induces an "Uncultivable" microorganism to grow in vitro. *Applied and Environmental Microbiology* 74(15): 4889–4897.
No, H.K., and S.P. Meyers. 1995. Preparation and characterization of chitin and chitosan – a review. *Journal of Aquatic Food Product Technology* 4: 27–52.
Oh, H.M., A. Choi, and T.I. Mheen. 2003. High-value materials from microalgae. *Korean Journal of Microbiology and Biotechnology* 31(2): 95–102.
Okamoto, Y., R. Yano, K. Miyatake, I. Tomohiro, Y. Shigemasa, and S. Minami. 2003. Effects of chitin and chitosan on blood coagulation. *Carbohydrate Polymers* 53: 337–342.
Onsoyen, E., and O. Skaugred. 1990. Metal recovery using chitosan. *Journal of Chemical Technology and Biotechnology* 49(4): 395–404.
Peter, M.G., G. Kegel, and R. Keller. 1986. Structural studies on sclerotized unsect cuticle. In *Chitin in nature and technology*, ed. R.A.A. Muzzarelli, C. Jeuniaux, and G.W. Gooday. New York: Plenum Press.
Piel, J. 2006. Bacterial symbionts: Prospects for the sustainable production of invertebrate-derived pharmaceuticals. *Current Medicinal Chemistry* 13: 39–50.
Polak, J.T., M. Balaban, A. Peplow, and A.J. Philips. 1989. Supercritical carbon dioxide extraction of lipids from algae. In *Supercritical fluid science and technology*, ACS Symposium Series, vol. 406, ed. K.P. Johnston and J.M.L. Penninger, 449–467. American Chemical Society: Washington, DC, USA.
Pomponi, S.A., and R. Willoughby. 1994. Sponge cell culture for production of bioactive metabolites. In *Sponges in time and space*, ed. R.W.M. van Soest, T.M.G. van Kempen, and J.C. Braekman, 395–400. Balkmana, Rotterdam, The Netherlands.
Pomponi, S.A., D.G. Baden, and Y. Zohar. 2007. Marine biotechnology: Realizing the potential. *Marine Technology Society Journal* 41(3): 24–31.
Prasher, D.C., V.K. Eckenrode, W.W. Ward, F.G. Prendergast, and M.J. Cormier. 1992. Primary structure of the *Aequorea victoria* green-fluorescent protein. *Gene* 111(2): 229–233.
Pulz, O., and W. Gross. 2004. Valuable products from biotechnology of microalgae. *Applied Microbiology and Biotechnology* 65: 635–648.
Qiang, H., Y. Zarmi, and A. Richmond. 1998. Combined effects of light intensity, light path and culture density on output rate of *Spirulina platensis* (cyanobacteria). *European Journal of Phycology* 33: 165–171.
Richmond, A. 1992. Open systems for the mass production of photoautotrophic microalgae outdoors: Physiological principles. *Journal of Applied Phycology* 4(3): 281–286.
Richmond, A. 2003. *Handbook of microalgal cultures*. Blackwell Publishers, Oxford, UK.
Richmond, A., Z. Cheng-Wu, and Y. Zarmi. 2003. Efficient use of strong light for high photosynthetic productivity: Interrelationships between the optical path, the optimal population density and cell growth inhibition. *Biomolecular Engineering* 20: 229–236.
Riguera, R. 1997. Isolating bioactive compounds from marine organisms. *Journal of Marine Biotechnology* 5: 187–193.
Rinkevich, B. 2005. Marine invertebrate cultures: New millennium trends. *Marine Biotechnology* 7: 429–439.
Rudrapatnam, N., T. Kittur, and F.S. Kittur. 2002. Chitin – the undisputed biomolecule of great potential. *Critical Reviews in Food Science and Nutrition* 43: 61–87.
Salazar, F.N., F. Zamora, J.M. Canals, and F. Lopez. 2008. Protein stabilization in sparkling base wine using zirconia and bentonite: Influence on the foam parameters and protein fractions. *Journal International des Sciences de la Vigne et du Vin* 42: 51–58.
Sandford, P.A. 1989. Chitosan: Commercial uses and potential applications. In *Chitin and chitosan: Sources, chemistry, biochemistry, physical properties and applications*, ed. G. Skjak-Braek, T. Anthonsen, and P. Sandford, 1–69. London: Elsevier Applied Sci.
San-Lang, W., I.-L. Shih, C.-H. Wang, K.-C. Tseng, W.-T. Chang, Y.-K. Twu, J.-J. Ro, and C.-L. Wang. 2002. Production of antifungal compounds from chitin by *Bacillus subtilis*. *Enzyme and Microbial Technology* 31: 321–328.
Schirmer, A., R. Gadkari, C.D. Reeves, F. Ibrahim, E.F. DeLong, and C.R. Hutchinson. 2005. Metagenomic analysis reveals diverse polyketide synthase gene clusters in microorganisms

associated with the marine sponge *Discodermia dissoluta. Applied and Environmental Microbiology* 71: 4840–4849.

Schmidt, E.W., A.Y. Obratsova, S.K. Davidson, D.J. Faulkner, and M.G. Haygood. 2000. Identification of the antifungal peptide-containing symbiont of the marine sponge Theonella swinhoei as a novel -proteobacterium, '*CandidatusEntotheonella palauensis*'. *Marine Biology* 136: 969–977.

Sfanos, K., D. Harmody, P. Dang, A. Ledger, S. Pomponi, P. McCarthy, and J. Lopez. 2005. A molecular systematic survey of cultured microbial associates of deep-water marine invertebrates. *Systematic and Applied Microbiology* 28: 242–264.

Shahidi, F., and J. Synowiecki. 1991. Isolation and characterization of nutrients and value-added products from snow crab (*Chinoecetes opilio*) and shrimp (*Pandalus borealis*) processing discards. *Journal of Agricultural and Food Chemistry* 39: 1527–1532.

Shelef, G., A. Sukenik, and M. Green. 1984. Microalgae harvesting and processing: A literature review: A subcontract report. http://www.nrel.gov/docs/legosti/old/2396.pdf.

Singer, N., and J. Wooten. 2003. Method of extracting chitin from the shells of exoskeletal animals. US Patent No. US2003/0060610 A1.

Sipkema, D., R. Osinga, W. Schatton, D. Mendola, J. Tramper, and R.H. Wijffels. 2005. Large-scale production of pharmaceuticals by marine sponges: Sea, cell or synthesis? *Biotechnology and Bioengineering* 90: 201–222.

Subhadraa, B., and G. Grinson. 2011. Algal biorefinery-based industry: An approach to address fuel and food insecurity for a carbon-smart world. *Journal of the Science of Food and Agriculture* 91: 2–13.

Subra, P., R. Tufeu, Y. Garrabos, and P. Boissinot. 1991. Supercritical fluid extraction from a Mediterranean brown alga. *The Journal of Supercritical Fluids* 4(4): 244–249.

Suryanarayana Rao, S.V., K.P. Yashodha, N.S. Mahendrakar, and P. Puttarajappa. 1987. Deactylation of chitin at low temperature by novel alkali impregnation technique. *Indian Journal of Technology* 25: 194–196.

Synowiecki, J. 2007. Production, properties and some new applications of chitin and its derivatives. *Critical Reviews in Food Science and Nutrition* 43(2): 145–171.

Tang, W.L., and H. Zhao. 2009. Industrial biotechnology: Tools and applications. *Biotechnology Journal* 4(12): 1725–1739.

Taylor, M.W., R. Radax, D. Steger, and M. Wagner. 2007. Sponge-associated microorganisms: Evolution, ecology, and biotechnological potential. *Microbiology and Molecular Biology Reviews* 71: 295–347.

Tharanathan, R.N., and F.S. Kittur. 2003. Chitin – the undisputed biomolecule of great potential. *Critical Reviews in Food Science and Nutrition* 43: 61–87.

Tsai, G.J., and W.H. Su. 1999. Antibacterial activity of shrimp chitosan against *Escherichia coli*. *Journal of Food Protection* 62: 239–243.

Tsai, G.-T., W.-H. Su, H.-C. Chen, and C.-L. Pan. 2002. Antimicrobial activity of shrimp chitin and chitosan from different treatments and applications of fish preservation. *Fisheries Science* 68: 170–177.

Tsioptsias, C., I. Tsivintzelis, L. Papadopoulou, and C. Panayiotou. 2009. A novel method for producing tissue engineering scaffolds from chitin, chitin-hydroxyapatite, and cellulose. *Materials Science and Engineering: C* 29: 159–164.

Tsugita, T. 1990. Chitin/chitosan and their applications. In *Advances in fisheries technology and biotechnology for increased profitability*, ed. M.N. Voigt and R.J. Botta, 287–298. Lancaster: Technomic Publications.

Turkiewicz, M., E. Gromek, H. Kalinowska, and M. Zielinska. 1999. Biosynthesis and properties of an extracellular metalloprotease from the Antarctic marine bacterium *Sphingomonas paucimobilis*. *Journal of Biotechnology* 70(1–3): 53–60.

Turkiewicz, M., J. Kur, A. Biakowska, H. Cieslinski, H. Kalinowska, and S. Bielecki. 2003. Antarctic marine bacterium *Pseudoalteromonas* sp. 22b as a source of cold-adapted β-galactosidase. *Biomolecular Engineering* 20(4–6): 317–324.

Uchida, Y., M. Izume, and A. Ohtakara. 1989. Preparation of chitosan oligomers with purified chitosanase and its application. In *Chitin and chitosan*, ed. G. Skjak-Bræk, T. Anthonsen, and P. Sandford, 373–382. Amsterdam: Elsevier.
Venter, J.C., K. Remington, J.F. Heidelberg, A.L. Halpern, D. Rusch, J.A. Eisen, D. Wu, I. Paulsen, K.E. Nelson, W. Nelson, D.E. Fouts, S. Levy, A.H. Knap, M.W. Lomas, K. Nealson, O. White, J. Peterson, J. Hoffman, R. Parsons, H. Baden-Tillson, C. Pfannkoch, Y.-H. Rogers, and H.O. Smith. 2004. Environmental genome shotgun sequencing of the Sargasso Sea. *Science* 304: 66–74.
Wagner-Döbler, I., W. Beil, S. Lang, M. Meiner, and H. Laatsch. 2002. Integrated approach to explore the potential of marine microorganisms for the production of bioactive metabolites. *Advances in Biochemical Engineering/Biotechnology* 74: 207–238.
Walls, D., and S.T. Loughran. 2011. Tagging recombinant proteins to enhance solubility and aid purification. *Methods in Molecular Biology* 681: 151–175.
Wang, S.L., and S.H. Chio. 1998. Deproteinization of shrimp and crabshell with the protease of *Pseudomonas aeruginosa* K-187. *Enzyme and Microbial Technology* 22: 629–633.
Wang, S.-L., T.-Y. Lin, Y.-H. Yen, H.-F. Liao, and Y.-J. Chen. 2006. Bioconversion of shellfish chitin wastes for the production of *Bacillus subtilis* W-118 chitinase. *Carbohydrate Research* 341: 2507–2515.
Wang, W., B. Huang, Z.G. Feng, X.P. Chen, W.X. Tang, and Z.W. Chen. 2010. Expression, purification, and characterization of functional recombinant human daintain/AIF-1 in *Escherichia coli*. *Zeitschrift für Naturforschung C* 65: 726–732.
Whistler, R.S., and J.N. BeMiller. 1962. Chitin. *Journal of Organic Chemistry* 27: 1161–1164.
Wijffels, R.H. 2008. Potential of sponges and microalgae for marine biotechnology. *Trends Biotechnology* 26(1): 26–31.
Winter, J.M., and B.S. Moore. 2009. Exploring the chemistry and biology of vanadium-dependent haloperoxidases. *Journal of Biological Chemistry* 284(28): 18577–18581.
Zheng, W., M.L. Wise, A. Wyrick, J.G. Metz, L. Yuan, and W.H. Gerwick. 2002. Polyenoic fatty acid isomerase from the marine alga *Ptilota filicina*: Protein characterization and functional expression of the cloned cDNA. *Archives of Biochemistry and Biophysics* 401(1): 11–20.

Chapter 6
Extraction and Characterization of Bioactive Carbohydrates with Health Benefits from Marine Resources: Macro- and Microalgae, Cyanobacteria, and Invertebrates

Rita M. Hickey

6.1 Introduction

This chapter addresses extraction of bioactive carbohydrates from sustainable marine resources. It describes an integrated approach and workflow of extraction, purification, and characterization methods from the marine targets both in their unmodified form and following suitable processing such as freeze-drying. A major output of this chapter is technologies, processes, and protocols at pilot and analytical scale designed to isolate the bioactive compounds and the optimization of these processes to allow for commercial development of the bioactive compounds by SMEs. This chapter also addresses chosen bioactive carbohydrates that may be isolated from the waste streams of macroalgae and marine fisheries by-products as this is an underutilized source for marine bioactive compounds and an economic problem for the fish processing and seaweed harvesting industries.

Marine organisms such as macro-and microalgae, sponges, fish, and bacteria have all developed diverse and unique characteristics that allow them to survive under conditions with varying degrees of salinity, pressure, temperature, and illumination (Rasmussen and Morrissey 2007). Despite the fact that the biodiversity of the marine environment far exceeds that of the terrestrial environment, research into the use of marine natural products as pharmaceutical and cosmeceutical agents as well as functional food ingredients still offers a great deal of opportunity. A number of molecules isolated from various marine organisms are currently under study at an advanced stage of clinical trials; some of them have already been marketed as drugs (Bourguet-Kondracki and Kornprobst 2005). In particular, carbohydrates that can be derived from the marine environment represent a wide diversity of biomolecules with potential applications in therapeutics and biotechnology. Such molecules possess marked immunological properties ranging from nonspecific stimulation of the

R.M. Hickey (✉)
Teagasc Food Research Centre, Moorepark Fermoy, Co.Cork, Ireland
e-mail: rita.hickey@teagasc.ie

M. Hayes (ed.), *Marine Bioactive Compounds: Sources, Characterization and Applications*, DOI 10.1007/978-1-4614-1247-2_6,

host immune system, resulting in antitumor, antiviral, and anti-infective effects, to antioxidant, antimutagenic, or hematopoietic activity (Zhang et al. 2010). Nucleic acids and proteins are linear assemblies, however, carbohydrates are the most complex and diverse class of biopolymers commonly found in nature. Although monosaccharides are comprised of a single saccharide unit, a wide array of available monosaccharide building blocks, as well as the possibility of different stereochemical linkages between each pair, results in tremendous complexity. In addition, the chain length of oligosaccharides can also vary widely from monosaccharides up to branched oligosaccharides with more than 30 building blocks, or in the case of polysaccharides to several thousand building blocks (Werz and Seeberger, 2005).

In simple terms, carbohydrates can be divided into two categories: the linear sugars or polysaccharides, which consist of repeating pyranose monosaccharide rings; and the branched sugars, which are saccharide structures based on multiple linkages between monosaccharide rings (Shiver et al. 2004). Moreover, carbohydrates can exist as glycoconjugates (glycoproteins and glycolipids). In proteins, glycans can be attached at *N*-linked (Asn), *O*-linked (Ser/Thr), and *C*-linked (Trp) sites. Additionally, they are found in glycolipids including globosides and glycosphingolipids (Krishnamoorthy and Mahal, 2009). Our understanding of carbohydrates is also complicated by the fact that synthesis of glycan motifs is not template driven; that is, there is no genetic code for this complex modification. Several factors including levels of glycosyltransferases, glycosidases, nucleotide sugar transporters, and protein and lipid trafficking contribute significantly to the glycome of an organism or cell (Krishnamoorthy and Mahal, 2009). Therefore, to study carbohydrates comprehensively, particularly from marine sources, technology development is crucial. This chapter therefore focuses on the role that marine-based carbohydrates have on human health and the various technologies currently used to extract, characterize, and produce such carbohydrates at both analytical and pilot scales. The isolation of a number of key bioactive carbohydrates is reviewed.

6.2 Health-Promoting Properties of Marine Carbohydrates

6.2.1 Macroalgae

Marine macroalgae, or seaweed as they are more commonly known, are one of nature's most biologically active resources, as they hold a wealth of bioactive compounds. For example, compounds isolated from marine macroalgae have demonstrated various biological activities, such as antibacterial activity, antioxidant potential, anti-inflammatory properties, anticoagulant activities, antiviral activities, apoptotic activities, and prebiotic activity (O'Sullivan et al. 2010). Seaweed is commonly classified into three main groups based on their pigmentation; Phaeophyta (brown seaweed), Chlorophyta (green seaweed), and Rhodophyta (red seaweed). In addition to lipids and proteins, macroalgae are a rich source of polysaccharides (d'Ayala et al. 2008). The primary polysaccharides in brown seaweed include alginates, laminarins, fucans, and cellulose (Haugan and Liaaenjensen, 1994; Goni et al. 2002) whereas in green macroalgae, ulvans are the major polysaccharide component (Robic et al. 2009). The primary polysaccharides in red seaweed are agars and carrageenans (McHugh 2003).

Alginate, a high molecular weight polysaccharide, forms hydrogels under relatively mild pH and temperature conditions and is generally regarded as nontoxic, biocompatible, biodegradable, inexpensive, and abundantly available. All these properties make alginates very useful materials for biomedical applications, especially for controlled delivery of drugs and other biologically active compounds and for the encapsulation of cells. Calcium alginate is a natural hemostat, so alginate-based dressings are often used in the treatment of bleeding wounds. The gel-forming alginate property helps in removing the dressing without much difficulty (d'Ayala et al. 2008). Another polysaccharide produced from brown seaweed, laminaran, is a small glucan, approximately 5 kDa in size, with a degree of polymerization (DP) between 20 and 25 (Chizhov et al. 1998). In terms of bioactivity, laminaran modulates the immune response (Neyrinck et al. 2007), but also has other activities including antitumor (Jolles et al. 1963), and antiapoptosis (Kim et al. 2006). Fucoidan, a sulfated polysaccharide also from brown seaweed displays antibacterial (Zapopozhets et al. 1995), antiviral (Yasuhara-Bell and Lu 2010), anticoagulant (Cumashi et al. 2007), antioxidant (Wang et al. 2008), anti-inflammatory (Cumashi et al. 2007), and immunomodulatory effects (Zapopozhets et al. 1995). It also has inhibitory action against the growth of Lewis lung carcinoma and B16 melanoma in mice (Koyanagi et al. 2003) and induces apoptosis in HT-29 and HCT116 human colon cancer cells (Kim et al. 2010).

Ulvan, the major water-soluble polysaccharide found in green seaweed has been shown to possess antioxidant properties that appear to be related to the sulfate content of the polysaccharide (Redoucan et al. 2011). Sulfated polysaccharides from *Ulva pertusa* were also reported to affect levels of low- and high-density cholesterols and triglyceride in the plasma and serum of mice, with the resulting conclusion that they have potential for use in preventing ischemic cardiovascular and cerebrovascular diseases (Pengzhan et al. 2003). *Chondrus crispus* and *Gelidium cartilagineum*, the well-known sources of carrageenan and agar, respectively, have been found to possess antiviral properties attributed to the galactan units in the polysaccharides of both. The antiviral activity was active against influenza B and mumps virus in embryonated eggs even after 24 h incubation (Bhakuni and Rawat 2005). Carrageenan also exhibits several other types of antiviral properties. It is cointernalized into infected cells with the *Herpes simplex* virus (HSV), and inhibits the growth of this DNA-containing virus. Carrageenan interferes with fusion between cells infected with human immunodeficiency virus (HIV), an RNA-containing virus, and inhibits the retroviral enzyme (reverse transcriptase; Bhakuni and Rawat 2005). Carrageenan was also shown to act as an anticoagulant (Caceres et al. 2000; Zúñiga et al. 2006) and antihyperlipidemic agent (Panlasigui et al. 2003). The use of carrageenan in ulcer therapy has been studied extensively and in many cases of ulcer formation, carrageenan proves an effective cure (Bhakuni and Rawat 2005). Antitumor and immunomodulation activities are also found in different molecular weight α-carrageenans from *Chondrus ocellatus* (Zhou et al. 2004). Agar also has medicinal or pharmaceutical industrial applications including use as a suspending agent for radiological solutions (barium sulfate), as a bulk laxative as it gives a smooth and nonirritating hydrated bulk in the digestive tract, and as a formative ingredient for tablets and capsules to carry and release drugs (Laurienzio 2010).

6.2.2 Microalgae

Microalgae are the primary producers for food chain ecosystems and are considered as a low utility group. However, as land resources become depleted worldwide, microalgae are fast becoming substitute resources or new sources for various bioactive compounds. The high variety of species in various habitats is one of the reasons why microalgae attract attention from scientists and industry. Microalgae can be used to produce a wide range of primary and secondary metabolites, including carbohydrates (Kim and Lee, 2005). Not only does the presence of a particular compound make these organisms interesting but also their huge diversity and the possibility of culturing them in different controlled environments, means that they may act as natural reactors, enabling enrichment for certain, specific bioactive compounds (Plaza et al. 2009). However, to date only a few species have been identified as useful for commercial application. These include *Spirulina*, *Chlorella*, *Haematococcus*, *Dunaliella*, *Botryococcus*, *Phaeodactylum,* and *Porphyridium* (Raja et al. 2008; Milledge 2011). Nevertheless, Guzman-Murillo and Ascencio (2000) demonstrated that most of the 21 strains of microalgae used in their study produced extracellular polysaccharides highlighting the potential of these organisms as sources of bioactive polysaccharides. Among the strains used in this study, the largest amounts of polysaccharides produced were associated with *Porphyridium cruentum* and *Chlorella* sp., (Milledge 2011).

Extracellular carbohydrates in *Chlorella pyrenoidosa* cultures are represented mainly by water-soluble polysaccharides containing galactose, mannose, arabinose, xylose, ribose, fucose, and rhamnose (Maksimova et al. 2004; Milledge 2011). In general, for *Chlorella*, it is clear that it is essential to control the particular growing conditions, because those factors can significantly determine the whole and final chemical composition of the microalga (Plaza et al. 2009; Milledge 2011). A number of studies have indicated that *Chlorella* has desirable immunostimulatory properties, both in vitro and in vivo (Suárez et al. 2010). The immunostimulatory effects of proprietary aqueous extracts of the green microalga *Chlorella pyrenoidosa* have been demonstrated by studies performed in mice, in human blood cells, and in a human clinical trial, and are due to the presence of polysaccharides, the main components of the aqueous extract (Suárez et al. 2010).

Porphyridium spp. cells are encapsulated within a sulfated polysaccharide, which has a wide range of promising industrial and medicinal applications (Geresh et al. 2002; Milledge 2011). It is composed of about ten different sugars, the main ones being xylose, glucose, and galactose, and the minor ones being mannose, methylated galactose, and pentose. The polysaccharide is negatively charged, due to the presence of glucuronic acid and half-sulfate ester groups (Li et al. 2000; Milledge 2011). Huheihel et al. (2002) demonstrated that the cell wall-sulfated polysaccharide exhibited significant antiviral activity against herpes simplex virus types 1 and 2 both in vitro (cell culture) and in vivo (rats and rabbits; Milledge 2011). Moreover, the in vitro anti-inflammatory and antiproliferative activity of a sulfoglycolipidic fraction isolated from *P. cruentum* has also been studied (Bergé et al. 2002). This sulfoquinovosylacylglycerol fraction exhibited in vitro antioxidant, anti-inflammatory, and antiproliferative effects and a chemopreventive potential was also indicated (Bergé et al. 2002; Plaza et al. 2009). Sulfated exopolysaccharides of microalgae have also been found to possess antiadhesive properties against certain bacteria. For instance, adhesion of the human pathogen *Helicobacter pylori* to the HeLa S3 cell line and adhesion of the fish pathogens *Vibrio campbellii*, *V. ordalii*, *Streptococcus saprophyticus*, and *Aeromonas veronii* to spotted sand bass primary tissue culture cells, can be effectively blocked with sulfated exopolysaccharides isolated from various microalgae (Guzman-Murillo and Ascencio, 2000; Milledge 2011). Moreover, a sulfated polysaccharide from *Gyrodinium impudicum* was shown in vivo to possess immunostimulating activities. The polysaccharide stimulates various functions of macrophages and NK cells, such as tumoricidal activities and the production of IL-1β, TNF-α, NO, and H_2O_2 (Yim et al. 2005; Milledge 2011).

6.2.3 Marine Waste Streams and By-Products

Every year a considerable amount of total fishing catch is discarded as processing leftovers, including trimmings, fins, frames, heads, skin, and viscera. In addition to fish processing, a large quantity of processing by-products are accumulated as shells of crustaceans and shellfish from marine bioprocessing plants. Therefore, there is a great potential in the marine bioprocess industry to convert and utilize more of these by-products as valuable bioactives. Recent studies have identified a number of bioactive compounds from remaining fish muscle proteins, collagen and gelatin, fish oil, fish bone, internal organs, and shellfish and crustacean shells. Chitin is one of the major structural components of these shell wastes and as discussed possesses many bioactive properties (reviewed by Hayes et al. 2008).

6.3 Processing of Marine Carbohydrates

6.3.1 Extraction

For nearly all marine bioactive products, the starting point is a dilute suspension that ultimately should become a highly purified dry product. Therefore, the main aim of the first separation step is to achieve a volume reduction and a first-stage purification of the bioactive by removing as many other components from the sample as possible. Sample disruption is one of the first steps in processing. However, if the wrong disruption process is used, the bioactive can be destroyed. Therefore, short effective procedures are required (Muffler and Ulber 2005). The method used depends on the marine source and on the nature of the carbohydrate of interest and are either based on mechanical action (e.g., cell homogenizers, bead mills, ultrasounds, autoclave, and spray drying) or nonmechanical action (e.g., freezing, organic solvents, and osmotic shock; and acid, base, and enzyme reactions). Homogenizers are easy to clean and to sterilize and they do not require much maintenance, however, they do produce aerosols, which can be harmful. By using bead mills, cell disruption can be achieved in a single run with better temperature distribution and temperature control than homogenizers. Aerosol generation is minimized. However, bead mills are difficult to clean and sterilize. Microwaves or ultrasound can be combined with sample extraction by organic solvents. Microwave techniques are widely used in acid digestion of solid samples. Cell lysis using mild conditions and appropriate hydrolyzing enzymes are also an alternative for intracellular release of carbohydrates. Moreover, enzymes offer selectivity during bioactive release (reviewed by Muffler and Ulber 2005).

In the extraction of seaweed polysaccharides many of the above disruption methods are employed. For instance, for the extraction of alginate, brown algae are washed, macerated, and then extracted by sodium carbonate. Fucoidans are extracted from algae in dilute acid conditions, and agar which is soluble in boiling water but insoluble in cold water is extracted from boiled seaweed; the extract is then frozen and thawed. During this last step, water separates from the agar, carrying with it soluble impurities. Washing, bleaching, sterilization, and drying are performed depending on the applications. Extraction of carrageenans is normally performed under strong alkaline conditions at a temperature near the boiling point for several hours; the extract is filtered to get a clear solution and directly precipitated by addition of isopropanol (reviewed by Rinaudo 2008). There are two most commonly used techniques to cultivate microalgae. These are the open raceway pond system and closed photobioreactor system. Methods such as flocculation, centrifugation, and filtration are used to dewater the microalgae (Harun et al. 2010) whereas polysaccharide extraction from microalgae generally involves liquid extraction. In contrast, chitin is extracted from the exoskeleton of marine organisms, mainly crabs and shrimp, as described by Burrows et al. (2007) and Hayes et al. (2008). Briefly, the exoskeletons are crushed and washed, then treated with boiling sodium hydroxide to dissolve the proteins and sugars, thus isolating the crude chitin. In cyanobacteria strains, exocellular polysaccharides are present as outermost investments forming sheaths, capsules, and slimes that protect the bacterial cells from the environment. Moreover, most polysaccharide-producing cyanobacteria release aliquots of capsules and slimes as soluble polymers in the culture medium. Therefore, cyanobacteria strains that possess abundant capsules and slimes (and so release a large amount of soluble polysaccharides) allow for the easy recovery of the polysaccharides from liquid cultures (Laurienzo 2010).

The traditional extraction methods described above can have several drawbacks: they can be time consuming, laborious, have low selectivity, and/or low extraction yields. Moreover, many of these techniques employ large amounts of toxic solvents.

At present, extraction methods able to overcome such drawbacks are being researched, among them, supercritical fluid extraction which was discussed in Chapter 2 of this book. These extraction techniques provide higher selectivities and shorter extraction times. When a fluid is forced to a pressure and temperature above its critical point, it becomes a supercritical fluid. Under these conditions, various properties of the fluid are placed between those of a gas and those of a liquid (Chapter 2 of this book). As a consequence extraction becomes faster than for conventional liquid extraction and extraction conditions are easily controlled. Solubility of carbohydrates in supercritical CO_2 is low but it shows a marked increase when alcohols are added as polar solvents (Sanz and Martínez-Castro 2007). Despite this, few studies have been performed using SFE to fractionate marine carbohydrates.

6.3.2 *Fractionation*

Before any sample fractionation can take place, a number of properties of the carbohydrate of interest must be considered (Sanz and Martínez-Castro 2007):

- Molecular weight and whether the carbohydrate consists of straight or branched chains
- Solubility in water and polar solvents
- Thermal stability
- Chemical reactivity

Once all available information on the carbohydrate has been compiled, certain clean-up steps must be initiated. These procedures are generally performed to remove insoluble material, lipids, and proteins, desalt the sample, and remove impurities. The most important separation techniques include filtration and centrifugation. For example, harvesting microalgae often means concentrating the biomass from a concentration of <1 g dry weight L^{-1} in the photobioreactor to as much as 250 g dry weight L^{-1}. Here centrifugation is the method of choice (Olaizola 2003). Filtration separates solids from a liquid by forcing the liquid through a solid support or filter medium. Two different designs of filtration can be used:

- Dead-end filtration
- Cross-flow filtration

Both types are comprehensively reviewed by Muffler and Ulber (2005). In dead-end filtration, the total process fluid stream flows through the membrane. The retained solids collect on the membrane and form a buildup. When the membrane pores get clogged by the solids it must be changed. Cross-flow filtration is the term used to describe when the feed flow is directed parallel to the membrane surface. The tangential flow of liquid removes any retained particles from the membrane surface, which results in a stable flux for a longer time period.

The cost of dead-end filtration is low relative to cross-flow filtration. However, costs for filter aids can be very high depending on the filtration medium. The pressure drop and the shear stress increase with filtration time while the flow rate through the filter decreases. By using cross-flow filtration, relatively low shear stress is possible and filter aids are not required. In addition, scale-up is simple. However, a common problem in cross-flow filtration is membrane fouling. Therefore, membrane replacement is often frequently necessary.

Membrane separation is based mainly on molecular size, but to a lesser extent on shape and charge. The *permeate* (filtrate) is the stream that passes through the membrane, whereas the material remaining on the membrane is termed *retentate* (concentrate). Membrane processes are easy to scale up and the possibility of using the same materials and conditions in different sizes from laboratory to process scale reduces method validation hugely. The function of membranes has been greatly expanded in recent years and a number of membrane processes have been developed for molecular separation. These filtration techniques can be divided into four major groups: reverse osmosis (hyperfiltration; RO), nanofiltration (NF), ultrafiltration (UF), and microfiltration (MF). The dimensions of the components involved in these separations are <0.0001, 0.0001–0.001, 0.001–0.02, and 0.02–10 μm, respectively (Muffler and Ulber 2005). As an example, Lignot et al. (2003) developed a downstream process for the recovery of chondroitin sulfate (CS), a fishery by-product. CS is a glycosaminoglycan well known for its chondroprotective effect. After enzymatic extraction, cross-flow filtration was used to concentrate and purify CS. The researchers demonstrated that CS solution can be partially concentrated effectively by ultrafiltration up to four times.

Membranes can be converted into efficient adsorbers by attaching functional groups to the inner surface of synthetic microporous membranes. Affinity adsorption, ion exchange, or immobilized metal affinity chromatography can be carried out by these membranes. Whereas other separation techniques rely primarily on molecular size, charge, or solubility, affinity adsorption relies on highly specific binding interactions. Affinity adsorption offers the advantages of high purification factors (up to 1,000-fold) and high recovery rates. It is based on the formation of a reversible complex between the target and the ligand. Often the ligand is used in an immobilized form (insoluble matrix). Membrane ion exchangers of strong acidic (sulfonic acid), strongly basic (quarternary ammonium), weakly acid (carboxylic acid), and weakly basic (diethylamine) types are commercially available. These membranes are available in products for laboratory and process scale. For process applications the modules and systems can be adapted to the special needs of the specific separation process to achieve optimal conditions (Muffler and Ulber 2005).

6.4 Detection Methods

Carbohydrates do not have chromogenic or fluorogenic group(s) and therefore do not absorb visible or ultraviolet radiation (sugars generally absorb <200 nm; Rocklin and Pohl 1983). Moreover, they are not volatile enough for direct analysis by gas chromatography. For these reasons, most assays rely on derivatization by a chromogen or other molecules such as trimethylsilyl ethers (Me3Si) or trifluocacetate (TFA) esters that make the sugars volatile. The name of the chromogen used is generally attributed to the method name. It should be noted that the use of pulsed amperometry (PAD) is performed without prior derivatization (Johnson and LaCourse 1990), unlike other methods. There are a number of colorimetric methods that have been used for carbohydrate determination (reviewed comprehensively by Panagiotopoulos and Sempéré 2005). One of the most commonly used is the phenol–sulfuric acid (PSA) or Dubois method which involves dehydration of carbohydrates in the presence of concentrated H_2SO_4 at high temperature, forming furfurals or hydroxymethylfurfurals. Condensation of the latter compounds with a phenol group produces orange–yellow substances that absorb at 480–490 nm. The color produced at a constant phenol concentration is proportional to the amount of sugar originally present. This method has good precision (<20% at the 50 μM level) and a detection limit of 25–50 μM (Dubois et al. 1956; Gerchakov and Hatcher 1972). However, when used for environmental samples, reproducibility may be low. Despite this and other drawbacks (reviewed by Panagiotopoulos and Sempéré 2005), the method is still in use for polysaccharides from algae and seaweed (Siddhanta et al. 2001; Rocha de Souza et al. 2007).

The lack of specificity and the poor detection limit of the Dubois method may be overcome by the MBTH method that combines three well-known reactions of monosaccharides to analyze one final compound (formaldehyde). The assay relies on reduction of monosaccharides to the corresponding alditols with potassium borohydride (KBH_4) (Abdel-Akher et al. 1951). The resulting alditols are treated with periodic acid (HIO_4) and their terminal alditol glycol groups ($-CH_2OH$) produce two moles of formaldehyde (Sawicki et al. 1961). The latter compound further reacts with the chromogen 3-methyl-2-benzo thiazoline hydrazone hydrochloride (MBTH), producing a blue complex that absorbs at 635 nm. Despite the fact that the MBTH method is laborious given its three chemical reactions, it has been used extensively in the determination of mono-and polysaccharides from marine samples (Aluwihare and Repeta 1999). The TPTZ method was developed (Myklestad et al. 1997), to analyze marine carbohydrates. It combines the low detection limits and precision of the MBTH method with the rapidity and simplicity of the PSA method. This method uses the reducing properties of sugars and is based on the oxidation of sugars by Fe^{3+} ($K_3[Fe(CN)_6]$) in alkaline media (Avigad 1968). The resultant Fe^{2+} reacts with the chromogen TPTZ, producing a violet complex that absorbs at 595 nm. This method is also widely used for carbohydrate analysis of marine samples (van Oijen et al. 1951).

6.5 Purification and Characterization

The detection methods discussed above provide an estimate of the total carbohydrates present in marine samples. Compositional analyses or determination of individual sugars in marine samples require many additional techniques. Any detailed investigation into the structure of carbohydrates can involve the application of many combined techniques. Carbohydrates can be so complex that a single technique is usually not capable of sequencing these molecules. Such techniques can include enzymatic or chemical digests, high-performance liquid chromatography (HPLC), capillary electrophoresis (CE), mass spectrometry (MS), nuclear magnetic resonance (NMR), or a combination of any of the above (Behr and Sasisekharan 2007).

6.6 Industrial Industrial-Scale Production of Marine Carbohydrates: Case Study

6.6.1 Fucoidan

Fucoidans from brown seaweed have several food, pharmaceutical, and medicinal applications. These sulfated polysaccharides consist of L-fucose with molecular weights ranging from 13 to 950 kDa (Hahn et al. 2011). Their heterogeneous structures can vary considerably from alga to alga and this leads to difficulty in determining the structure and branching of the fucopyranosyl residues that can vary with sulfation (Hahn et al. 2011).

Figure 6.1 describes an industrial method for extraction of fucoidan from brown seaweed. Seaweed is pretreated overnight with an ethanol/formaldehyde/water solution (80:5:15 volume/volume) and the dried algae is subsequently extracted using a 0.01 M HCL solution supplemented with calcium chloride ($CaCl_2$) at 4% using Soxhlet at 70°C for approximately 3 h. This removes alginates, cellulose, and proteins (Hahn et al. 2011). Ethanol is then used to precipitate the carbohydrate. Fucoidan and laminaran are separated from each other using ion exchange chromatography and gel filtration methods (Hahn et al. 2011).

6.7 Perspectives and Future Directions

Marine resources, specifically carbohydrates, have potential for use as products targeted at the functional food and pharmaceutical markets. Patent activity in this area has increased recently. For example, the Kabushiki Kaisha Yakult Honsha Company, Japan, has patented a polysaccharide derivative known to contain fucoidan and rhamnan or rhamnan sulfate polysaccharides, extracted from the marine macroalgae, *Phaeophyceae* species, *Cladosiphon okamuranus*, *Chordaricles nemacystus*, *Hydrilla* sp., *Fucus* sp., and *Monostroma nitidum*. The purpose of this compound is as a therapeutic agent for the prevention and treatment of gastric ulcers caused by *Helicobacter pylori*. Furthermore, Takara Shuzo Company, Kyoto, Japan developed a medicinal composition exemplified by viscous polysaccharides isolated from the red algae, *Gelidium amansil*, *G. japonicum*, *G. pacificum*, *G. subcostatum*,

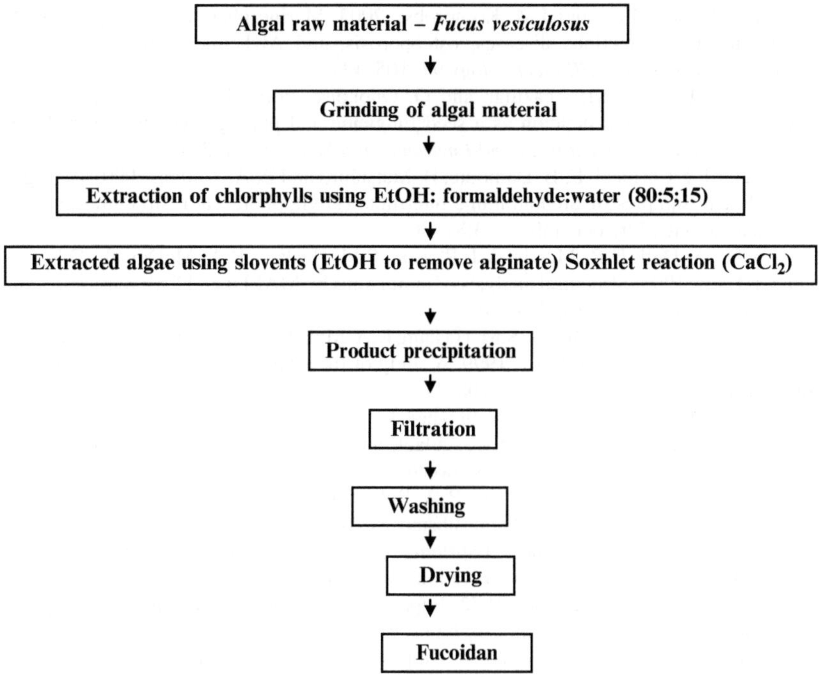

Fig. 6.1 Preparation of fucoidan from seaweed (adapted from Hahn et al. 2011)

Pteocladia tennis, and *Hypneaceae* species consisting of at least one 3,6-anhydrogalactopyranose. This compound is proposed for the treatment or prevention of diabetes, rheumatism, and cancer, and contains various inhibitory factors. From these examples, it is clear that marine polysaccharides possess a multitude of bioactivities with potential for use in functional foods.

References

Abdel-Akher, M., J. K. Hamilton and F. Smith. 1951. The reduction of sugars with sodium borohydride. *Journal of American Chemical Society* 73: 4691–4692.

Aluwihare, L. I. and D. J. Repeta. 1999. A comparison of the Chemical Characteristics of Oceanic DOM and Extracellular DOM Produced by Marine Algae. Marine Ecology Progress Series 186: 105–117.

Avigad, G. 1968. A modified procedure for the colorimetric, ultramicro determination of reducing sugars with the alkaline ferricyanide reagent. *Carbohydrate Research* 7: 94–97.

Behr J. and R. Sasisekharan. 2007. Chapter 8 in *Glycobiology* (C. Sansom and O. Markman, eds., Scion Publishing Ltd, UK).

Bergé, J. P., E. Debiton, J. Dumay, P. Durand and C. Barthomeuf. 2002. *In vitro* anti-inflamatory and anti-proliferative activity of sulfolipids from the red alga *Porphyridium cruentum. Journal of Agricultural Food Chemistry* 50: 6227–6232.

Bhakuni, D.S., and D.S. Rawat. 2005. *Bioactive marine natural products*. New Delhi: Springer.

Bourguet-Kondracki, M.L., and J.M. Kornprobst. 2005. Marine pharmacology: potentialities in the treatment of infectious diseases, osteoporosis, and Alzheimer's disease. *Advances in Biochemical Engineering/Biotechnology* 97: 105–131.

Burrows, F., C. Loumie, M. Abazinge and O. Onokpise. 2007. Extraction and evaluation of Chitosan from crab exoskeleton as a seed fungicide and plant growth enhancer. *American-Eurasian Journal of Agricultural and Environmental Science* 2: 103–111.

Caceres, P. J., M. J. Carlucci, E. B. Damonte, B. Matsuhiro and E. A. Zuniga. 2000.. Carrageenans from Chilean samples of Stenogramme interrupta (Phyllophoraceae): structural analysis and biological activity. Phytochemistry 53:81–86.

Chizhov, A.O., A. Dell, H.R. Morris, A.J. Reason, S.M. Haslam, R.A. McDowell, O.S. Chizhov and A.I. Usov. 1998. Structural analysis of laminarans by MALDI and FAB mass spectrometry. *Carbohydrate Research* 310: 203–210.

Chizhov, A.O., A. Dell, H.R. Morris, S.M. Haslam, R.A. McDowell, A.S. Shashkov, N.E. Nifant'ev, E.A. Khatuntseva, and A.I. Usov. 1999. A study of fucoidan from the brown seaweed Chorda filum. *Carbohydrate Research* 320: 108–119.

Cumashi, A., N.A. Ushakova, M.E. Preobrazhenskaya, A. D'Incecco, et al. 2007. A comparative study of the anti-inflammatory, anticoagulant, antiangiogenic, and antiadhesive activities of nine different fucoidans from brown seaweeds. *Glycobiology* 17(5): 541–552.

Dubois, M., K. A. Gilles, J. K. Hamilton, P. A. Rebers and F. Smith. 1956. Colorimetric method for determination of sugar and related substances. *Analytical Chemistry* 28: 350–356.

d'Ayala, G.G., M. Malinconico, and P. Laurienzo. 2008. Marine derived polysaccharides for biomedical applications: chemical modification approaches. *Molecules* 13: 2069–2106.

Gerchakov, S. M. and P. G. Hatcher. 1972. Improved technique for analysis of carbohydrates in sediments. *Limnology and Oceanography* 17: 938–943.

Geresh, S., I. Adin, E. Yarmolinsky and M. Karpasas, 2002. Characterization of the extracellular polysaccharides of *Porphyridium* sp.: Molecular weight determination and rheological properties. *Carbohydrate Polymers* 50: 183–189.

Goni, I., L. Valdivieso, and M. Gudiel-Urbano. 2002. Capacity of edible seaweeds to modify *in vitro* starch digestibility of wheat bread. *Nahrung* 46: 18–20.

Guzman-Murillo, A. and F. Ascencio. 2000. Anti-adhesive activity of sulphated exopolysaccharides of microalgae on attachment of red sore disease-associated bacteria and *Helicobacter pylori* to tissue culture cells. *Letters in Applied Microbiology* 30: 473–478.

Hahn, T., S. Kelly, K. Muffler, N. Tippkotter, and R. Ulber. 2011. Extraction of lignocellulose and algae for the production of bulk and fine chemicals. In *Industrial scale natural products extractions*, ed. B. Hans-Jorg and P. Stephan, 221–245. Weinheim: Wiley-VCH.

Haugan, J.A. and S. Liaaenjensen. 1994. Algal Carotenoids 54. Carotenoids of Brown-Algae (Phaeophyceae). *Biochemical cystematics and Ecology* 22: 31–41.

Hayes, M., B. Carney, J. Slater, and W. Bruck. 2008. Mining marine shellfish waste for bioactive molecules: chitin and chitosan – Part A: Extraction methods. *Journal of Biotechnology* 3: 871–877.

Huheihel, M., V. Ishanu, J. Tal and S. M. Arad. 2002. Activity of *Porphyridium* sp. polisaccharide against herpes simplex viruses *in vitro* and *in vivo*. *Journal of Biochemical and Biophysical methods* 50: 189–200.

Johnson D.C and W.R. LaCourse. 1990. LC with pulsed ECD at gold and platinum. *Analytical Chemistry* 62: 589A–597A.

Jolles, B., M. Remington, and P. S. Andrews. 1963. Effects of sulphated degraded laminarin on experimental tumour grouth. *British Journal of Cancer* 17: 109–115.

Harun, R., M. K. Danquah and G. M. Forde. 2010. Microalgal biomass as a fermentation feedstock for bioethanol production. *Journal of Chemical Technology and Biotechnology* 85(2): 199–203.

Kim, J. D. and C. G. Lee. 2005. Systemic optimization of microalgae for bioactive compound production. *Biotechnology and Bioprocessing Engineering* 10: 418–424.

Kim, K.-H., Y.-W Kim, H.B. Kim, B.J. Lee and D.S. Lee. 2006. Anti-apoptotic activity of laminarin polysaccharides and their enzymatically hydrolyzed oligosaccharides from *Laminaria japonica*. *Biotechnology Letters* 28: 439–446.

Kim, E.J., S. Y. Park, J. Y, Lee and J. H. Park. 2010. Fucoidan present in brown algae induces apoptosis of human colon cancer cells. *BMC Gastroenterology* 10: 96.

Krishnamoorthy L. and L. K. Mahal. 2009. Glycomic Analysis: An Array of Technologies. *ACS Chemical Biology* 4(9): 715–732.

Koyanagi, S., H. Nakagawa, Y. Kuramoto, S. Ohdo, S. Soeda, and H. Shimeno. 2003. Optimizing the dosing schedule of TNP-470 [*O*-(chloroacetylcarbamoyl)fumagillol] enhances its antitumor and antiangiogenic efficacies. *The Journal of Pharmacology and Experimental Therapeutics* 304: 669–674.

Krishnamoorthy, L., J.W. Bess, A.B. Preston, K. Nagashima, and L.K. Mahal. 2009. HIV-1 and microvesicles from T-cells share a common glycome. Arguing for a common origin. *Nature Chemical Biology* 5(4): 244–250.

Laurienzio, P. 2010. Marine polysaccharides in pharmaceutical applications: an overview. *Marine Drugs* 8: 2435–2456.

Li, S. Y., Y. Shabtai and S. M. Arad. 2000. Production and composition of the sulphated cell wall polysaccharide of *Porphyridium* (Rhodophyta) as affected by CO_2 concentration. *Phycologia* 39: 332–336.

Lignot, B., V. Lahogue and P. Bourseau. 2003. Enzymatic extraction of Chondroitin Sulfate from skate cartilage and concentration-desalting by ultrafiltration. *Journal of Biotechnology* 103: 281–284.

Maksimova, I. V., L. B. Bratkovskaya and S. E. Plekhanov. 2004. Extracellular carbohydrates and polysaccharides of the alga *Chlorella pyrenoidosa* Chick S-39. *The Biological Bulletin* 31: 175–181.

McHugh, D. 2003. A guide to the seaweed industry. FAO Fisheries Technical Paper No. 441, Rome.

Milledge, J.J. 2011. Commercial applications of microalgae other than as biofuels: a brief overview. *Reviews in Environmental Science and Biotechnology* 10(1): 31–41.

Muffler, K., and R. Ulber. 2005. Downstream processing in marine biotechnology. *Advances in Biochemical Engineering/Biotechnology* 97: 63–103.

Myklestad, S. V., E. Skånøy and S. Hestmann. 1997. A sensitive method for analysis of dissolved mono- and polysaccharides in seawater. *Marine Chemistry* 56: 279–286.

Neyrinck, A.M., A. Mouson and N.M. Delzenne. 2007. Dietary supplementation with laminarin, a fermentable marine beta (1–3) glucan, protects against hepatotoxicity induced by LPS in rat by modulating immune response in the hepatic tissue. *International Immunopharmacology* 7: 1497–1506.

O' Sullivan, L., B. Murphy, P. McLoughlin, P. Duggan, P.G. Lawlor, H. Hughes, and G.E. Gardiner. 2010. Prebiotics from marine macroalgae for human and animal health applications. *Marine Drugs* 8(7): 2038–2064.

Olaizola, M. 2003. Commercial development of microalgal biotechnology: from the test tube to the marketplace. *Journal of Biomolecular Engineering* 20: 459–466.

Panagiotopoulos, C., and R. Sempéré. 2005. Analytical methods for the determination of sugars in marine samples: a historical perspective and future directions. *Limnology and Oceanography: Methods* 3: 419–454.

Panlasigui, L. N., O. Q. Baello, J. M. Dimatangal and B. D. Dumelod. 2003. Blood cholesterol and lipid-lowering effects of carrageenan on human volunteers. *Asia Pacific Journal of Clinical Nutrition* 12: 209–214.

Pengzhan, Y., Z. Quanbin, L. Ning, X. Zuhong, W. Yanmei, L. Zhi'en, 2003. Polysaccharides from *Ulva pertusa* (Chlorophyta) and preliminary studies on their antihyperlipidemia activity. *Journal of Applied Phycology* 15: 21–27.

Plaza, M., M. Herrero, A. Cifuentes and E. Ibáñez. 2009. Innovative natural functional ingredients from microalgae. *Journal of Agricultural Food Chemistry* 57: 7159–7170.

Raja, R., Hemaiswarya, S., Ashok Kumar, N., Sridhar, S, and R. Rengasamy. 2008. A Perspective on the Biotechnological Potential of Microalgae. *Critical Reviews in Microbiology* 34(2): 77–88.

Rasmussen, R.S., and M.T. Morrissey. 2007. Marine biotechnology for production of food ingredients. *Advances in Food and Nutrition Research* 52: 237–292.

Robic, A., C. Gaillard, J.F. Sassi, Y. Lerat, and M. Lahaye. 2009. Ultrastructure of ulvan: a polysaccharide from green seaweeds. *Biopolymers* 91: 652–664.

Rocha de Souza, M.C., C. Texeira-Masques, C.M. Guerra-Dore, F.R. Ferreira da Silva, H.A. Olivera-Rocha and E. Lisboa-Leite. 2007. Antioxidant activity of sulfated polysaccharides from brown and red seaweeds. *Journal of Applied Phycology* 19: 153–160.

Rinaudo, M. 2008. Main properties and current applications of some polysaccharides as biomaterials. *Polymer International* 57(3): 397–430.

Rocklin, R. D. and C. A. Pohl. 1983. Determination of carbohydrates by anion exchange chromatography with pulsed amperometric detection. *Journal of Liquid Chromatography* 6: 1577–1590.

Sanz, M.L. and I. Martínez-Castro. 2007. Recent developments in sample preparation for chromatographic analysis of carbohydrates. *Journal of Chromatography* A 1153: 74–89.

Sawicki, E., T. R. Hauser, T. W. Stanley and W. Elbert. 1961. The methyl-2-benzothiazolone hydrazone test. *Analytical Chemistry* 33: 93–96.

Shriver, Z., S. Raguram and R. Sasisekharan. 2004. Glycomics: a pathway to a class of new and improved therapeutics, *Nature Reviews Drug Discovery* 3: 863–873.

Siddhanta, A.K., A.M. Goswami, B.K. Ramavat, K. H. Mody and O.P. Mairh. 2001. Water soluble polysaccharides of marine algal species Ulva (Ulvales, Chlorophyta) of Indian waters. *Indian Journal of Marine Sciences* 30: 166–172.

Suárez E. R., J. A. Kralovec and T. B. Grindley. 2010. Isolation of phosphorylated polysaccharides from algae: the immunostimulatory principle of *Chlorella pyrenoidosa*. *Carbohydrate Research* 345(9): 1190–1204.

Van Oijen, T., M. J. W. Veldhuis, M. Y. Gorbunov, J. Nishioka, M. A. Van Leeuwe and H. J. W. de Baar. 2005. Enhanced carbohydrate production by Southern Ocean phytoplankton in response to in situ fertilization. *Marine Chemistry* 93: 33–52.

Wang, J., Q. Zhang, Z. Zhang and Z. Li. 2008. Antioxidant activity of sulfated polysaccharide fractions extracted from *Laminaria japonica*. *International Journal of Biological Macromolecules* 42(2): 127–132.

Werz D. B. and P. H. Seeberger. 2005. Carbohydrates as the Next Frontier in Pharmaceutical Research. *Chemistry - A European Journal* 11(11): 3194–3206.

Werz, P., H. Seeberger, A. Matthias, B. Oberli, and B.P. Daniel. 2008. Synthesis of a hexasaccharide repeating unit from bacillus anthracis vegetative cell walls. *Organic Letters* 10: 905–908.

Yasuhara-Bell, J., and Y. Lu. 2010. Marine compounds and their antiviral activities. *Antiviral Research* 86(3): 231–240.

Yim, J. H., E. Son, S. Pyo and H. K. Lee. 2005. Novel sulfated polysaccharide derived from red-tide microalga *Gyrodinium impudicum* strain KG03 with immunostimulating activity in vivo. *Marine Biotechnology* (NY) 7: 331–338.

Zapopozhets T. S., N.N. Besednova and IuN Loenko. 1995. Antibacterial and immunomodulating activity of fucoidan. *Antibiot Khimioter* 40(2): 9–13.

Zhang, Z., F. Wang, X. Wang, X. Liu, and Q. Zhang. 2010. Extraction of the polysaccharides from five algae and their potential antioxidant activity in vitro. *Carbohydrate Polymers* 82(1): 118–121.

Zhou, G., Y. Sun, H. Xin, Y. Zhang, Z. Li and Z. Xu. 2004. *In vivo* antitumor and immunomodulation activities of different molecular weight lambda-carrageenans from *Chondrus ocellatus*. *Pharmacological Ressearch* 50: 47–53.

Zúñiga, E.A., B. Matsuhiro, and E. Mejías. 2006. Preparation of a low-molecular weight fraction by free radical depolymerization of the sulfated galactan from Schizymenia binderi (Gigartinales, Rhodophyta) and its anticoagulant activity. *Carbohydrate Polymers* 66(2): 208–215.

Chapter 7
Medicinal Chemistry and Ligand Profiling for Evaluation of Promising Marine Bioactive Molecules

A.K. Croft, W. Groenewald, and M.S. Tierney

7.1 Introduction

Many marine natural products exhibit a range of bioactivities, including anticancer, antiviral, antifungal, and antihypertensive properties. As such, they are excellent lead compounds for further drug discovery. In recent years, due to the more accessible cost of computing in terms of both money and time, the complex and expensive process of drug discovery has been significantly enhanced through the use of computational approaches. Here we describe key aspects of the process where computation has helped, including lead validation, optimization, profiling, and discovery, as well as in silico ADME (absorption, distribution, metabolism, and elimination) and toxicological methods, giving relevant examples of the uses of such approaches from the marine natural product world.

Traditionally, pharmaceutical companies have considered natural products as bioactive goldmines from which they have sourced a significant number of commercial drugs (Shu 1998; Eisenhauer et al. 1994). For instance, in 1997 approximately 60% of the antitumor and anti-infective agents that were either commercially available or in late stages of clinical trials at the time were of natural product origin. However, interest in naturally sourced compounds has waned, resulting in a 30% decrease in the number of development projects within this area between 2001 and 2008 (Schuster 2010). This is due to the challenging nature of natural product drug discovery, exemplified by supply issues of raw material (Cragg et al. 1993), lengthy extraction, isolation, and characterization procedures (Rollinger 2009), and the

A.K. Croft (✉) • W. Groenewald
School of Chemistry, University of Wales Bangor, Bangor, UK
e-mail: a.k.croft@bangor.ac.uk

M.S. Tierney
School of Chemistry, University of Wales Bangor, Bangor, UK

Food BioSciences Department, Teagasc Food Research Centre, Ashtown, Dublin 15, Ireland

M. Hayes (ed.), *Marine Bioactive Compounds: Sources, Characterization and Applications*, DOI 10.1007/978-1-4614-1247-2_7,

inherent complexity of natural product chemistry, which has resulted in many medicinal chemists viewing such drug-leads as "ugly ducklings" (Strohl 2000).

Marine plants and animals experience harsh ecological conditions, and therefore possess an arsenal of secondary metabolites for protection that often differ structurally from terrestrial compounds, making them ideal potential leads for new drug discovery. The relatively recent FDA approval of the marine-derived drug Prialt™, derived from a peptide toxin found in the marine snail *Conus magus* (Belden 2005; McIntosh, et al. 1982), has renewed interest in marine lead discovery. However, this discovery came following a 28-year gap after the FDA approval of the marine-derived antiviral drug vidarabine from the sponge *Tethya crypta* (Glaser and Mayer 2009). In 2009, the antitumoral drug Yondelis®, derived from the tunicate *Ecteinascidia turbinata* (Rinehart et al. 1990) achieved E.U. approval for the treatment of ovarian carcinoma in addition to its earlier approval for the treatment of soft tissue sarcoma (Yap et al. 2009). However, marine lead discoveries are few and far between, often due to the hurdles met in the early stages of the drug discovery process. Issues relating to the supply of raw material are particularly prominent in marine drug discovery because of the cost of specialist collection equipment and the extremity of supply locations. These issues can be addressed to a certain extent by a concept known as reverse pharmacognosy, which seeks to find biological targets for known natural compounds by virtual screening and identifying the natural resources that contain the active molecules (Do et al. 2007). Thereby, only marine plants and organisms identified as producing bioactive compounds are collected, which lessens the cost and time involved, relative to the traditional approach of identifying leads. The establishment of various marine-specific databases (Babu et al. 2008; Davis and Vasanthi 2011; Lei and Zhou 2002; MarinLit) means that there are many thousands of marine compounds available for virtual screening (VS), based on the reverse pharmacognosy principle. One of the most comprehensive marine databases is the Dictionary of Marine Natural Products (DMNP), a subset of the Dictionary of Natural Products (DNP) database, which contains over 30,000 compounds (Dictionary of Natural Products).

Computer-aided drug discovery and development (CADDD) encompasses a wide range of in silico tools designed to modernize the drug discovery and development process and has experienced increased implementation over the last number of years. The increased popularity of CADDD coincides with advancements made in software and hardware computational power, establishment of large compound libraries and the availability of accurate protein target structures. Theoretical models also provide a way of expanding the molecule space by virtually creating and screening all possible and viable small molecules (Reymond et al. 2010). Predictions claim that molecular modeling and other in silico approaches will account for approximately 20% of pharmaceutical research and development spending by 2016 (van de Waterbeemd and Gifford 2003). Some current computational resources that have been utilized with a variety of natural products, including marine natural products, are collected in Table 7.1, along with examples of their use.

With the increased demand for new pharmaceutical compounds, high-throughput screening (HTS) methods have been developed to raise the number of substances

Table 7.1 Examples of virtual screening (VS) tools used for natural and marine product profiling

	VS Tool	Characteristics	Function in NP screening
Pharmacophore modeling	*RI-4D QSAR*	Allows the construction of optimized spatial QSAR models in the form of 3-D pharmacophores, which are dependent on conformation, alignment, and pharmacophore grouping (Hopfinger et al. 1999)	A training set of 26 lamellarins (**17a** and **17b**) was used to build pharmacophore models to investigate for cytotoxicity against human breast cancer cells (Thipnate et al. 2009)
	LigandScout (Ligand Scout, Europe)	Automatic construction and visualization of pharmacophore models derived from protein structures (Wolber and Langer 2005)	Pharmacophore model generation for the screening of natural AChE inhibitors (Rollinger et al. 2004)
	DISCOtech in SYBYL 8.0 (SYBYL 8.0) Martin et al. (1993)	DISCOtech™ works in distance space and can carry out clique detection to create pharmacophore hypotheses (Khanfar et al. 2009)	Generated a pharmacophore model based on four triterpenoid sipholanes (**13–16**) compounds from the sponge *Callyspongia siphonella* (Jain et al. 2009)
	Catalyst	Maps pharmacophore features and searches databases using the fast flexible search method (Kahnberg et al. 2004)	Identified the coumarins scopoletin (**18**) and its glucoside scopolin (**19**), which are found in *Scopolia carniolica*, as potential ACE inhibitors (Rollinger et al. 2004)

(continued)

Table 7.1 (continued)

	VS Tool	Characteristics	Function in NP screening
Molecular docking	*Alpha Triangle in MOE* (Molecular Operating Environment, Canada)	Produces poses by superposition of ligand atom triplets and triplet points within the target site (Indarte et al. 2007)	Modeling of the interactions between a novel ACE-inhibitory peptide, Lys-Val-Leu-Ile-Leu-Ala, from milk casein and the ACE active site (Wang et al. 2011)
	AutoDock 4.0 (Morris et al. 1998)	AutoDock 4.0 provides fast prediction of bound conformations with predicted free energies of association (Morris et al. 1998)	To investigate the binding modes of marine phenolic inhibitors of ARL2, an enzyme implicated in diabetic complications (Wang et al. 2009)
	AutoDock 4.0 (Morris et al. 1998) *& FRED 2.0* (Jain et al. 2009)	FRED docks molecules using an extensive search algorithm that searches rotations and translations of each ligand conformer within the active site at a specified resolution (McGann 2011)	To confirm the inhibition mode of BACE1 inhibitor phlorotannins, eckol (**9**), phlorofurofucoeckol-A (**10**), and dieckol (**11**) extracted from the brown algae *Eisenia bicyclis*
	AutoDock Vina	Updated version of AutoDock 4.0, with improved speed and prediction accuracy (Trott and Olson 2010)	Investigate the capability of marine-sulfated sterols (**1–8**) to act as FXR modulators (Sepe et al. 2011)
	Molegro Virtual Docker (Thomsen and Christensen 2006*)*	Based on a hybrid search algorithm, called guided-differential evolution (DE), for fast and accurate identification of potential binding modes (Thomsen and Christensen 2006)	Investigate the binding mode of an ACE-I pentapeptide, derived from peanut protein, with tACE (Jimsheena and Gowda 2011)
	Surflex-Dock (Surflex-Dock 2.0, Tripos International, Missouri, USA) *with SYBYL 8.0* (SYBYL 8.0)	Combines the scoring function from the Hammerhead docking system (Welch et al. 1996) with a search engine that relies on a surface-based molecular similarity method to rapidly generate suitable poses (Jain 2003)	Docking of sipholenone E (**14**), sipholenol L (**15**), siphonellinol D (**16**), and sipholenol J (**20**) from a marine sponge into binding sites of a P-gp receptor to investigate their potential to reverse P-gp-mediated MDR in cancer cells (Abraham et al. 2010)
	GOLD 3.1.1 with ChemScore scoring function	Employs a genetic algorithm to explore the full range of ligand conformational flexibility with partial flexibility of the protein (Jones et al. 1997)	To predict the optimal binding positions and relative binding propensities of manzamine A (**21**) from the marine sponge *Haliclona* sp. (Sakai et al. 1986) and two of its analogs to glycogen synthase kinase GSK-3β (Ibrahim et al. 2008)
	Selnergy™	An inverse docking tool that provides precious estimations on the selectivity and/or the synergy that molecules may have for a set of protein targets (Do et al. 2007)	Identified three targets, from 400 proteins that were screened, for the coumarin meranzin (**12**) from the plant *Limnocitrus littoralis*

ACE-I angiotensin-converting enzyme inhibitor, *tACE* testis angiotensin-converting enzyme, *AChE* acetylcholine esterase, *ALR2* aldose reductase, *BACE1* β-site APP cleaving enzyme 1, *FXR* farnesoid-X-receptor, *MDR* multidrug resistance, *P-gp*, P-glycoprotein that can be screened, and increase the probability of hit identification. However, even HTS has not shown the desired increase in success rate, and is very costly due to its large scale. In this approach, the principal theme of medicinal chemistry is to find compounds that are potent and selective. This often leads to high attrition rates in the later stages of development and low success rates in the drug discovery process due to toxicity and unfavorable physicochemical properties. Therefore, this approach has undergone remodeling to one in which potency is no longer the main focus. Now factors that influence later development (such as absorption, distribution, metabolism, elimination, and toxicity, ADMET for short) are considered at all stages of discovery because it has been found that potency is more readily targeted later, and does not cause a bottleneck (Keseru and Makara 2006). The introduction of various computational methods that aim to predict ADMET properties, potency, and molecular mechanism of action has also brought a new dimension into drug discovery. Such theoretical methods have the ability to reduce time and cost by screening and filtering compounds at all or various stages of the discovery process. In this way, only compounds with the highest potential for success need to be synthesized and tested, reducing costs dramatically.

7.2 Overview of the Drug Discovery Process: From Hits to Leads to Drug Candidates

7.2.1 Screening Approaches

In pharmaceutical companies HTS will typically commence once the target and a suitable biological assay for activity have been validated. HTS centers around small reaction volumes carried out in an automated fashion in multiple-well microplates so that thousands of assays can be carried out in a day. The compounds that are found to show good activity in HTS are called "hits" and are defined as nonreactive compounds with known structure, purity, and good minimal potency (<20 μM) from the HTS in vivo assay (Wunberg et al. 2006). The hits are then validated to confirm that the observed signal was due to a desired mechanism and not due to nonspecific binding, aggregate formation, precipitation, or other physical processes. A variety of techniques has been employed in this verification process, as summarized by Keseru and Makara (2006) and it is imperative to eliminate compounds that are false-positives as early as possible before continuing to spend more resources on optimizing them. Hits are prioritized based on their biological profile (potency, selectivity, and specificity) as well as ADMET and physicochemical properties in order to identify compounds with the least number of liabilities for further optimization (Wunberg et al. 2006). Other factors that may influence the prioritization include chemical tractability, the binding mechanism, and the patentability (Keseru and Makara 2006).

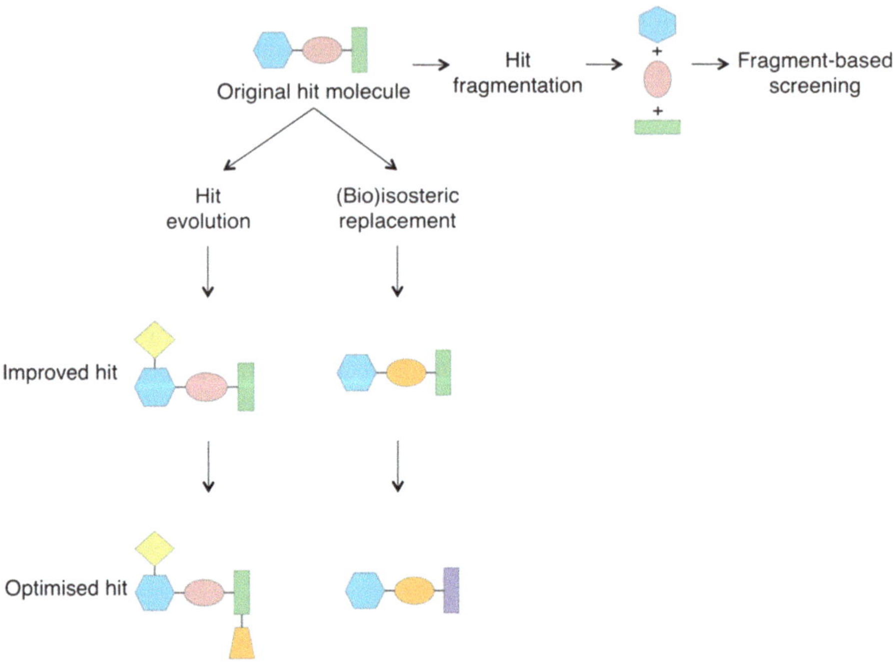

Fig. 7.1 A schematic representation of the modifications performed on hits to establish initial SAR in the hit-to-lead process (Keseru and Makara 2006)

Generally, clusters of hits are favored over single compounds (Keseru and Makara 2006) Next, the hits are filtered and assessed in the hit-to-lead process (Wunberg et al. 2006). Limited chemical modifications are also performed on these preliminary hits in order to establish initial structure–activity relationships (SAR; Wunberg et al. 2006). The chemical modifications performed in the hit-to-lead process may include (a) hit evolution (analogs of the original hit are synthesized), (b) (bio)isosteric replacement (the same as hit evolution, but has greater dependence on the structure of the target), (c) hit fragmentation (fragmentation of large molecules to identify fragments or minimum pharmacophores), or (d) a combination of these (Keseru and Makara 2006). These modifications are shown schematically in Fig. 7.1. Fragments from HTS hits or other fragment libraries can in turn be used in fragment-based drug discovery. This approach has developed tremendously in the past few years and can now be used as a concrete alternative to search for hits. Fragment hits are smaller than conventional hits, and therefore show lower affinity of binding to their targets (in the micromolar to millimolar range) and sensitive assays such as NMR, protein X-ray crystallography, and surface plasmon resonance are required for their detection (Congreve et al. 2008). For fragment hits it is key to know the binding mode and there should also be scope to produce structural analogs readily (Keseru and Makara 2006). Fragments, in turn, can be chemically modified by (a) fragment expansion/evolution (similar to the usual optimization in medicinal chemistry, functionality is added to include additional protein–ligand interactions),

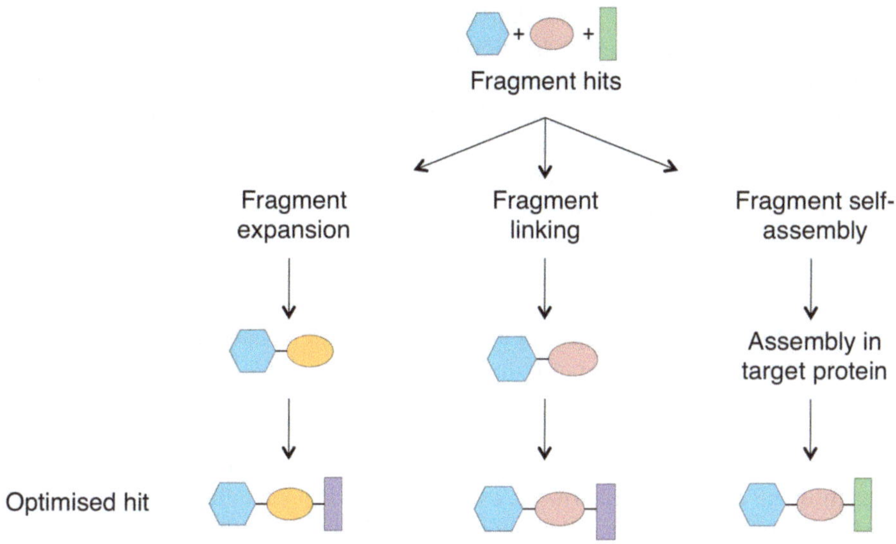

Fig. 7.2 A schematic representation of the ways in which fragments can be modified to form druglike molecules (Keseru and Makara 2006)

(b) fragment linking/merging (the linking of two fragments, or often linking a fragment to a part of a known active molecule in order to improve its properties overall), (c) fragment self-assembly (the protein effectively assembles its own inhibitor by bringing together fragments that can cross-link), or (d) a combination of these, as shown in Fig. 7.2 (Erlanson et al. 2004; Blundell et al. 2002; Murray and Blundell 2010).

The hit-to-lead process could take up to a year and is an important point for deciding which compounds to continue with in the lead optimization stage. The leads that enter this stage are compounds that show potency, selectivity for the target, and confirmed specific binding. They also show promising SAR and favorable biophysical and ADMET properties for their optimization into drugs (Wunberg et al. 2006). In vivo experiments are also important, and at this stage the in vivo and in vitro data are brought together to assess undesirable pharmacodynamic effects in humans prior to clinical trials. Large-scale ADMET and safety assays are also performed at this point. The lead optimization process can take up to 3 years. The leads are ranked according to their pharmacological profile, affinity for the target, and their ADMET properties in order to make the very important decision of which compounds to put forward as drug candidates.

7.3 Marine Natural Products as Leads

The chemical space covered by natural products (NPs) is more varied and more druglike than standard combinatorial collections (Feher and Schmidt 2003; Clardy and Walsh 2004) despite the large percentages of drugs that are either naturally occurring compounds or were derived from a NP or a NP pharmacophore. Traditionally, bioassay-guided fractionation – a process consisting of recurring cycles of separation and assessment of biological activity until a pure compound is obtained – would be performed in order to extract the active NPs. NP purification and identification is typically too slow to be part of mainstream HTS. However, there are ways in which NPs such as marine compounds can be made available for screening (Harvey 2007). Great improvements have been made in the purification and identification of NPs with enhanced HPLC, NMR, and MS techniques, and their automation (Harvey 2007; Bugni et al. 2008). For example, a marine NP (MNP) library has been constructed using HPLC-MS purification techniques to prepare the MNPs for HTS (Bugni et al. 2008). Most wells had three or fewer compounds, providing sufficient purity for the identification of lead compounds and for identifying their structures. This is an excellent example of how MNP libraries can accelerate drug discovery.

Another inroad for NPs into screening is to prepare collections of purified samples from NP extracts. Furthermore, extracts of NPs can be included in screening as early as possible in order to allow more time to follow up on hits. An alternative is to use NP extracts in niches outside of HTS, where HTS could not provide hits with conventional libraries. NPs have also been used as scaffolds for combinatorial approaches (Boldi 2004; Haustedt et al. 2006) and for starting points for creating libraries of diverse compounds or analogs (Pelish et al. 2001; Mang et al. 2006) with the rationale that there will be improved chances for creating biologically active compounds. These approaches produce compounds that are suitable for use in HTS. Bryostatin, which was extracted from the marine bryozoan *Bugula nerita*, showed potent activation of protein kinase C. The structure has been used to create a pharmacophore model from which simpler analogs, which also show activity, have been generated (Wender et al. 1988) and research in this area is still ongoing (Trindade-Silva et al. 2010).

Another innovative approach has been the use of similarity searching to identify target protein structures related to that of a protein with a known natural ligand, coupled with the synthesis of variations of the natural ligand (Koch et al. 2004). This has resulted in both identified targets to screen, and a screening library that is enriched for those particular targets. Alternatively, *combinatorial biosynthesis*, the nonnatural combination of biosynthetic enzymes to transform polyketides and peptides, has been utilized to obtain analogs of original NP with improved properties (Harvey 2007). With the development of a natural product-likeness score (Ertl et al. 2008), which measures the overall similarity of a compound with the currently known NP structural space, the NP-likeness score can be used to select structures that support combinatorial synthesis. The score can also be used in virtual screening to determine the NP-likeness of compounds, in addition to drug-likeness, novelty, and absence of undesirable structures in order to prioritize compounds for purchase for HTS. Pharmaceutical companies rely more and more on HTS to find leads but, as has been discussed, NPs can be good lead compounds for further investigation and optimization, and therefore including NPs in virtual screening is yet another route for NPs to play their role in the discovery process (Harvey 2007).

Pharmaceutical companies have large collections of compounds for screening that have been developed over many years and can be expanded to hundreds of thousands of compounds. However, the expansion of the libraries through the development of lead compounds is thought to bias the chemical diversity in these libraries. Computational tools provide ways of measuring the diversity in such collections (Mishra et al. 2008). In silico screening for drug-likeness or lead-likeness can also be applied from the initial stages to filter and prioritize compounds for further development. Such screens are based on scoring functions. Drug-likeness was defined by Lipinski's rule of five (Lipinski et al. 2001) and states that a compound is more likely to show poor absorption or permeation when:

- There are more than five hydrogen bond donors.
- The molecular weight is over 500 Da.
- The logP is over five.
- There are more than ten hydrogen bond acceptors.

Compounds that are substrates for biological transporters are exceptions to this rule. Lead-likeness, also referred to as the rule of three (Congreve et al. 2003), implies that compounds are high-quality leads if they:

- Have a molecular weight less than 300 or 350 Da.
- Have logP lower than three.
- Number of hydrogen bond donors is three or less.
- Number of hydrogen bond acceptors is three or less.

Computational methods are valuable to screen for favorable ADMET properties because this filtering plays such an important role in preventing attrition in later stages of drug development. Normally hits won't be discarded based on in silico screening only, but these results provide quick ways of scoring and prioritizing compounds in easily visualized ways (Wunberg et al. 2006). In silico methods are also an attractive option for prioritizing structures for focused screening, for example, to screen analogs of natural compounds with known activity or to use the natural compound as a scaffold for combinatorial synthesis, as previously highlighted. Virtual screening technologies are also one of the main influences of in silico techniques in contributing to the discovery of novel drug compounds.

7.4 Virtual Screening in Drug Discovery

Virtual screening, together with traditional HTS, has proved to be an essential in silico tool when screening for new bioactive compounds in the pharmaceutical industry (Stahl et al. 2006; Oprea and Matter 2004). The initial concept of VS was that a set of compounds could be initially screened computationally, to reduce and enrich the set with the most bioactive ligands to be screened experimentally. In silico VS can be employed following in vivo/in vitro experiments to gain information on binding interactions responsible for bioactivities observed (Sepe et al. 2011) or in vitro/in vivo experiments follow in silico investigations to confirm or falsify predictions (Boldi 2004). Ligand profiling using VS is generally carried out by ligand-based virtual screening (LBVS) and/or structure-based virtual screening (SBVS). LBVS refers to methods based on evaluating resemblance among ligands, such as the use of pharmacophore models or molecular-similarity analysis for carrying out screening. SBVS primarily entails the use of molecular docking methods.

Of late, VS tools have become more extensively available through grid computing (von Korff et al. 2011), which enables "sharing resources on a worldwide scale, across different institutions to run computationally intensive, scientific applications without the need for a centralized supercomputer" (Levesque et al. 2009), and more recently also through cloud computing (Foster et al. 2009). These distribution systems are comparable in their architecture and technology, but differ in areas such as security, programming model, computational model, and applications (Foster et al. 2009). Distribution systems, such as grid and cloud computing, offer researchers the opportunity to carry out computationally exhaustive tasks, such as molecular docking and molecular dynamics that may otherwise not be feasible. Another concept, open notebook science, which aims to make the full record of scientific research available also has the potential to positively influence VS studies by introducing more transparency in relation to results, by encouraging researcher collaborations, and improving the availability of data (Wald 2010).

7.5 Molecular Docking and Reverse-Docking

Structure-based VS involves the docking of molecules into X-ray or NMR structures, into homology models, and the use of target structures based on three-dimensional information for database searching (Ripphausen et al. 2010). Molecular docking is the computational prediction of noncovalent binding between macromolecules, and a macromolecule with either a small molecule or ligand (Trott and Olson 2010). Docking methods consist of a sampling algorithm, commonly based on Monte Carlo simulation, distance geometry, simulated annealing or genetic algorithm, which seeks to find the best ligand–target conformational fit. When choosing a docking method for a particular screening study there is always an innate dilemma as to whether to sacrifice accuracy for computational power or vice versa. In an attempt to overcome this problem, many docking algorithms are becoming more specialized to focus on specific areas of application (Kirchmair et al. 2008).

Regardless of the docking algorithm preferred, it must have the capacity to accurately predict the protein-bound ligand conformation. Binding enthalpies of the docked ligands are estimated by assessing the degree of compatibility to the macromolecular target through "scoring" methods. Therefore, the scoring function associated with the docking algorithm must be considered as a ligand "score" used as a means to rank the complementarity of related compounds to the biological target of interest (Stahl and Rarey 2001). Scoring functions are fast approximate mathematical approaches that measure the strength of the noncovalent interactions (Samantray and Sahu 2010). It is generally assumed that lower energy scores represent better protein–ligand bindings compared to higher energy values and as a result, molecular docking may be viewed as an optimization problem, where the task is to find the ligand-binding mode, also known as pose prediction, with the lowest energy (Thomsen and Christensen 2006).

Scoring functions have been defined, based on how they predict binding affinity, into three main classes: force-field-based, knowledge-based, and empirical scoring functions. Knowledge-based scoring functions use structural information stored in databases of protein–ligand complexes to derive atom pair interaction potentials (Muegge 2000); force-field-based scoring functions employ classical molecular mechanics energy functions, and empirical scoring functions sum interaction terms derived from weighted structural parameters obtained by fitting the scoring function to experimental binding constants of a training set of protein–ligand complexes (Ferrara et al. 2004). The significance of scoring accuracy on the quality of docking results is reflected by the number of studies that compare the various scoring functions (Wang et al. 2003; Bissantz et al. 2000; Warren et al. 2006; Zhou et al. 2007). A commonly used scoring function is based on the root-mean-square deviation (RMSD) between a generated docking pose and the experimental ligand conformation (Kirchmair et al. 2008). Despite the widespread use of the RMSD, which generates a measure of structural similarity, inconsistencies cause misclassification of both correct and incorrect poses, and it has been advised that this measure be combined with others for assessing pose prediction accuracy (Kroemer et al. 2004). Consensus scoring attempts to combine multiple scoring functions to reassess the docked conformer of every compound and only compounds ranked highly in all scoring functions applied are considered for further validation (Wang and Wang 2001).

The capacity of docking algorithms and their associated scoring functions to correctly rank the most active ligands in a test set of compounds can be assessed using the Directory of Useful Decoys (DUDS; Huang et al. 2006). DUDS has been devised to minimize bias in docking algorithms by providing challenging decoys and to benchmark virtual screening methods. DUDS contain 2,950 active compounds, active against a total of 40 target proteins. There are 36 decoys for every active agent present within the directory that has similar physical properties but dissimilar topology to the active agent (Huang et al. 2006). Other decoy sets have been established prior to DUDS, but many have inherent bias, as the sets often consist of decoy compounds that are dissimilar structurally from test set compounds, making them more likely to be identified as decoys on this basis rather than on predicted chemical interaction with the receptor (Wallach and Lilien 2011). The use of DUDS is also restrictive, as it represents only a single decoy set and therefore there is a risk that all algorithms may be optimized to this single benchmark. It would be more favorable if multiple decoy sets of similar quality were available. In addition, DUDS only spans a small synthetically feasible subset of small-molecule space (Wallach and Lilien 2011). Wallach and colleagues have shown that the removal of the criteria for synthetic feasibility of the molecules when building a decoy set results in unbiased virtual decoy sets (VDS) that reduce the risk of overfitting of docking algorithms and yield in silico-produced sets that more closely match the physical properties of the active ligands (Wallach and Lilien 2011).

Proteins of plant (Segura Campos et al. 2010; Jang et al. 2011; Jimsheena and Gowda 2011) and marine origin (Wijesekara and Kim 2010; Sato et al. 2002) are rich natural sources of angiotensin-converting enzyme inhibitor (ACE-I) peptides, which are currently of great interest to the pharmaceutical and functional food industries (Wijesekara and Kim 2010; Li et al. 2004; Tierney et al. 2010). The use of molecular docking techniques allows information to be gained on peptide/ACE receptor interactions and to identify critical characteristics of natural, potent ACE-I peptides. Jimsheena and Gowda (2011) investigated the ACE-I inhibitory action of the pentapeptide NAQRP (IC_{50} of 32 ± 2 μM) isolated from arachin, a storage protein from peanut, through molecular docking simulations. A model of human testis ACE (tACE) bound to lisinopril, a marketed antihypertensive drug, was derived from the Brookhaven Protein Data Bank (PDB: 1086). PepBuild, a public domain Web server, was employed for the construction and optimization of the ligand structure. As in a previous study by this group (Jimsheena and Gowda 2010), the binding potential, location of binding sites, and mode of binding (pose) for the ligand were carried out using the docking software Molegro Virtual Docker that employs a force-field based screening function (Thomsen and Christensen 2006).

Following the observed binding of NAQRP to tACE and the interactions as given by the best pose (Jimsheena and Gowda 2011) it was suggested that the binding mode of the pentapeptide differed from the mode observed previously for arachin-derived tripeptide ACE inhibitors (Jimsheena and Gowda 2010). It was suggested that the lower ACE inhibition of NAQRP, compared to those observed previously in the tripeptides, may be due to its larger size. Therefore, the tripeptide QRP was modeled and docked at the ACE active site and appeared to be an ideal fit and was also twofold (IC_{50} of 15 ± 2 μM) more potent than the original pentapeptide (Jimsheena and Gowda 2011). Similar docking approaches have the potential to be applied to various marine peptides that possess in vitro ACE-I inhibitory activity to confirm their bioactivities, such as those isolated from shrimp (Hai-Lun et al. 2006), seaweed (Sato et al. 2002), and salmon (Ono et al. 2006).

Nuclear receptors (NRs) are one of the largest groups of transcription factors and are viewed as master controllers of genes involved in metabolic control (Francis et al. 2003). Therefore identification of novel ligands for these receptors may be used to combat diseases such as atherosclerosis, diabetes, and obesity (Lizcano et al. 2011; Viennois et al. 2011). An adopted member of the metabolic NR family is the farnesoid-X-receptor (FXR), also known as the bile acid receptor (BAR). Bile acids (BAs) are endogenous ligands of the FXR receptor (Wang et al. 1999), which regulates bile acid synthesis, conjugation, and transport, as well as various aspects of lipid and glucose metabolism (Claudel et al. 2005). Marine-sulfated polyhydroxysterols from ophiuroids (brittle stars), illustrated in Fig. 7.3, share some common structural characteristics with BAs (Sepe et al. 2011), thereby suggesting that they may have the ability to act as FXR modulators. Molecular docking studies were carried out using AutoDock Vina software on eight selected sterols (**1–8**) isolated from four different *Ophiocoma* species to investigate the potential effects they have on the FXR at the atomic level (Sepe et al. 2011). AutoDock Vina is an improved version of AutoDock, showing two orders of magnitude improvement in speed and

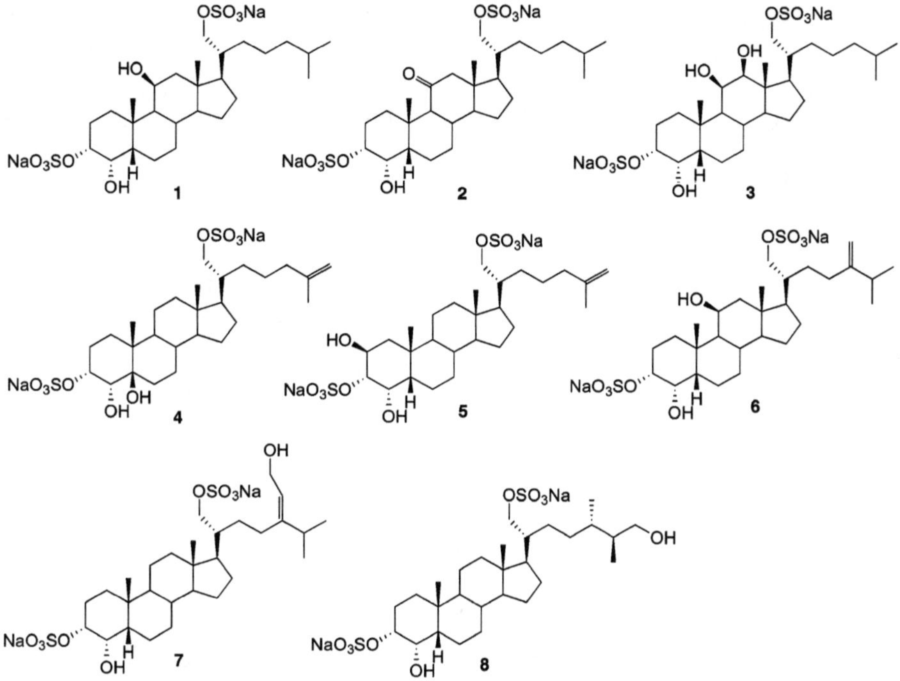

Fig. 7.3 Selected natural and marine products involved in VS studies, including the ophiuoid sterols **1–8**, phlorotannins **9–11**, merazin **12**, the sipolanes **13–16**, the lamellarins **17a** and **17b**, the scopoletins **18** and **19**, sipholenol J **20,** and manzamine A **21**

significantly better accuracy of the binding mode predictions compared to AutoDock 4.0. This improved performance is achieved by employing a gradient optimization technique and a multithreading process that takes advantage of multiple CPUs or CPU cores (Trott and Olson 2010). For all docking complexes relative to the eight sterols, a common sulfate group was seen to establish H-bonds with three amino acids essential for receptor activation and the steroid nucleus formed H-bonds with the cavity pocket to accommodate the binding site. However, one of the sterols (**8**), isolated from *Ophiocoma superba*, that possessed the most potent antagonistic activity on the expression of FXR-regulated genes in liver cells also exhibited additional interactions with the FXR receptor. The bent shape of this molecule and the presence of an hydroxyl group on the molecule's side-chain are thought to be the crucial elements that increase this molecule's potency relative to the other seven ophiuroid sterols investigated (Sepe et al. 2011).

Phlorotannins, polyphenolic compounds abundant in brown macroalgae, have been the center of much attention due to their promising diverse bioactivities, which, in addition to their well-documented antioxidant capabilities (Kim et al. 2009; Shibata et al. 2008) include antidiabetic, antiproliferative (Nwosu et al. 2011), prevention of carcinogenesis (Parys et al. 2010), and ACE-I-inhibitory activity

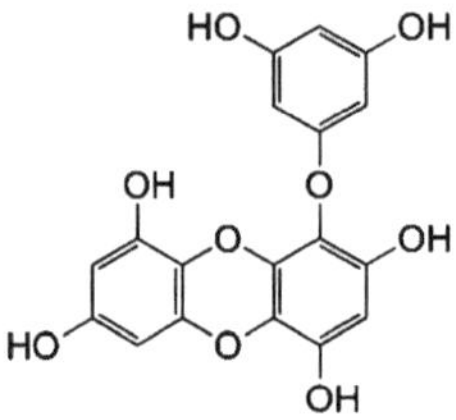

9

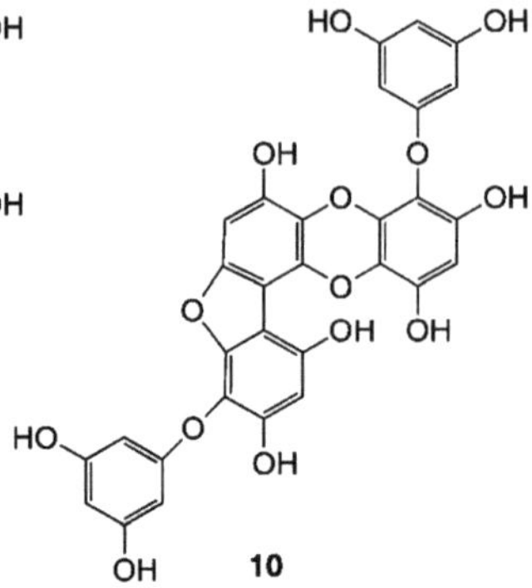

10

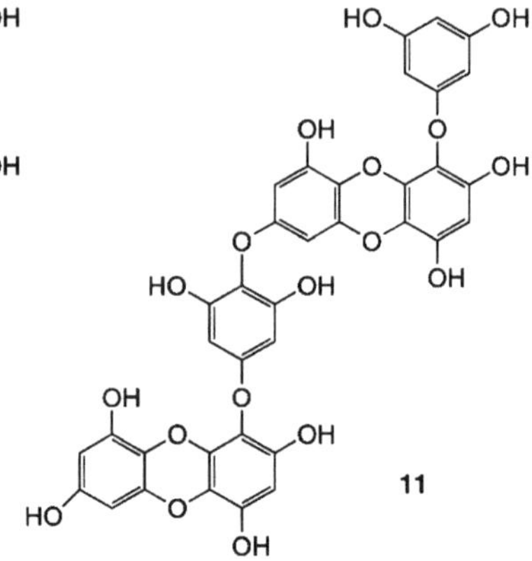

11

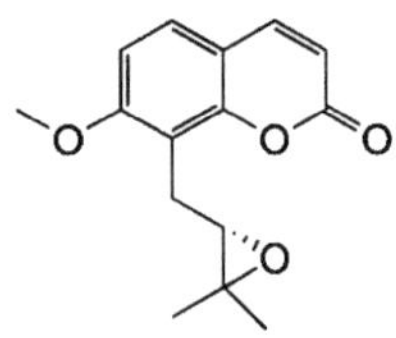

12

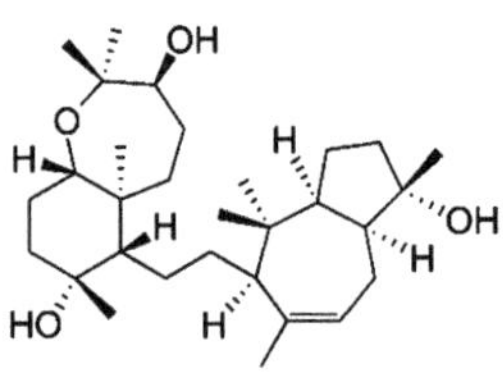

13

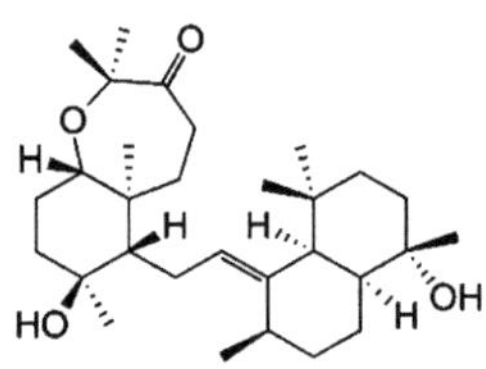

14

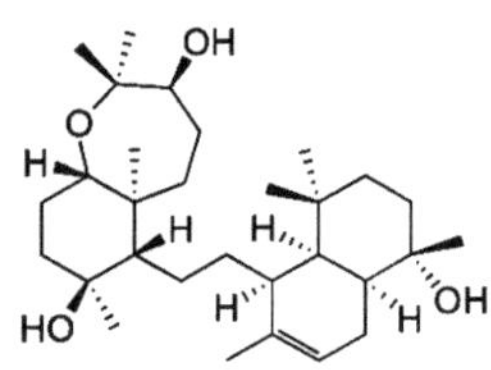

15

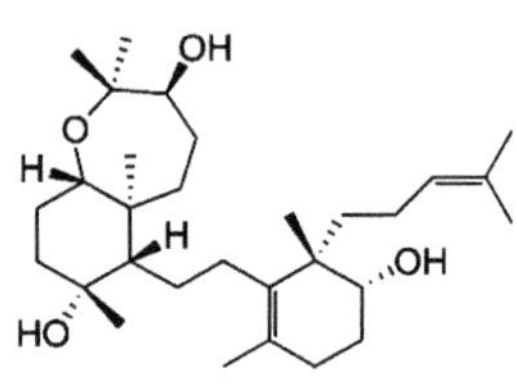

16

Fig. 7.3 (continued)

(Jung et al. 2006). Phlorotannins have also been investigated as potential preventive agents of the neurodegenerative Alzheimer's disease (AD; Jung et al. 2010). The development of AD is caused partially by the formation of amyloid plaques, which are composed of a β-amyloid peptide (Cole and Vassar 2007). The β-site APP cleaving enzyme 1 (BACE1) is essential for the generation of β-amyloid plaques and agents that have the ability to block the action of this enzyme may have potential roles in future AD treatment. Several phlorotannins isolated from *Eisenia bicyclis* have displayed significant in vitro BACE1 inhibition and to confirm the inhibition mode of

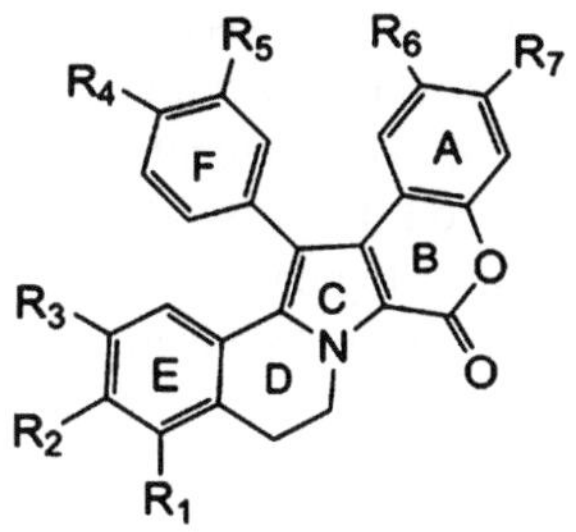

17a

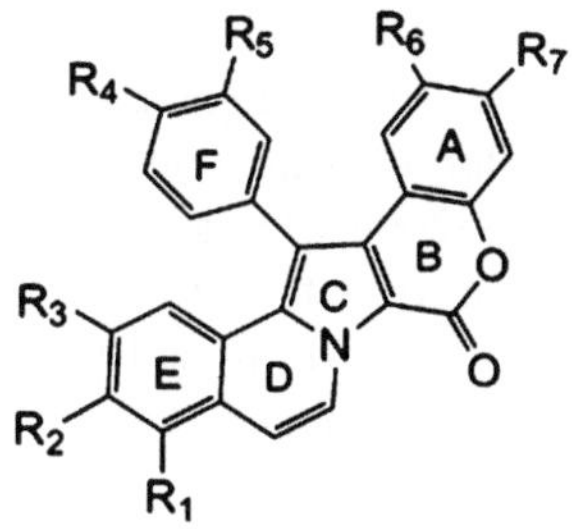

17b

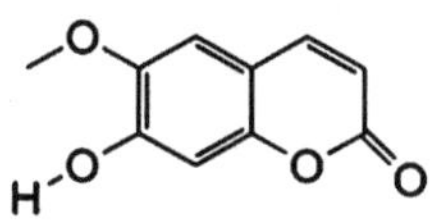

18

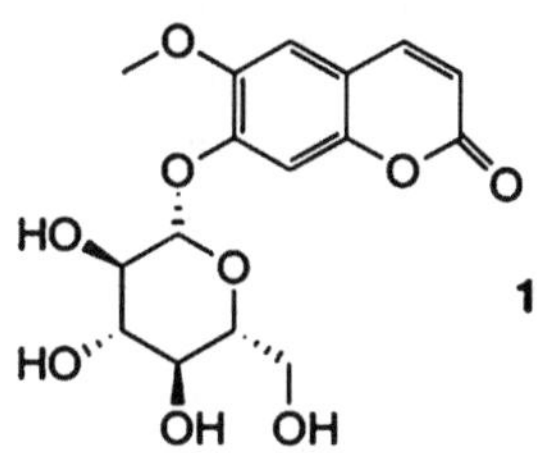

19

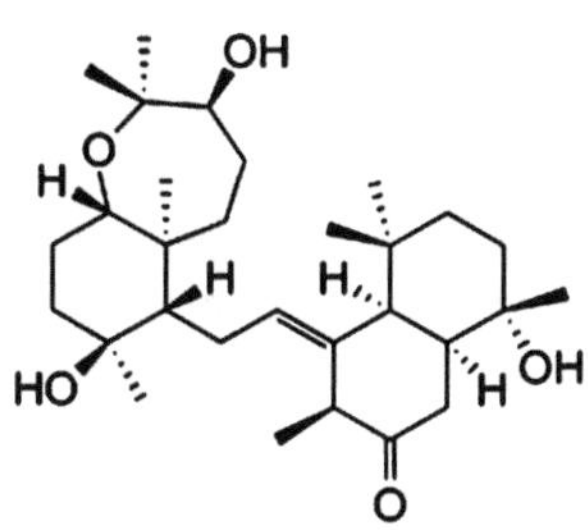

20

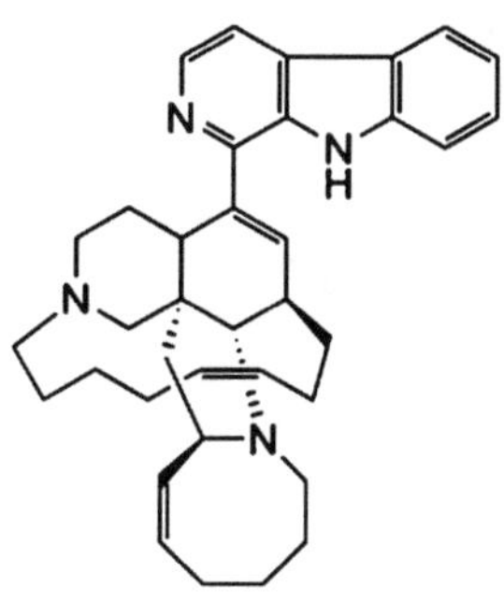

21

Fig. 7.3 (continued)

the selective BACE1 inhibitors a molecular docking study of three of the phlorotannins, namely eckol (**9**), phlorofurofucoeckol (**10**), and dieckol (**11**; Fig. 7.3), was carried out. In order to gain increased accuracy, repeatability, and reliability, Jung et al. (2006) used two docking programs, AutoDock 4.0 and Fast Rigid Exhaustive Docking (FRED) 2.0 (FRED), to predict the protein–ligand conformation. The success of the docking analysis was reflected by statistically significant scores and it highlighted a common binding mode of the phlorotannins to the Tyr132 and Thr133 residues of the site responsible for BACE1 inhibition (Jung et al. 2010).

Reverse, or inverse, VS is a valuable tool that can be employed to reposition an existing pharmacological drug in order to cut the costs associated with new drug development as repeat toxicological and pharmacokinetic assessments would not be necessary (Chong and Sullivan 2007). Reverse docking attempts to position a flexible receptor around a rigid ligand, as opposed to regular molecular docking which involves insertion of a flexible ligand into a mainly rigid receptor (Harriman and Deslongehamps 2004). Do and colleagues employed the software tool Selnergy™ to gain more information regarding the pharmacological activity of the plant *Limnocitrus littoralis,* from the Rutaceae family, from which the coumarin meranzin (**12**; Fig. 7.3) was isolated (Do et al. 2007). Selnergy™ is a recently developed inverse docking tool that consists of a database of 7,000 protein structures with annotated biological properties, such as antibacterial, anti-inflammatory, and anti-aging (Selnergy). Of the 400 proteins that were screened, three targets, namely COX1, COX2, and PPARγ were selected. Selnergy predictions were validated by comparison of the docking results with available experimental results (Do et al. 2007). This study highlighted the importance of inverse docking tools as a component of reverse pharmacognosy for finding multiple targets of natural products. Inverse docking also paves the way for the development of natural multitarget-directed ligands (MTDLS) that have the potential to reveal new approaches for existing drugs to tackle various serious diseases such as neurodegenerative syndromes, (Cavalli et al. 2008) cardiovascular disease, inflammatory conditions, and metabolic disease, which tend to have multiple pathogenic manifestations (Morphy and Rankovic 2005).

7.6 Ligand-Based VS

Ligand-based VS can be subdivided into two- and three-dimensional methods, and combinations of both. However, three-dimensional techniques, such as pharmacophore-based VS, are more popular than two-dimensional LDVS methods, for example, fingerprint similarity searching (Ripphausen et al. 2010). The advantage of pharmacophore-based screening, when compared to other ligand similarity screening approaches, lies in the capacity of this screening method to detect a diverse set of supposed active compounds with totally different chemical scaffolds (Dror et al. 2009). Pharmacophore modeling involves the identification of a common functional group(s) responsible for a ligand–receptor interaction associated with a group of bioactive compounds, and mapping of these interactions in three-dimensional space (Guner 2002). Pharmacophoric features, such as hydrogen bonding and electrostatic and hydrophobic interactions form the basis of database searches during virtual screens (Rella et al. 2006).

There are two general types of pharmacophore models: target-based models, which use structural information from either experimental data, for example, NMR, or from theoretical studies, for example, homology modeling; and ligand-based models, where the common chemical elements of a known ligand cluster with a similar binding mechanism are aligned in three-dimensional space to create a model (Rollinger 2009). Similarly to reverse docking, reverse pharmacophore mapping provides a means of finding new mechanisms of action activities of pharmacologically active ligands. This approach can be employed when the molecular targets of a biologically active small molecule are detected in a cell- or animal-based bioassay screen, and a corresponding natural product or an existing drug has (have) not been identified. Liu et al. (2010) recently developed the first Web-based tool, PharmMapper, to carry out drug target identification.

Research carried out on the Red Sea sponge *Callyspongia siphonella* has identified a number of sipholane triterpenoid compounds with the potential to reverse P-glycoprotein (P-gp)-mediated multidrug resistance (MDR) in human epidermoid cancer cells (Jain et al. 2009). A ligand-based pharmacophore model was generated of the most potent compounds in the sipholane group, namely sipholenol A (**13**), sipholenone E (**14**), sipholenol L (**15**), and siphonellinol D (**16**; Fig. 7.3). The program SYBYL 8.0 was utilized for the molecular modeling and a module, DISCOtech, within this program was used to build the pharmacophore models based on common chemical features. DISCOtech generated ten models with superimposed conformations of the active sipholanes and the best model, selected based on maximum structural overlap, comprised of three hydrophobic points and two hydrogen bond acceptors. Therefore, it was concluded that these four sipholane compounds possess a common pharmacophore that enables them to reverse P-gp-mediated MDR (Jain et al. 2009). The introduction of pharmacophore modeling is a positive step forward in the development of P-gp modulators which, until recently, had been hampered due to the lack of a high-resolution 3-D structure of P-gp (Srinivas et al. 2006).

These pharmacophore models, along with numerous others representing a host of pharmacological targets, are employed in pharmacophore-based activity profiling. The predictive power of each individual model must be taken into account, as this will ultimately affect the integrity of the final ligand profiling method. Enrichment factor, goodness of hit (GH) score, and the receiver operating characteristic curve-area under the curve (ROC-AUC) are important criteria that can be assessed in order to sufficiently evaluate a model (John et al. 2011; Klabunde and Evers 2005). The enrichment factor is the most widely applied validation method and it quantifies the number of active compounds found in the hit list, in relation to the fraction of inactives (Dror et al. 2009).

The marine lamellarins have been the focus of much attention pertaining to their reported bioactivities, including HIV-I integrase inhibition (Taha et al. 2010; Reddy et al. 1999), antitumor properties (Baunbaek et al. 2008), antioxidant activity (Krishnaiah et al. 2004), and ability to reverse multidrug resistance (Vanhuyse et al. 2005). They are pyrrole hexacyclic alkaloids that were initially isolated from the marine mollusc *Lamellaria* species (Andersen et al. 1985). Thipnate et al. (2009) used a training set of 26 lamellarins, where compounds were based on either scaffold **17a** or **17b**, to build a 4-D quantitative structure–activity relationship (QSAR) and pharmacophore models to investigate for cytotoxicity against human hormone-dependent breast cancer cells. QSAR, as mentioned earlier, involves the prediction of the biological activities of new compounds using mathematical relationships based on various properties of previously tested potential agents. Relative to the commonly used 3-D QSAR, 4-D QSAR has been improved by the inclusion of molecular state ensemble averaging, which adds conformational and alignment freedom (Martins et al. 2009) allowing the assembly of optimized dynamic spatial

QSAR models in the form of 3-D pharmacophores (Thipnate et al. 2009). In this study, the receptor geometry was not known and therefore receptor-independent (RI-) 4-D QSAR models were built from eight possible receptor-binding alignments for the entire training set.

The final step of RI-4-D QSAR modeling involves identifying the potentially most active conformers of each compound. This modeling approach ensures that the quantity of 3-D pharmacophore information is maximized from such a small set of structurally complex lamellarins. Subsequently, 4-D fingerprint virtual screening analysis was applied to the lamellarin dataset. The derivation and validation of a potential set of universal descriptors produce the 4-D fingerprints and are derived from 4-D-molecular similarity analysis (4-D-MS; Senese et al. 2004). The 4-D fingerprint virtual screens were constructed using a trial descriptor pool, which was compiled from the sets of 4-D fingerprints across each of the molecules in the training set. A consensus set of 4-D QSAR models proposed that the degree of cytotoxicity against the breast cancer cells is controlled by the ability of substituents on the lamellarin E-ring to form a ligand–receptor intermolecular hydrogen bond and a hydrophobic interaction. The high degree of consistency observed between the 3-D pharmacophore from the RI-4-D QSAR model and the 4-D fingerprint VS model (Thipnate et al. 2009) shows that, even with a small training set of compounds, pharmacophores responsible for specific bioactivities can be identified and advance in the drug development process.

The recent advancements in free Web-based VS tools, in addition to the establishing of online marine databases and distribution systems, means that drug discovery will no longer be an activity restricted to those with large amounts of research funds and CPU time, allowing the wider research community to initiate drug discovery projects. Schneidman-Duhovny et al. (2008) have developed the first Web server, called PharmaGist, for elucidating 3-D pharmacophores from a set of druglike molecules. In PharmaGist pharmacophores are detected by multiple flexible alignment of the input ligands and this program has the ability to detect pharmacophores common to subsets of input ligands, which makes PharmaGist tolerant to outliers and to several binding modes (Schneidman-Duhovny et al. 2008). PharmaGist is a dual-module screening tool; it detects pharmacophore candidates from a set of input ligands and can carry out pharmacophore-based screening (Dror et al. 2009). Although the DUDS was primarily developed for assessing docking algorithms it can also be employed for gauging the performance of pharmacophore-based screening. By employing the DUDS, a large-scale evaluation of PharmaGist was carried out and was found to perform comparably to other frequently used screening tools such as Catalyst, MTree, and DOCK (Ewing et al. 2001).

7.7 Quantum Mechanics Within VS

The use of quantum mechanics (QM) to improve VS methods has slowly increased over the past number of years (Queiroz et al. 2009; Tawari and Degani 2010). Until recently, QM calculations were considered quite arduous due to the lack of access of many researchers to the large amounts of computational power and storage required to carry out such calculations. This hindrance has been removed to a certain extent due to a fall in the cost of CPU time and with the introduction of grid and cloud-based computing.

Structural knowledge for the investigations of receptor–ligand electrostatic interactions is often calculated using molecular mechanical (MM) forcefields. However, one of the drawbacks of MM for such calculations is that the Coulomb interactions between fixed charges, used to define the electrostatic effects, do not allow for fluctuations in charges resulting from changes in the environment and therefore, reliable quantitative comparison with experimental data is hampered (Halgren and Damm 2001; Patel and Brooks 2006). QM provides the drug discovery process with an array of new descriptors, that is, molecular representations, due to its ability to characterize molecules of all kinds. The availability of these descriptors, such as molecular electrostatic potential (ESP) maps, local hardness and softness, frontier orbital analysis, and density of states, allows interactions between targets and ligands to be viewed from a standpoint not possible with MM force-fields. Also, QM may be employed as an alternative method for the calculation of binding free energies, considering the deficiencies associated with scoring functions used in docking methods. Hybrid quantum mechanical/molecular mechanical (QM/MM) methods have the capability to calculate more accurate binding free energies when investigating binding modes of protein–ligand interaction (Raha et al. 2007). In QM/MM calculations, the residues involved in ligand binding are treated using QM whereas the remainder of the protein is modeled using MM calculations (Gleeson et al. 2010). QM/MM can be employed to prepare the structures of small molecules and proteins, for example, optimizing binding poses obtained from docking (Gentilucci et al. 2008), and refining the geometries of enzyme-active sites obtained with MM (Mukherjee et al. 2008) or X-ray structures (Fanfrlik et al. 2008). Within the drug discovery route, QM/MM is thought to be important for interpreting poorly resolved electron density (Fanfrlik et al. 2008), finding the details of the interaction enzymes active sites (Mladenovic et al. 2009), and analyzing to what extent substituent effects have on the binding mode (Gleeson and Gleeson 2009). Interest in the use of combined quantum mechanical/molecular mechanical (QM/MM) approaches as general physics-based scoring functions for docking has grown in recent years (Beierlein et al. 2003) A QM/MM scoring function model was found to more accurately identify the pose of six HIV-1 protease inhibitors compared with the widely used ChemScore (Eldridge et al. 1997) and GoldScore (Verdonk et al. 2003) scoring functions.

7.8 Combined Pharmacophore and Docking VS

Virtual screening approaches can produce large numbers of false positives (Jansen and Martin 2004) such that compounds that are ranked highly in silico do not exhibit the same potency when tested in vivo or in vitro, which can, in turn, result in false negatives (Peach and Nicklaus 2009). Recent comparative studies that weigh up the advantages and disadvantages of pharmacophore-based VS and docking-based VS are plentiful (Hein et al. 2010; Steindl and Langer 2005). Pharmacophore-based VS in some aspects is more efficient than using docking methods, as these screening approaches often take less computational time and the results are less problematic to interpret (Steindl and Langer 2005). Prefiltering of databases and postprocessing of primary hit lists before undertaking either VS tool has been suggested due to scoring deficiencies (Hein et al. 2010).

In an attempt to surmount the shortcomings of individual molecular docking and pharmacophore screening, a combination of structure-based and ligand-based methods has been introduced to get the best of both worlds. An example is where pharmacophore models have been used as both pre- and postscreening methods for various docking studies (Peach and Nicklaus 2009; Hein et al. 2010; Steindl and Langer 2005; Lyne et al. 2004; Deng et al. 2008; Khanfar et al. 2009). A comprehensive VS approach, carried out by Khanfar and colleagues (2010) searching for novel glycogen synthase-3β (GSK-3β) inhibitors highlights how pharmacophore modeling can be successfully employed as a prescreening tool for molecular docking and results in an efficient multistage VS protocol as outlined in Fig. 7.4. GSK-3β is considered to be a promising pharmacological target for the treatment of a range of diseases, such as Alzheimer's disease and chronic inflammatory diseases (Li et al. 2006; Martinez et al. 2002). Initially, GSK-3β inhibitor scaffolds were sourced from three databases, including an in-house database consisting of mainly natural products or their semisynthetic or microbial transformation derivatives, and filtered according to an extended Lipinski's rule of five.

A validated pharmacophore model, previously developed by the same research group (Khanfar et al. 2009), based on a marine-derived phenylmethylene hydantoin GSK-3β inhibitors was employed for the LBVS. For the docking studies a stringent criterion, relating to ligand–target hydrogen-binding interactions, ligand binding orientation, and ligand binding score, was implemented in an attempt to ensure the enrichment of the bioactive compounds and to reduce the number of false positives. An initial 14 validated hits exhibited in vitro GSK-3β inhibition (Fig. 7.4), but was reduced to two compounds, 2-anilino-1,3,4-oxadiazole and a phenylmethylene hydantoin, following relevant in vitro toxicity tests. The important outcome from this all-encompassing approach is that the identified hits possess only modest structural similarity to any other known GSK-3β inhibitors, and therefore VS makes it possible to identify receptor inhibitors that may not have been considered on a structural basis alone.

7.9 ADME/TOX and Selectivity

Attrition is common in the later stages of drug development – not due to low potency – but rather due to other unfavorable qualities. A realization of the importance of optimizing druglike properties as well as potency and selectivity in early stages has now caused ADMET assessment of compounds to be performed at nearly every stage of the discovery process (Kassel 2004).

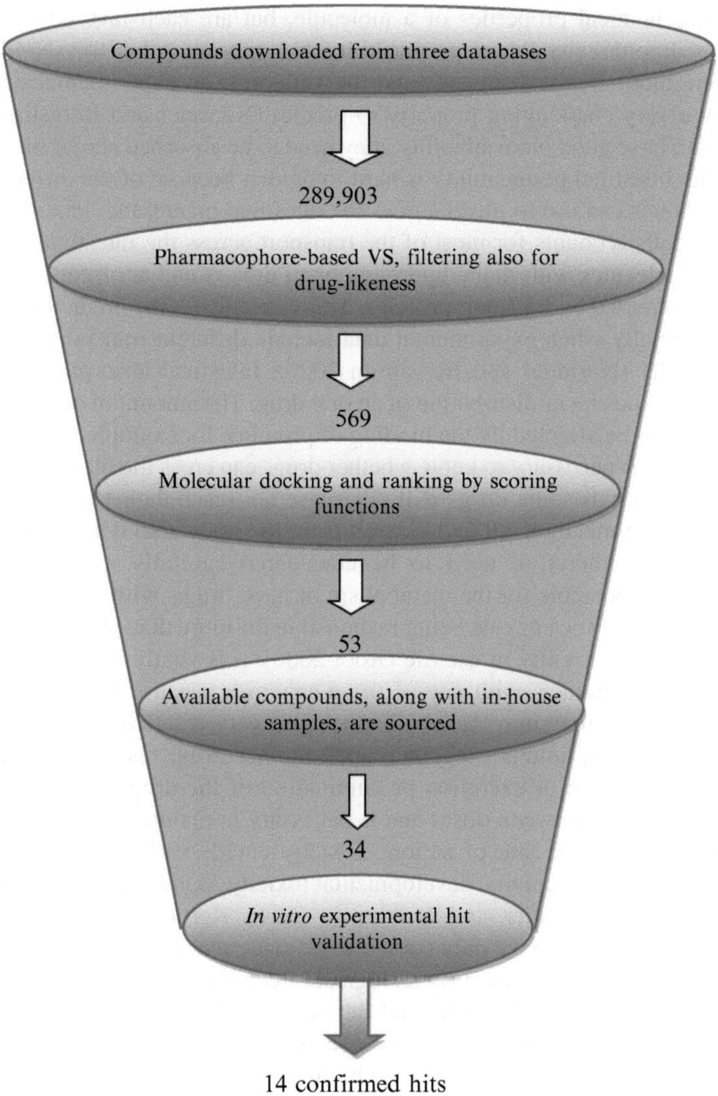

Fig. 7.4 Schematic of a multistage VS approach to experimental validation

Absorption, distribution, metabolism, elimination, and toxicity are each very important factors to consider in potential drugs. Absorption is a significant starting point, as the absorption in the intestine will determine whether a drug can be administered orally. An oral formulation is very convenient because it allows the patient to take the drug independently. Solubility in aqueous intestinal fluids and permeability across the intestinal membrane are vital for absorption in the intestine (Norinder and Bergstrom 2006). Solubility and permeability are both dependent on the physicochemical properties of a molecule, but are each favored by opposite properties; lipophilicity favors permeability, but opposes solubility (Norinder and Bergstrom 2006). These contrasts, and the variability in experimental data, make solubility a very challenging property to predict (Norinder and Bergstrom 2006). For drugs to have good bioavailability, they need to be absorbed across the intestinal membrane. Intestinal permeability is hard to predict because of the different routes that compounds can use to travel across the intestinal membrane. Passive diffusion across the cells accounts for most of the transport across the intestinal membrane, but larger molecules with many hydrogen bond donors and acceptors may also be actively transported by transport proteins. This complicates the prediction of permeability, especially when experimental data include different transport mechanisms unintentionally (Norinder and Bergstrom 2006). Intestinal absorption determines the systemic exposure or distribution of an oral drug. The amount of drug reaching its target may also be affected by the binding of proteins, for example, in the serum.

Similarly, it is necessary to know whether drugs can cross the blood–brain barrier (BBB), inasmuch as some drugs will otherwise not reach their target. The rates at which drugs are metabolized and cleared from the body also determine their bioavailability, and therefore need to be considered carefully. Cytochrome P450 enzymes are responsible for the metabolism of most drugs, with 7 of the 57 known human isoforms of the enzyme being responsible for more than 90% of the metabolism of all drugs currently in use (de Groot 2006). It is vitally important to predict whether a drug interacts with any of these enzymes because they may make the molecule inactive, they may metabolize it into a very reactive and hence harmful molecule, or it may interact together with another drug, resulting in drug–drug interactions. The rate of excretion or elimination of the drug from the body will determine intervals between doses and is necessary in maintaining acceptable concentration levels at the site of action. Toxicity entails various domains, such as mutagenicity, carcinogenicity, developmental toxicity, skin sensitization, and reproductive toxicity (Norinder and Bergstrom 2006; Kortagere and Ekins 2010). Predictions for toxicity have only started to be made recently (Kortagere and Ekins 2010), and are arduous because toxicological effects may arise by way of a number of different mechanisms (Norinder and Bergstrom 2006).

Innovations in analytical chemistry and developments in LC-MS technology have enabled ADME studies to become high-throughput (Kassel 2004). Predictive ADMET utilizes the large databases of collected ADMET data from experiments, with associated structures, and facilitates the building of computational models that link structural changes with changes in response, in order to predict which compounds will have improved properties (Davis and Riley 2004). The in silico ADME approach facilitates drug discovery by obtaining high-quality hits through simultaneous filtering for potency and ADME properties. The approach additionally allows rapid screening of molecules in libraries as viable options for continuation. The ideal situation would be to be able to predict the effect of proposed structural modifications in order to direct such continuation (Butina et al. 2002).

ADME is biologically complex and therefore hard to predict. Attempts to generate effective ADME models are currently limited by the quantity and quality of available

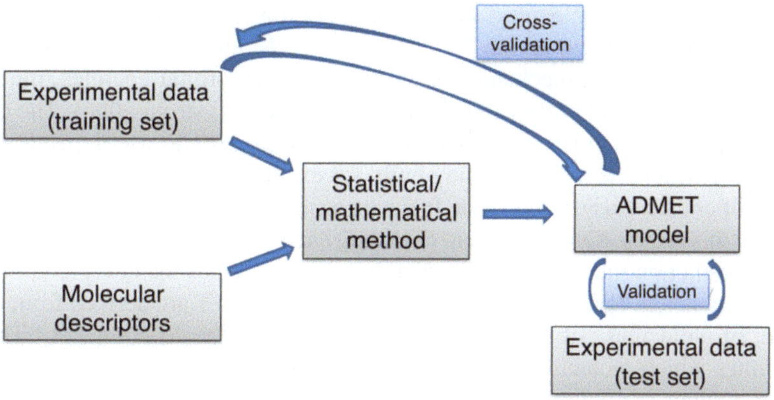

Fig. 7.5 A schematic representation of the components and processes in setting up an ADMET model

data to create these predictive models, limitations in modeling methods, and limitations in the understanding of the biological and chemical systems to be modeled (Davis and Riley 2004). The current methodological approaches include: QSAR, which is quick and is based on empirical data, and the first-principle methods of molecular dynamics (MD) and quantum mechanics (QM), which are comparably much slower methods but generally are more accurate, especially for systems for which there are few or no experimental data. Approaches that are commonly applied to the chemical data derived from these approaches include pattern recognition, statistical regression, classification methods, artificial neural networks, and genetic algorithms (Butina et al. 2002). The models selected can significantly affect the direction of future research and therefore it is imperative that they are rigorously validated (Davis and Riley 2004). However, strong links to the data in the training sets have been observed, affording biased outcomes, and the tendency of pharmaceutical companies to keep their experimental data private limits the rapid further development and improvement of these models (Davis and Riley 2004).

At the heart of the components of an ADMET model (Fig. 7.5) is the statistical or mathematical method that will be employed to discover QSAR that relates the set of descriptors (*X*) that describe the chemical structure to the observed ADMET property (*Y*; Davis and Riley 2004). There are a variety of methods that have been applied to ADMET models and they can generally be divided into linear and nonlinear multivariate models. Linear multivariate models include multiple linear regression (MLR) and partial least squares (PLS). For relationships for which linear models are not adequate, nonlinear PLS could be used, but other methods such as artificial neural networks, genetic algorithms, support vector machines, Bayesian probability, and clustering methods may be more appropriate to predict such relationships. Molecular descriptors for the molecules are also a fundamental part of the ADMET model. These can be 1-D, such as weight, molecular refractivity, number of atoms, and number of bonds; 2-D, where the properties are computed from the connectivity of the molecules such as electrotopological descriptors; and 3-D descriptors that are based on the 3-D structures of the molecules, such as surfaces, volumes, and quantum mechanics descriptors such as orbital energies, multipole moments, charges, and polarizabilities (Norinder and Bergstrom 2006). The choice of which descriptors, or which combination of descriptors to use depends on the purpose for which the model is being built, for example, in the screening of a large number of compounds or to provide focus on a smaller scale. Some factors that will influence this are the time to be spent obtaining each descriptor and the time to spend on each prediction made, and the predictive accuracy versus the interpretability.

The parameters for the ADMET model that relate *X* to *Y* are derived from fitting to a set of molecules with available experimental data, the "training set." This set determines which molecules can be accurately predicted by the model, because in general, reliable predictions can only be made for molecules that are similar to those in the training set (Butina et al. 2002). Therefore, the diversity of molecules for which experimental data are available is the cardinal limitation of the applicability of models that are based on empirical fitting methods (Butina et al. 2002) and this can represent a challenge for marine-derived materials, which can often be quite different in structure to those derived from terrestrial sources. It is ideal to have data from experiments that were performed uniformly to limit noise in the model, but this is often impossible, inasmuch as most experimental studies on large datasets are performed by pharmaceutical companies, and these data are not accessible. Another difficulty in fitting is to avoid overfitting of the model so that the model fits the training set better, but loses the capability to relate to molecules outside the training set in a general way (Butina et al. 2002), losing its predictive power. One way of preventing this is to use cross-correlation where one or more molecules are left out of the training set and the model is refitted with the remaining molecules. Predictions are made for the molecules that were left out, with the resulting model and the process repeated until predictions have been made for all the molecules in the training set. The correlation coefficient (Q^2) between the predicted and experimental values is calculated and compared with that for the model that was trained on all the molecules in the training set (R^2). If the model is not overfitted, no significant difference should be noted between these.

Another approach, which should preferably be used in conjunction with cross-correlation, is to have a separate test set that is never included in the training. The use of such a separate set is primarily governed by the availability of data. The correlation coefficient of the test set should be comparable to that of the training set to indicate that the model is not overfitted (Norinder and Bergstrom 2006; Butina et al. 2002). In addition, the dependent variable in question can be randomly redistributed among molecules and the model derived again on the redistributed values. This can be done numerous times in order to make certain that there is a definitive difference between the predictive power of the redistributed values in comparison to a model using the true dependent variables (Norinder and Bergstrom 2006).

No single approach can be used to perform a complete range of ADMET predictions (Butina et al. 2002). A combination of models that predict the same property but are based on different principles is a good method for gaining or losing certainty (Butina et al. 2002). Consensus or ensemble models have been proven to have high predictive accuracy and robustness, although they are more complex (Norinder and Bergstrom 2006).

The process of unraveling and solving these (often contradictory) requirements of activity and good ADMET properties is termed "multi-objective or multidimensional optimization" (Ekins et al. 2010). In contrast to the development of simple and uncomplicated rules to filter compounds (Gleeson 2008), multidimensional optimization methods, such as Pareto optimization, offer a way of finding the optimal combination of all the requirements without any weighting of these requirements. A curve is generated with points on it that represent the best possible trade-off between the different requirements. A great advantage to this method is that it immediately rules out the majority of molecules that are not acceptable and can give a plot or surface as a visual way of investigating the trade-offs (Gleeson 2008).

With many requirements to take into account at once, a visual representation is an extremely useful way to summarize a large amount of data. Various approaches to graphically representing ADME properties have been developed in order to aid in decision making during the drug discovery process: various plots, traffic-light red, yellow and green representations, and even Chernoff faces and radar plots (Ritchie et al. 2011).

7.10 Data and Datasets

ADME testing can now be done on a large scale, providing substantial amounts of data on a wide array of compounds (Kassel 2004), and concomitantly leading to a large quantity of data being generated by the drug discovery process. The availability and quality of data are important for use in building accurate theoretical models, as these models do show a dependency on the data on which they are based (Davis and Riley 2004).

The availability of experimental data on large datasets performed uniformly will greatly assist in areas such as the prediction of ADMET properties of drugs. Some attempts at collecting such data in open repositories include the E.U.-funded OpenTox website (Hardy 2011), the U.S. EPA EcoTox database (U.S. Environment Protection Agency), the ONS solubility challenge (Bradley et al. 2010), OpenQSAR (Watson et al. 2010), and QSAR World (QSAR World 2007) as well as other attempts to collect models and descriptors such as those based on the chemistry development kit (Willighagen et al. 2011), Mold2 (U.S. Food and Drug Administration 2010), and repositories such as Chemspider (Williams 2011). High-performance computing networks, such as the world community grid (World Community Grid 2011), provide computational power for large calculations and assist both by the provision of service and in favoring open access to the data that are generated with these network resources.

7.11 Perspective

A confluence of recent events has placed computational methodology at the forefront of pharmaceutical development, including an increase in computational power available through new technology such as grid and cloud computing; increased availability to biochemical and chemical data through large-scale genome sequencing; proteomic and metabolomic projects; improved information distribution and accessibility through the Internet; and the development of new bioinformatic and visualization methodology to handle large datasets. This has led to screening methodologies that complement traditional approaches, along with reverse methodologies to probe the drug-space of existing molecules. These methods are able to better tackle the challenging aspects of bringing new drugs to market, including increased speed and reduced costs and waste in drug-candidate assessment. These developments are particularly relevant for marine bioactives, whereby rapid target identification, toxicity screening, and chemical modification can pose special problems due to their often unique structural nature, relative to the better-explored terrestrial natural products. The different structural aspects of marine natural products offer an exceptional opportunity to develop highly novel and potent drug candidates, for which the various computational methods have already had a significant impact, and it is likely that, as these approaches improve with further technological and methodological advances, many more marine compounds and their derivatives will be brought to market in the future.

References

Abraham, I., S. Jain, C.P. Wu, M.A. Khanfar, Y. Kuang, C.L. Dai, Z. Shi, X. Chen, L. Fu, S.V. Ambudkar, K. El Sayed, and Z.S. Chen. 2010. Marine sponge-derived sipholane triterpenoids reverse P-glycoprotein (ABCB1)-mediated multidrug resistance in cancer cells. *Biochemical Pharmacology* 80: 1497–1506.

Andersen, R.J., D.J. Faulkner, C.H. He, G.D. Van Duyne, and J. Clardy. 1985. Metabolites of the marine prosobranch mollusk *Lamellaria* sp. *Journal of the American Chemical Society* 107: 5492–5495.

Babu, P.A., S.S. Puppala, S.L. Aswini, M.R. Vani, C.N. Kumar, and T. Prasanna. 2008. A database of natural products and chemical entities from marine habitat. *Bioinformation* 3: 142–143.

Baunbaek, D., N. Trinkler, Y. Ferandin, O. Lozach, P. Ploypradith, S. Rucirawat, F. Ishibashi, M. Iwao, and L. Meijer. 2008. Anticancer alkaloid lamellarins inhibit protein kinases. *Marine Drugs* 6: 514–527.

Beierlein, F., H. Lanig, G. Schurer, A.H.C. Horn, and T. Clark. 2003. Quantum mechanical/molecular mechanical (QM/MM) docking: An evaluation for known test systems. *Molecular Physics* 101: 2469–2480.

Belden, H. 2005. First pain drug in new class comes from snail. *Drug Topics* 149: 8.

Bissantz, C., G. Folkers, and D. Rognan. 2000. Protein-based virtual screening of chemical databases. 1. Evaluation of different docking/scoring combinations. *Journal of Medicinal Chemistry* 43: 4759–4767.

Blundell, T.L., H. Jhoti, and C. Abell. 2002. High-throughput crystallography for lead discovery in drug design. *Nature Reviews Drug Discovery* 1: 45–54.

Boldi, A.M. 2004. Libraries from natural product-like scaffolds. *Current Opinion in Chemical Biology* 8: 281–286.
Bradley, J.C., C. Neylon, R. Guha, A.J. Williams, B. Hooker, A.S.I.D. Lang, B. Friesen, T. Bohinski, D. Bulger, M. Federici, J. Hale, J. Mancinelli, K.B. Mirza, M.J. Moritz, D. Rein, C. Tchakounte, and H.T. Truong. 2010. Open notebook science challenge: Solubilities of organic compounds in organic solvents. *Nature Precedings*. doi:10.1038/npre.2010.4243.1033.
Bugni, T.S., B. Richards, L. Bhoite, D. Cimbora, M.K. Harper, and C.M. Ireland. 2008. Marine natural product libraries for high-throughput screening and rapid drug discovery. *Journal of Natural Products* 71: 1095–1098.
Butina, D., M.D. Segall, and K. Frankcombe. 2002. Predicting ADME properties in silico: Methods and models. *Drug Discovery Today* 7: S83–S88.
Cavalli, A., M.L. Bolognesi, A. Minarini, M. Rosini, V. Tumiatti, M. Recanatini, and C. Melchiorre. 2008. Multi-target-directed ligands to combat neurodegenerative diseases. *Journal of Medicinal Chemistry* 51: 347–372.
Chong, C.R., and D.J. Sullivan. 2007. New uses for old drugs. *Nature* 448: 645–646.
Clardy, J., and C. Walsh. 2004. Lessons from natural molecules. *Nature* 432: 829–837.
Claudel, T., B. Staels, and F. Kuipers. 2005. The Farnesoid X receptor: A molecular link between bile acid and lipid and glucose metabolism. *Arteriosclerosis, Thrombosis, and Vascular Biology* 25: 2020–2030.
Cole, S.L., and R. Vassar. 2007. The Alzheimer's disease β-secretase enzyme, BACE1. *Molecular Neurodegeneration* 2: 22.
Congreve, M., R. Carr, C. Murray, and H. Jhoti. 2003. A rule of three for fragment-based lead discovery? *Drug Discovery Today* 8: 876–877.
Congreve, M., G. Chessari, D. Tisi, and A.J. Woodhead. 2008. Recent developments in fragment-based drug discovery. *Journal of Medicinal Chemistry* 51: 3661–3680.
Cragg, G.M., S.A. Schepartz, M. Suffness, and M.R. Grever. 1993. The taxol supply crisis. New NCI policies for handling the large-scale production of novel natural product anticancer and anti-HIV agents. *Journal of Natural Products* 56: 1657–1668.
Davis, A.M., and R.J. Riley. 2004. Predictice ADMET studies, the challenges and the opportunities. *Current Opinion in Chemical Biology* 8: 378–386.
Davis, G.D., and A.H. Vasanthi. 2011. Seaweed metabolite database (SWMD): A database of natural compounds from marine algae. *Bioinformation* 5: 361–364.
de Groot, M.J. 2006. Designing better drugs: Predicting cytochrome P450 metabolism. *Drug Discovery Today* 11: 601–606.
Deng, X.Q., H.Y. Wang, Y.L. Zhao, M.L. Xiang, P.D. Jiang, Z.X. Cao, Y.Z. Zheng, S.D. Luo, L.T. Yu, Y.Q. Wei, and S.Y. Yang. 2008. Pharmacophore modelling and virtual screening for identification of new Aurora-A kinase inhibitors. *Chemical Biology & Drug Design* 71: 533–539.
Dictionary of Natural Products. London: Chapman & Hall/CRC Informa.
Do, Q.T., C. Lamy, I. Renimel, N. Sauvan, P. Andre, F. Himbert, L. Morin-Allory, and P. Bernard. 2007. Reverse pharmacognosy: Identifying biological properties for plants by means of their molecule constituents: Application to meranzin. *Planta Medica* 73: 1235–1240.
Dror, O., D. Schneidman-Duhovny, Y. Inbar, R. Nussinov, and H.J. Wolfson. 2009. Novel approach for efficient pharmacophore-based virtual screening: Method and applications. *Journal of Chemical Information and Modeling* 49: 2333–2343.
Eisenhauer, E.A., W.W. ten Bokkel Huinink, K.D. Swenerton, L. Gianni, J. Myles, M.E. van der Burg, I. Kerr, J.B. Vermorken, K. Buser, and N. Colombo. 1994. European-Canadian randomized trial of paclitaxel in relapsed ovarian cancer: High-dose versus low-dose and long versus short infusion. *Journal of Clinical Oncology* 12: 2654–2666.
Ekins, S., J.D. Honeycutt, and J.T. Metz. 2010. Evolving molecules using multi-objective optimization: Applying to ADME/Tox. *Drug Discovery Today* 15: 451–460.
Eldridge, M.D., C.W. Murray, T.R. Auton, G.V. Paolini, and R.P. Mee. 1997. Empirical scoring functions: I. The development of a fast empirical scoring function to estimate the binding affinity of ligands in receptor complexes. *Journal of Computer-Aided Molecular Design* 11: 425–445.

Erlanson, D.A., R.S. McDowell, and T. O'Brien. 2004. Fragment-based drug discovery. *Journal of Medicinal Chemistry* 47: 3463–3482.
Ertl, P., S. Roggo, and A. Schuffenhauer. 2008. Natural product-likeness score and its application for prioritization of compound libraries. *Journal of Chemical Information and Modeling* 48: 68–74.
Ewing, T.J., S. Makino, A.G. Skillman, and I.D. Kuntz. 2001. DOCK 4.0: Search strategies for automated molecular docking of flexible molecule databases. *Journal of Computer-Aided Molecular Design* 15: 411–428.
Fanfrlik, J., J. Brynda, J. Rezac, P. Hobza, and M. Lepsik. 2008. Interpretation of protein/ligand crystal structure using QM/MM calculations: Case of HIV-1 protease/metallacarborane complex. *Journal of Physical Chemistry B* 112: 15094–15102.
Feher, M., and J.M. Schmidt. 2003. Property distributions: Differences between drugs, natural products, and molecules from combinatorial chemistry. *Journal of Chemical Information and Computer Sciences* 43: 218–227.
Ferrara, P., H. Gohlke, D.J. Price, G. Klebe, and C.L. Brooks III. 2004. Assessing scoring functions for protein. *Journal of Medicinal Chemistry* 47: 3032–3047.
Foster, I., Zhao, Y., Raicu, I., and Lu, S. 2009. Cloud computing and grid computing 360-degree compared. ArXiv e-prints: 0901.0131v0901.
Francis, G.A., E. Fayard, F. Picard, and J. Auwerx. 2003. Nuclear receptors and the control of metabolism. *Annual Review of Physiology* 65: 261–311.
FRED. 2008. OpenEye Scientific Software. New Mexico, United States.
Gentilucci, L., F. Squassabia, R. Demarco, R. Artali, G. Cardillo, A. Tolomelli, S. Spampinato, and A. Bedini. 2008. Investigation of the interaction between the atypical agonist c[YpwFG] and MOR. *Febs Journal* 275: 2315–2337.
Glaser, K.B., and A.M. Mayer. 2009. A renaissance in marine pharmacology: From preclinical curiosity to clinical reality. *Biochemical Pharmacology* 78: 440–448.
Gleeson, M.P. 2008. Generation of a set of simple, interpretable ADMET rules of thumb. *Journal of Medicinal Chemistry* 51: 817–834.
Gleeson, M.P., and D. Gleeson. 2009. QM/MM calculations in drug discovery: A useful method for studying binding phenomena? *Journal of Chemical Information and Modeling* 49: 670–677.
Gleeson, M.P., S. Hannongbua, and D. Gleeson. 2010. QM methods in structure based design: Utility in probing protein-ligand interactions. *Journal of Molecular Graphics and Modelling* 29: 507–517.
Guner, O.F. 2002. History and evolution of the pharmacophore concept in computer-aided drug design. *Current Topics in Medicinal Chemistry* 2: 1321–1332.
Hai-Lun, H., C. Xiu-Lan, S. Cai-Yun, Z. Yu-Zhong, and Z. Bai-Cheng. 2006. Analysis of novel angiotensin-I-converting enzyme inhibitory peptides from protease-hydrolyzed marine shrimp *Acetes chinensis*. *Journal of Peptide Science* 12: 726–733.
Halgren, T.A., and W. Damm. 2001. Polarizable force fields. *Current Opinion in Structural Biology* 11: 236–242.
Hardy, B., OpenTox. 2011. Barry Hardy. http://www.opentox.org/. Accessed 3 May 2011.
Harriman, D.J., and G. Deslongehamps. 2004. Reverse-docking as a computational tool for the study of asymmetric organocatalysis. *Journal of Computer-Aided Molecular Design* 18: 303–308.
Harvey, A.L. 2007. Natural products as a screening resource. *Current Opinion in Chemical Biology* 11: 480–484.
Haustedt, L.O., C. Mang, K. Siems, and H. Schiewe. 2006. Rational approaches to natural-product-based drug design. *Current Opinion in Drug Discovery & Development* 9: 445–462.
Hein, M., D. Zilian, and C.A. Sotriffer. 2010. Docking compared to 3D-pharmacophores: The scoring function challenge. *Drug Discovery Today: Technologies* 7: e229–e236.
Hopfinger, A.J., A. Reaka, P. Venkatarangan, J.S. Duca, and S. Wang. 1999. Construction of a virtual high throughput screen by 4D-QSAR analysis: Application to a combinatorial library of glucose inhibitors of glycogen phosphorylase b. *Journal of Chemical Information and Computer Sciences* 39: 1151–1160.

Huang, N., B.K. Shoichet, and J.J. Irwin. 2006. Benchmarking sets for molecular docking. *Journal of Medicinal Chemistry* 49: 6789–6801.

Ibrahim, M.A., A.G. Shilabin, S. Prasanna, M. Jacob, S.I. Khan, R.J. Doerksen, and M.T. Hamann. 2008. 2-N-methyl modifications and SAR studies of manzamine A. *Bioorganic and Medicinal Chemistry* 16: 6702–6706.

Indarte, M., J.D. Madura, and C.K. Surratt. 2007. Dopamine transporter comparative molecular modeling and binding site prediction using the LeuTAa leucine transporter as a template. *Proteins: Structure, Function, and Bioinformatics* 70: 1033–1046.

Jain, A.N. 2003. Surflex: Fully automatic flexible molecular docking using a molecular similarity-based search engine. *Journal of Medicinal Chemistry* 46: 499–511.

Jain, S., I. Abraham, P. Carvalho, Y.H. Kuang, L.A. Shaala, D.T. Youssef, M.A. Avery, Z.S. Chen, and K.A. El Sayed. 2009. Sipholane triterpenoids: Chemistry, reversal of ABCB1/P-glycoprotein-mediated multidrug resistance, and pharmacophore modeling. *Journal of Natural Products* 72: 1291–1298.

Jang, J.-H., S.-C. Jeong, J.-H. Kim, Y.-H. Lee, Y.-C. Ju, and J.-S. Lee. 2011. Characterisation of a new antihypertensive angiotensin I-converting enzyme inhibitory peptide from *Pleurotus cornucopiae*. *Food Chemistry* 127: 412–418.

Jansen, J.M., and E.J. Martin. 2004. Target-biased scoring approaches and expert systems in structure-based virtual screening. *Current Opinion in Chemical Biology* 8: 359–364.

Jimsheena, V.K., and L.R. Gowda. 2010. Arachin derived peptides as selective angiotensin I-converting enzyme (ACE) inhibitors: Structure-activity relationship. *Peptides* 31: 1165–1176.

Jimsheena, V.K., and L.R. Gowda. 2011. Angiotensin I-converting enzyme (ACE) inhibitory peptides derived from arachin by simulated gastric digestion. *Food Chemistry* 125: 561–569.

John, S., S. Thangapandian, S. Sakkiah, and K.W. Lee. 2011. Potent bace-1 inhibitor design using pharmacophore modeling, in silico screening and molecular docking studies. *BMC Bioinformatics* 12: S1–S28.

Jones, G., P. Willett, R.C. Glen, A.R. Leach, and R. Taylor. 1997. Development and validation of a genetic algorithm for flexible docking. *Journal of Molecular Biology* 267: 727–748.

Jung, H.A., S.K. Hyun, H.R. Kim, and J.S. Choi. 2006. Angiotensin-converting enzyme I inhibitory activity of phlorotannins from *Ecklonia stolonifera*. *Fisheries Science* 72: 1292–1299.

Jung, H.A., S.H. Oh, and J.S. Choi. 2010. Molecular docking studies of phlorotannins from *Eisenia bicyclis* with BACE1 inhibitory activity. *Bioorganic and Medicinal Chemistry Letters* 20: 3211–3215.

Kahnberg, P., M.H. Howard, T. Liljefors, M. Nielsen, E.O. Nielsen, O. Sterner, and I. Pettersson. 2004. The use of a pharmacophore model for identification of novel ligands for the benzodiazepine binding site of the GABAA receptor. *Journal of Molecular Graphics and Modelling* 23: 253–261.

Kassel, D.B. 2004. Applications of high-throughput ADME in drug discovery. *Current Opinion in Chemical Biology* 8: 339–345.

Keseru, G.M., and G.M. Makara. 2006. Hit discovery and hit-to-lead approaches. *Drug Discovery Today* 11: 741–748.

Khanfar, M.A., B.A. Asal, M. Mudit, A. Kaddoumi, and K.A. El Sayed. 2009. The marine natural-derived inhibitors of glycogen synthase kinase-3β phenylmethylene hydantoins: *In vitro* and *in vivo* activities and pharmacophore modeling. *Bioorganic and Medicinal Chemistry* 17: 6032–6039.

Khanfar, M.A., R.A. Hill, A. Kaddoumi, and K.A. El Sayed. 2010. Discovery of novel GSK-3beta inhibitors with potent in vitro and in vivo activities and excellent brain permeability using combined ligand- and structure-based virtual screening. *Journal of Medicinal Chemistry* 53: 8534–8545.

Kim, A., T. Shin, M. Lee, J. Park, K. Park, N. Yoon, J. Kim, J. Choi, B. Jang, D. Byun, N. Park, and H. Kim. 2009. Isolation and identification of phlorotannins from *Ecklonia stolonifera* with antioxidant and anti-inflammatory properties. *Journal of Agricultural and Food Chemistry* 57: 3483–3489.

Kirchmair, J., P. Markt, S. Distinto, G. Wolber, and T. Langer. 2008. Evaluation of the performance of 3D virtual screening protocols: RMSD comparisons, enrichment assessments, and decoy selection–what can we learn from earlier mistakes? *Journal of Computer-Aided Molecular Design* 22: 213–228.

Klabunde, T., and A. Evers. 2005. GPCR antitarget modeling: Pharmacophore models for biogenic amine binding GPCRs to avoid GPCR-mediated side effects. *ChemBioChem: A European Journal of Chemical Biology* 6: 876–889.

Koch, M.A., L.O. Wittenberg, S. Basu, D.A. Jeyaraj, E. Gourzoulidou, K. Reinecke, A. Odermatt, and H. Waldmann. 2004. Compound library development guided by protein structure similarity clustering and natural product structure. *Proceedings of the National Academy of Sciences of the United States of America* 101: 16721–16726.

Kortagere, S., and S. Ekins. 2010. Troubleshooting computational methods in drug discovery. *Journal of Pharmacological and Toxicological Methods* 61: 67–75.

Krishnaiah, P., V.L. Reddy, G. Venkataramana, K. Ravinder, M. Srinivasulu, T.V. Raju, K. Ravikumar, D. Chandrasekar, S. Ramakrishna, and Y. Venkateswarlu. 2004. New lamellarin alkaloids from the Indian ascidian *Didemnum obscurum* and their antioxidant properties. *Journal of Natural Products* 67: 1168–1171.

Kroemer, R.T., A. Vulpetti, J.J. McDonald, D.C. Rohrer, J.Y. Trosset, F. Giordanetto, S. Cotesta, C. McMartin, M. Kihlen, and P.F. Stouten. 2004. Assessment of docking poses: Interactions-based accuracy classification (IBAC) versus crystal structure deviations. *Journal of Chemical Information and Computer Sciences* 44: 871–881.

Lei, J., and J. Zhou. 2002. A marine natural product database. *Journal of Chemical Information and Computer Sciences* 42: 742–748.

Levesque, M.J., K. Ichikawa, S. Date, and J.H. Haga. 2009. Design of a grid service-based platform for *in silico* protein-ligand screenings. *Computer Methods and Programs in Biomedicine* 93: 73–82.

Li, G.-H., G.-W. Le, Y.-H. Shi, and S. Shrestha. 2004. Angiotensin I–converting enzyme inhibitory peptides derived from food proteins and their physiological and pharmacological effects. *Nutrition Research* 24: 469–486.

Li, X., F. Lu, Q. Tian, Y. Yang, Q. Wang, and J.Z. Wang. 2006. Activation of glycogen synthase kinase-3 induces Alzheimer-like tau hyperphosphorylation in rat hippocampus slices in culture. *Journal of Neural Transmission* 113: 93–102.

LigandScout, Inte: Ligand GmbH, Vienna, Austria, Europe.

Lipinski, C.A., F. Lombardo, B.W. Dominy, and P.J. Feeney. 2001. Experimental and computational approaches to estimate solubility and permeability in drug discovery and development settings. *Advanced Drug Delivery Reviews* 46: 3–26.

Liu, X., S. Ouyang, B. Yu, Y. Liu, K. Huang, J. Gong, S. Zheng, Z. Li, H. Li, and H. Jiang. 2010. PharmMapper server: D web server for potential drug target identification using pharmacophore mapping approach. *Nucleic Acids Research* 38: W609–614.

Lizcano, F., C. Romero, and D. Vargas. 2011. Regulation of adipogenesis by nuclear receptor PPAR gamma is modulated by the histone demethylase JMJD2C. *Genetics and Molecular Biology* 34: 19–24.

Lyne, P.D., P.W. Kenny, D.A. Cosgrove, C. Deng, S. Zabludoff, J.J. Wendoloski, and S. Ashwell. 2004. Identification of compounds with nanomolar binding affinity for checkpoint kinase-1 using knowledge-based virtual screening. *Journal of Medicinal Chemistry* 47: 1962–1968.

Mang, C., S. Jakupovic, S. Schunk, H.D. Ambrosi, O. Schwarz, and J. Jakupovic. 2006. Natural products in combinatorial chemistry: An andrographolide-based library. *Journal of Combinatorial Chemistry* 8: 268–274.

MarinLit. Marine Natural Product Bibliography Software. Christchurch, New Zealand: University of Cantebury.

Martin, Y.C., M.G. Bures, E.A. Danaher, J. DeLazzer, I. Lico, and P.A. Pavlik. 1993. A fast new approach to pharmacophore mapping and its application to dopaminergic and benzodiazepine agonists. *Journal of Computer-Aided Molecular Design* 7: 83–102.

Martinez, A., A. Castro, I. Dorronsoro, and M. Alonso. 2002. Glycogen synthase kinase 3 (GSK-3) inhibitors as new promising drugs for diabetes, neurodegeneration, cancer, and inflammation. *Medicinal Research Reviews* 22: 373–384.

Martins, J.P., E.G. Barbosa, K.F. Pasqualoto, and M.M. Ferreira. 2009. LQTA-QSAR: A new 4D-QSAR methodology. *Journal of Chemical Information and Modeling* 49: 1428–1436.

McGann, M. 2011. FRED pose prediction and virtual screening accuracy. *Journal of Chemical Information and Modeling* 51: 578–596.

McIntosh, M., L.J. Cruz, M.W. Hunkapiller, W.R. Gray, and B.M. Olivera. 1982. Isolation and structure of a peptide toxin from the marine snail *Conus magus*. *Archives of Biochemistry and Biophysics* 218: 329–334.

Mishra, K.P., L. Ganju, M. Sairam, P.K. Banerjee, and R.C. Sawhney. 2008. A review of high throughput technology for the screening of natural products. *Biomedicine & Pharmacotherapy* 62: 94–98.

Mladenovic, M., M. Arnone, R.F. Fink, and B. Engels. 2009. Environmental effects on charge densities of biologically active molecules: Do molecule crystal environments indeed approximate protein surroundings? *Journal of Physical Chemistry B* 13: 5072–5082.

Molecular Operating Environment, Chemical Computing Group (CCG), Montreal, Canada.

Morphy, R., and Z. Rankovic. 2005. Designed multiple ligands. An emerging drug discovery paradigm. *Journal of Medicinal Chemistry* 48: 6523–6543.

Morris, G.M., D.S. Goodsell, R.S. Halliday, R. Huey, W.E. Hart, R.K. Belew, and A.J. Olson. 1998. Automated docking using a Lamarckian genetic algorithm and an empirical binding free energy function. *Journal of Computational Chemistry* 19: 1639–1662.

MTree, BioSolveIT Gmbh, Sankt Augustin, Germany.

Muegge, I. 2000. A knowledge-based scoring function for protein-ligand interactions: Probing the reference state. *Perspectives in Drug Discovery and Design* 20: 99–114.

Mukherjee, P., P. Desai, A. Srivastava, B. Tekwani, and M. Avery. 2008. Probing the structures of leishmanial farnesyl pyrophosphate synthases: Homology modeling and docking studies. *Journal of Chemical Information and Modeling* 48: 1026–1040.

Murray, C.W., and T.L. Blundell. 2010. Structural biology in fragment-based drug design. *Current Opinion in Structural Biology* 20: 497–507.

Norinder, U., and C.A.S. Bergstrom. 2006. Prediction of ADMET properties. *ChemMedChem* 1: 920–937.

Nwosu, F., J. Morris, V.A. Lund, D. Stewart, H.A. Ross, and G.J. McDougall. 2011. Anti-proliferative and potential anti-diabetic effects of phenolic-rich extracts from edible marine algae. *Food Chemistry* 126: 1006–1012.

Ono, S., M. Hosokawa, K. Miyashita, and K. Takahashi. 2006. Isolation of peptides with angiotensin I-converting enzyme inhibitory effect derived from hydrolysate of upstream Chum Salmon Muscle. *Journal of Food Science* 68: 1611–1614.

Oprea, T.I., and H. Matter. 2004. Integrating virtual screening in lead discovery. *Current Opinion in Chemical Biology* 8: 349–358.

Parys, S., S. Kehraus, A. Krick, K.W. Glombitza, S. Carmeli, K. Klimo, C. Gerhauser, and G.M. Konig. 2010. In vitro chemopreventive potential of fucophlorethols from the brown alga *Fucus vesiculosus* L. by anti-oxidant activity and inhibition of selected cytochrome P450 enzymes. *Phytochemistry* 71: 221–229.

Patel, S., and C.L. Brooks III. 2006. Fluctuating charge force fields: Recent developments and applications from small molecules to macromolecular biological systems. *Molecular Simulation* 32: 231–249.

Peach, M.L., and M.C. Nicklaus. 2009. Combining docking with pharmacophore filtering for improved virtual screening. *Journal of Cheminformatics* 1: 6.

Pelish, H.E., N.J. Westwood, Y. Feng, T. Kirchhausen, and M.D. Shair. 2001. Use of biomimetic diversity-oriented synthesis to discover galanthamine-like molecules with biological properties beyond those of the natural product. *Journal of the American Chemical Society* 123: 6740–6741.

QSAR World. 2007. Strand Life Sciences Pvt. Ltd., http://www.qsarworld.com/. Accessed 3 May 2011.

Queiroz, A.N., B.A.Q. Gomes, W.M. Moraes Jr., and R.S. Borges. 2009. A theoretical antioxidant pharmacophore for resveratrol. *European Journal of Medicinal Chemistry* 44: 1644–1649.

Raha, K., M.B. Peters, B. Wang, N. Yu, A.M. Wollacott, L.M. Westerhoff, and K.M. Merz Jr. 2007. The role of quantum mechanics in structure-based drug design. *Drug Discovery Today* 12: 725–731.

Reddy, M.V., M.R. Rao, D. Rhodes, M.S. Hansen, K. Rubins, F.D. Bushman, Y. Venkateswarlu, and D.J. Faulkner. 1999. Lamellarin alpha 20-sulfate, an inhibitor of HIV-1 integrase active against HIV-1 virus in cell culture. *Journal of Medicinal Chemistry* 42: 1901–1907.

Rella, M., C.A. Rushworth, J.L. Guy, A.J. Turner, T. Langer, and R.M. Jackson. 2006. Structure-based pharmacophore design and virtual screening for novel angiotensin converting enzyme 2 inhibitors. *Journal of Chemical Information and Modeling* 46: 708–716.

Reymond, J.L., R. van Deursen, L.C. Blum, and L. Ruddigkeit. 2010. Chemical space as a source for new drugs. *Medicinal Chemistry Communications* 1: 30–38.

Rinehart, K.L., T.G. Holt, N.L. Fregeau, J.G. Stroh, P.A. Keifer, F. Sun, L.H. Li, and D.G. Martin. 1990. Ecteinascidins 729, 743, 745, 759A, 759B, and 770: Potent antitumor agents from the Caribbean tunicate *Ecteinascidia turbinata*. *Journal of Organic Chemistry* 55: 4512–4515.

Ripphausen, P., B. Nisius, L. Peltason, and J. Bajorath. 2010. Quo vadis, virtual screening? A comprehensive survey of prospective applications. *Journal of Medicinal Chemistry* 53: 8461–8467.

Ritchie, T.J., P. Ertl, and R. Lewis. 2011. The graphical representation of ADME-related molecule properties for medicinal chemists. *Drug Discovery Today* 16: 65–72.

Rollinger, J.M. 2009. Accessing target information by virtual parallel screening – The impact on natural product research. *Phytochemistry Letters* 2: 53–58.

Rollinger, J.M., A. Hornick, T. Langer, H. Stuppner, and H. Prast. 2004. Acetylcholinesterase inhibitory activity of scopolin and scopoletin discovered by virtual screening of natural products. *Journal of Medicinal Chemistry* 47: 6248–6254.

Sakai, R., T. Higa, C.W. Jefford, and G. Bernardinelli. 1986. Manzamine A, a novel antitumor alkaloid from a sponge. *Journal of the American Chemical Society* 108: 6404–6405.

Samantray, D., and R.K. Sahu. 2010. Drug designing and docking efficacy assessment of halogen substituted aspirin. *Researcher* 2: 17–23.

Sato, M., T. Hosokawa, T. Yamaguchi, N. Toshiki, K. Muramoto, T. Kahara, K. Funayama, A. Kobayashi, and T. Nakano. 2002. Angiotensin I-converting enzyme inhibitory peptides derived from Wakame (*Undaria pinnatifida*) and their antihypertensive effect in spontaneously hypertensive Rats. *Journal of Agricultural and Food Chemistry* 50: 6245–6252.

Schneidman-Duhovny, D., O. Dror, Y. Inbar, R. Nussinov, and H.J. Wolfson. 2008. PharmaGist: A webserver for ligand-based pharmacophore detection. *Nucleic Acids Research* 36: W223–W228. Catalyst, Accelrys Inc., San Diego, CA.

Schuster, D. 2010. 3D pharmacophores as tools for activity profiling. *Drug Discovery Today: Technologies* 7: e205–e211.

Segura Campos, M.R., L.A. Chel Guerrero, and D.A. Betancur Ancona. 2010. Angiotensin-I converting enzyme inhibitory and antioxidant activities of peptide fractions extracted by ultrafiltration of cowpea *Vigna unguiculata* hydrolysates. *Journal of the Science of Food and Agriculture* 90: 2512–2518.

Selnergy, Greenpharma S.A.S, Orléans, France.

Senese, C.L., J. Duca, D. Pan, A.J. Hopfinger, and Y.J. Tseng. 2004. 4D-fingerprints, universal QSAR and QSPR descriptors. *Journal of Chemical Information and Computer Sciences* 44: 1526–1539.

Sepe, V., G. Bifulco, B. Renga, C. D'Amore, S. Fiorucci, and A. Zampella. 2011. Discovery of sulfated sterols from marine invertebrates as a new class of marine natural antagonists of farnesoid-x-receptor. *Journal of Medicinal Chemistry* 54: 1314–1320.

Shibata, T., K. Ishimaru, S. Kawaguchi, H. Yoshikawa, and Y. Hama. 2008. Antioxidant activities of phlorotannins isolated from Japanese Laminariaceae. *Journal of Applied Phycology* 20: 705–711.

Shu, Y.Z. 1998. Recent natural products based drug development: A pharmaceutical industry perspective. *Journal of Natural Products* 61: 1053–1071.

Srinivas, E., J.N. Murthy, A.R.R. Rao, and G.N. Sastry. 2006. Recent advances in molecular modeling and medicinal chemistry aspects of phospho-glycoprotein. *Current Drug Metabolism* 7: 205–217.

Stahl, M., and D. Rarey. 2001. Detailed analysis of scoring functions for virtual screening. *Journal of Medicinal Chemistry* 44: 1035–1042.

Stahl, M., W. Guba, and M. Kansy. 2006. Integrating molecular design resources within modern drug discovery research: The Roche experience. *Drug Discovery Today* 11: 326–333.

Steindl, T., and T. Langer. 2005. Docking versus pharmacophore model generation: A comparison of high-throughput virtual screening strategies for the search of human rhinovirus coat protein inhibitors. *QSAR & Combinatorial Science* 24: 470–479.

Strohl, W.R. 2000. The role of natural products in a modern drug discovery program. *Drug Discovery Today* 5: 39–41.

Surflex-Dock 2.0, Tripos International, Missouri, USA.

SYBYL 8.0, Tripos International, Missouri, USA.

Taha, M.O., M. Tarairah, H. Zalloum, and G. Abu-Sheikha. 2010. Pharmacophore and QSAR modeling of estrogen receptor beta ligands and subsequent validation and in silico search for new hits. *Journal of Molecular Graphics and Modelling* 28: 383–400.

Tawari, N.R., and M.S. Degani. 2010. Pharmacophore mapping and electronic feature analysis for a series of nitroaromatic compounds with antitubercular activity. *Journal of Computational Chemistry* 31: 739–751.

Thipnate, P., J. Liu, S. Hannongbua, and A.J. Hopfinger. 2009. 3D pharmacophore mapping using 4D QSAR analysis for the cytotoxicity of lamellarins against human hormone-dependent T47D breast cancer cells. *Journal of Chemical Information and Modeling* 49: 2312–2322.

Thomsen, R., and M.H. Christensen. 2006. MolDock: A new technique for high-accuracy molecular docking. *Journal of Medicinal Chemistry* 49: 3315–3321.

Tierney, M.S., A.K. Croft, and M. Hayes. 2010. A review of antihypertensive and antioxidant activities in macroalgae. *Botanica Marina* 53: 387–408.

Trindade-Silva, A.E., G.E. Lim-Fong, K.H. Sharp, and M.G. Haygood. 2010. Bryostatins: Biological context and biotechnological prospects. *Current Opinion in Biotechnology* 21: 834–842.

Trott, O., and A.J. Olson. 2010. Software news and update AutoDock Vina: Improving the speed and accuracy of docking with a new scoring function, efficient optimization, and multithreading. *Journal of Computational Chemistry* 31: 455–461.

US Environment Protection Agency, ECOTOX (ECOTOXicology) Database, Release 4.0. 2011. http://cfpub.epa.gov/ecotox/ecotox_home.cfm. Accessed 3 May 2011.

US Food and Drug Administration, Mold2, Descriptors Generator Software. 2010. US Department of Health and Human Services. http://www.fda.gov/ScienceResearch/BioinformaticsTools/Mold2/default.htm. Accessed 3 May 2011.

van de Waterbeemd, H., and E. Gifford. 2003. ADMET *in silico* modelling: Towards prediction paradise? *Nature Reviews Drug Discovery* 2: 192–204.

Vanhuyse, M., J. Kluza, C. Tardy, G. Otero, C. Cuevas, C. Bailly, and A. Lansiaux. 2005. Lamellarin D: A novel pro-apoptotic agent from marine origin insensitive to P-glycoprotein-mediated drug efflux. *Cancer Letters* 221: 165–175.

Verdonk, M.L., J.C. Cole, M.J. Hartshorn, C.W. Murray, and R.D. Taylor. 2003. Improved protein–ligand docking using GOLD. *Proteins* 52: 609–623.

Viennois, E., A.J.C. Pommier, K. Mouzat, A. Oumeddour, F.Z. El Hajjaji, J. Dufour, F. Caira, D.H. Volle, S. Baron, and J.M.A. Lobaccaro. 2011. Targeting liver X receptors in human health: Deadlock or promising trail? *Expert Opinion on Therapeutic Targets* 15: 219–232.

von Korff, M., C. Rufener, M. Stritt, J. Freyss, R. Bar, and T. Sander. 2011. Integration of distributed computing into the drug discovery process. *Expert Opinion on Drug Discovery* 6: 103–107.

Wald, C. 2010. Scientists Embrace Openness. *Science Careers*. doi:10.1126/science.caredit.a1000036.

Wallach, I., and R. Lilien. 2011. Virtual decoy sets for molecular docking benchmarks. *Journal of Chemical Information and Modeling* 51: 196–202.

Wang, R., and S. Wang. 2001. How does consensus scoring work for virtual library screening? An idealized computer experiment. *Journal of Chemical Information and Computer Sciences* 41: 1422–1426.

Wang, H.B., J. Chen, K. Hollister, L.C. Sowers, and B.M. Forman. 1999. Endogenous bile acids are ligands for the nuclear receptor FXR BAR. *Molecular Cell* 3: 543–553.

Wang, R., Y. Lu, and S. Wang. 2003. Comparative evaluation of 11 scoring functions for molecular docking. *Journal of Medicinal Chemistry* 46: 2287–2303.

Wang, Z., B. Ling, R. Zhang, Y. Suo, Y. Liu, Z. Yu, and C. Liu. 2009. Docking and molecular dynamics studies toward the binding of new natural phenolic marine inhibitors and aldose reductase. *Journal of Molecular Graphics and Modelling* 28: 162–169.

Wang, Z., S. Zhang, W. Wang, F. Feng, and W. Shan. 2011. A novel angiotensin I converting enzyme inhibitory peptide from the milk casein: Virtual screening and docking studies. *Agricultural Sciences in China* 10: 463–467.

Warren, G.L., C.W. Andrews, A.M. Capelli, B. Clarke, J. LaLonde, M.H. Lambert, M. Lindvall, N. Nevins, S.F. Semus, S. Senger, G. Tedesco, I.D. Wall, J.M. Woolven, C.E. Peishoff, and M.S. Head. 2006. A critical assessment of docking programs and scoring functions. *Journal of Medicinal Chemistry* 49: 5912–5931.

Watson, P., Leahy, D., Cala, J., Searson, D., Sykora, V., Taylor, M., Woodman, S., Hiden, H., OpenQSAR. 2010. School of Computing Science, Newcastle University. http://www.openqsar.com/, Accessed 3 May 2011.

Welch, W., J. Ruppert, and A.N. Jain. 1996. Hammerhead: Fast, fully automated docking of flexible ligands to protein binding sites. *Chemistry and Biology* 3: 449–462.

Wender, P.A., C.M. Cribbs, K.F. Koehler, N.A. Sharkey, C.L. Herald, Y. Kamano, G.R. Pettit, and P.M. Blumberg. 1988. Modeling of the Bryostatins to the phorbol ester pharmacophore on protein kinase-C. *Proceedings of the National Academy of Sciences of the United States of America* 85: 7197–7201.

Wijesekara, I., and S.K. Kim. 2010. Angiotensin-I-converting enzyme (ACE) inhibitors from marine resources: Prospects in the pharmaceutical industry. *Marine Drugs* 8: 1080–1093.

Williams, A., Chemspider. 2011. Royal Society of Chemistry. http://www.chemspider.com/. Accessed 3 May 2011.

Willighagen, E., Guha, R., Steinbeck, C., Chemistry Development Kit. 2011. Sourceforge. http://sourceforge.net/projects/cdk/. Accessed 3 May 2011.

Wolber, G., and T. Langer. 2005. LigandScout: 3-D pharmacophores derived from protein-bound ligands and their use as virtual screening filters. *Journal of Chemical Information and Modeling* 45: 160–169.

World Community Grid, World Community Grid, technology solving problems. 2011. IBM. http://www.worldcommunitygrid.org/. Accessed 3 May 11.

Wunberg, T., M. Hendrix, A. Hillisch, M. Lobell, H. Meier, C. Schmeck, H. Wild, and B. Hinzen. 2006. Improving the hit-to-lead process: Data-driven assessment of drug-like and lead-like screening hits. *Drug Discovery Today* 11: 175–180.

Yap, T.A., C.P. Carden, and S.B. Kaye. 2009. Beyond chemotherapy: Targeted therapies in ovarian cancer. *Nature Reviews Cancer* 9: 167–181.

Zhou, Z., A.K. Felts, R.A. Friesner, and R.M. Levy. 2007. Comparative performance of several flexible docking programs and scoring functions: Enrichment studies for a diverse set of pharmaceutically relevant targets. *Journal of Chemical Information and Modeling* 47: 1599–1608.

Chapter 8
Marine-Derived Functional Foods: Claims and Current Legislation

Maria Hayes

8.1 Introduction

Functional foods are defined as foods that have disease-preventing and/or health-promoting benefits in addition to their basic nutritive value (Southon 2000). Marine-derived bioactive compounds fit convincingly within this definition and they may be used as functional food ingredients. For example, marine-derived dietary fibers including oligosaccharides such as fucoidan, can be viewed as functional food ingredients as they can impart health benefits to the consumer beyond those associated with basic nutrition. However, consumers require clarification regarding the concept of functional foods. It is now necessary, in Europe, to define the actual health benefit of a particular functional food due to regulations introduced by the European Union (that are discussed in this chapter) and the formation of the European Food Safety Authority (EFSA). Marketing of food ingredients as "bioactives with health benefits" now requires substantial scientific substantiation in Europe and it is now necessary to provide evidence of any benefit before any health claims can be made on a functional food product. The European Commission Concerted Action on Functional Food Science in Europe (FUFOSE), coordinated by the International Life Sciences Institute – ILSI Europe, was established in 1986 to establish a science-based approach for concepts in functional foods. The aims of FUFOSE were:

- To assess critically the science base required to provide evidence that specific nutrients and food components positively affect target functions in the body.
- To examine the available science from a function-driven perspective rather than a product-driven one.
- To reach consensus on targeted modifications of food and food constituents, and options for their application (Diplock et al. 1999 and Diplock et al. 2000).

M. Hayes (✉)
Food BioSciences Department, Teagasc Food Research Centre, Ashtown, Dublin 15, Ireland
e-mail: maria.hayes@teagasc.ie

M. Hayes (ed.), *Marine Bioactive Compounds: Sources, Characterization and Applications*, DOI 10.1007/978-1-4614-1247-2_8,

Today, in Europe, EFSA are charged with governing novel food and health claims made on food products and ingredients. China, Japan, and the United States of America each have their own regulatory bodies and laws governing the health claims made on foods and food ingredients and beverages. This chapter highlights the scientific substantiation of functional food health and novel food claims made on foods and food ingredients in the United States, China, Japan, and Europe and highlights specific marine-derived ingredients that have obtained EFSA novel food and health claims.

8.2 China

China is the home of traditional medicine and is today one of the world's most important and developed markets for functional foods. According to "The provision for functional foods administration," which was promulgated by the Ministry of Health (MOH) in March 1996, a *functional food* is defined as a food that has special health functions (Ministry of Health 1996). In China, functional foods are deemed suitable for consumption by special groups of people and function to regulate human body functions but may not be used for therapeutic purposes (Yang 2008). In 2005, the State Food and Drugs Administration (SFDA) established "The guideline of registration for functional foods." The functional foods definition was extended to "Health (functional) food means that a food has special health functions or is able to supply vitamins or minerals. It is suitable for consumption by special groups of people and has the function of regulating human body functions but is not used for therapeutic purposes. And it will not cause any harm whether acute or sub-acute or chronic" (Yang 2008; State Food and Drug Administration 2005). Nutrient substances, which do not provide energy to the consumer but provide vitamins and minerals to assist dietary insufficiency and foods with functional claims, are included in this definition.

8.3 Administration and Regulation of Functional Foods in China

Since 2003, the SFDA direct and conduct all regulatory affairs associated with functional food marketing and use. Tests, including toxicity, functional, stability, and hygiene tests, must be conducted and carried out in accordance with the standard procedures at specialized agencies qualified by the MOH and the SFDA. Figure 8.1 shows the examination and approval steps required of functional foods in China.

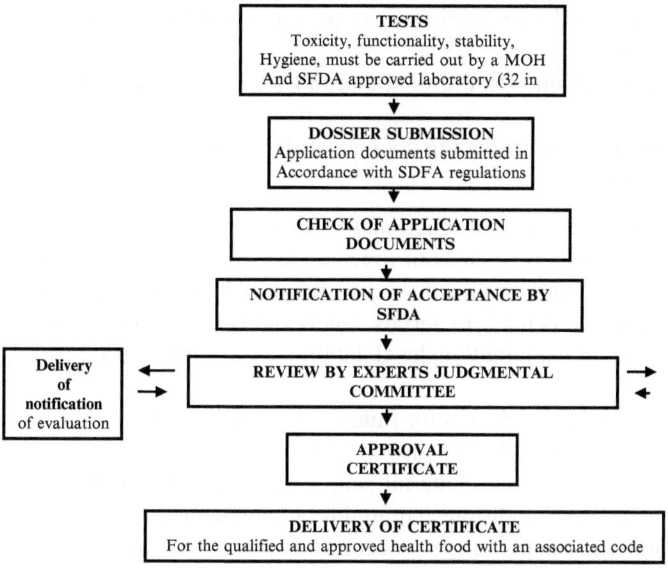

Fig. 8.1 The dossier of an application for a health food in China undergoes substantive examination prior to approval by the SFDA. As shown in Fig. 8.1, several steps are required and these include testing for toxicity, stability, functionality and hygiene, submission of data on tests, literature review data to SFDA, checking of formal application, acceptance, re-examination of application documents (in some instances), and approval certificate receipt by the manufacturer

8.4 Dossier Submission for Functional Food Claims in China

The dossier submitted in order to obtain a functional food claim in China should consist of the following five sections.

1. Information on the enterprise and the product for administrative data
2. Information specific to the food and its associated characteristics
3. Laboratory testing data obtained on the product
4. Consumer information
5. Scientific evidence from the literature regarding the functionality of the product

If the product is imported then additional information is required. This information should be provided in Chinese and should be consistent with the original language version of the documents. The information required includes:

1. A certificate confirming that the product complies with relevant local production quality control standards (QCS) issued by the country from which the product originated.
2. A certificate is required confirming that the products have been sold for greater than a year in the country where the product is produced, issued by the

certifying agency of the originating country and the Chinese embassy in the originating country.

3. Labels and the directions for use of the product in the originating country are required.
4. Product standards of the producing country of the product or international organizations are required.

The main directive in China for functional foods is "The guideline of registration for functional foods," which consists of nine chapters and 105 articles. It includes the general principles and applications of functional foods, examination and approval, raw and supplementary materials, labels and specifications, testing, registration, reregistration, re-examination, legal liabilities, and supplementary provisions (Yang 2008).

The "General Hygiene regulation for functional foods" and the "Good manufacture practice of functional foods" are also important documents and provide details of requirements concerning raw material supply and limits for heavy metals and microbial content, and the conditions required for manufacturing and processing the health foods along with details of the procedures regarding storage, transportation, and management (Yang 2008).

The "Regulation on the labelling of functional foods" was issued by the MOH in 1996 and was later revised in 2005 as the "Regulation on the advertisement of functional foods." This regulation stipulates what is allowed in terms of advertising functional foods to consumers (State Food and Drug Administration 2005).

8.5 Scientific Substantiation and Qualification Approval of Health Claims in China

Currently in China, there are 27 categories of product-specific health claims, which are related to reduction of disease risk or are function related. A claim for a functional food refers to any representation that states, suggests, or implies that a food or food component has particular characteristics relating to its origin, nutritional profile, health function, or any other quality. They refer to an improvement in health or a positive contribution to health.

If applying for a health claim in China, applicants should choose from the 27 permitted claims and produce a dossier as an application to the EJC for evaluation. Testing evidence and evidence from the literature that substantiate the health claim that is being suggested for the product must be produced and included in the claim.

With regard to the relevant strength of submitted testing evidence, there is no specification in the regulations that reflects what types of studies should be chosen. However, in general human intervention trials, safety studies and epidemiology or observation data should be included. Animal or model systems should only be included as supporting evidence and structural information regarding the food component or "bioactive" should be included especially if the food component is to be used for the first time in a functional food.

8.6 Future of Functional Foods in China

The future of functional foods in China will involve continued debate regarding substantiation of health claims and what evidence is required for this. The mechanisms of action of bioactive components will no doubt be of importance, as will evidence regarding the effects of the bioactive compounds when included, removed, increased, or decreased in food vehicles. At present, evidence-based mechanisms of action (EBM) of bioactive functional food ingredients are guided by the project "Process for the Assessment of Scientific Support for Claims on Foods" and this document offers practical information and guidance for preparation of scientific dossiers to support a health or novel food claim in China (Yang 2008).

8.7 Japan

Japan is the second largest market for functional food products in the world. During the 1980s, several Japanese scientists began to recognize the importance of diet in the prevention of age-related and geriatric diseases. In 1984, the Japanese Ministry of Education, Science and Culture, funded basic scientific research at Japanese universities to design and create functional physiologically active foods based on the tertiary functions of food ingredients. In 1992, a 3-year project entitled "Analysis and Molecular Design of Functional Foods" was chaired by Prof. S. Arai and sponsored by the Japanese Ministry of Education, Science and Culture (Arai 1996; Arai et al. 2001). It included aspects such as (1) analysis and design of body-regulating factors of foods, (2) analysis and (3) design of body-defending factors in foods, (4) factors involved in the immunological mechanisms of action of foods, (5) development of a technological basis for the design of functional foods, and (6) design of macroscopic structures.

8.8 Administration and Regulation of Functional Foods in Japan

The Japanese Ministry of Health, Labor and Welfare (MHLW) set up "Foods for Specified Health Use" (FOSHU) in 1991 as a regulatory system to approve the statements made on functional food labels concerning health claims made regarding the effect of food or food components on the human body (Shimizu 2003). FOSHU was enacted under the Nutrition Improvement Law. This law lists five categories of "food for special dietary uses."

1. Milk powder for infants
2. Formulated milk powder for infants

3. Foods for elderly individuals with difficulty in masticating or swallowing
4. Medical foods for the ill
5. FOSHU

FOSHU claims are health-related functions that can have positive effects on human physiological functions, and the related foods are intended to be consumed for the maintenance or promotion of health or special health uses by individuals who wish to control their health conditions (Shimizu et al. 2001). Food products applying for approval by FOSHU are evaluated by the Council of Pharmaceutical Affairs and Food Hygiene under the MHLW. In April 2001, FOSHU was expanded to include tablets and capsules along with functional food components of conventional foods. In 2001, MHLW brought in a new regulation, "Foods with health claims" which included existing FOSHU claims and "Foods with nutrient function claims" (Shimizu 2003).

8.9 Dossier Submission for Functional Food Claims in Japan

A dossier for FOSHU status should be submitted to the MHLW. This dossier should provide evidence regarding the effectiveness of the functional component or food product and this should be demonstrated by providing evidence of clinical studies in humans as well as biochemical and animal studies. Any human intervention studies should be carried out in accordance with the Helsinki Declaration and should have obtained approval from a committee on ethics (Morinaga et al. 2000). Furthermore, details regarding the safety of the product or functional component should be included and should also involve studies in human subjects. Finally, analytical determination of the functional component should be provided. Overall, the dossier should include the following.

- A sample of the entire package including labels and health claims
- Documentation of clinical and nutritional proof of the product and/or its functional components
- Documentation regarding the safety of the product or the functional food component and details of the eating experience
- Documentation regarding the stability of the product and the functional component
- Documentation regarding the physical attributes of the functional component and the product
- Details of methods of qualitative and quantitative analytical determination of its functional components and analytical results regarding the components of the product
- Details of the intake amount of the product and/or functional component
- Statement of the method and equipment used in the food's production, and an explanation of the quality control system used

8.10 Evaluation and Approval of Functional Food Health Claims in Japan

FOSHU applications are accepted every 3 months usually in March, June, September, and December. The approval process takes 6 months and starts with receipt of the dossier. In practice, approval of a health claim may take up to a year (Shimizu 2003).

8.11 Approved Health Claims in Japan Under (FOSHU)

Health claims approved by FOSHU can be categorized into eight groups:

1. Gastrointestinal conditions (G.I.)
2. Blood pressure
3. Serum cholesterol
4. Blood glucose
5. Absorption of minerals
6. Blood lipids
7. Dental health
8. Bone health

About half of the FOSHU products that have approval claim to improve GI conditions and the effective components are oligosaccharides, dietary fibers (Yamatoya 1995), and chitosan. With regard to a health claim related to blood pressure, peptide products containing either lacto tripeptides from fermented milks (functional components the peptides IPP and VPP; Hata et al. 1996), dodeca-peptide from casein (Sekiya et al. 1992), and a group of peptides from sardines (Fujita 2001) have been awarded FOSHU status and claim to reduce high blood pressure in cases where blood pressure is slightly elevated in the individual. Chitosan and soy proteins have FOSHU status related to reduction of serum cholesterol in Japan. Indigestive dextrin and wheat albumin can reduce blood glucose levels and were awarded FOSHU status in Japan. Globin digests have FOSHU status in relation to reduction of blood lipids and calcium and fucto-oligosaccharides have FOSHU status with regard to absorption of minerals in the gut. Others are detailed in a paper by Shimizu (2003).

8.12 United States of America

Bringing a new functional food product to market in the United States, such as one containing bioactive peptides, is a potentially complex process with many decision points. Functional foods are regulated by the Food and Drug Administration (FDA) in the United States and its regulatory mandate. Functional foods encompass a wide range of products and uses, including dietary supplements, medical foods, and food additives. Careful consideration of the potential market for a product must be weighed against the expense and time required determining the exact status of a product under the FDA laws and regulations. Regulatory status may form the basis for marketing a product so it is essential to ensure that any product making a claim is scientifically validated and may be legally defensible. Health claims for foods and dietary supplements are both subject to the Nutrition and Labeling Act (NLEA), which requires FDA premarket approval or authorization. Under the NLEA, the FDA must determine, based on "the totality of publicly available scientific evidence" that the claim is supported by "significant scientific agreement, among experts qualified by scientific training and experience." If the FDA finds that a claim can be made, it issues a regulation that allows any qualifying product to bear the claim. For example, all manufacturers of low-saturated fat products that also have low levels of total fat and cholesterol and that are not high in sodium may state: "Diets low in saturated fat and cholesterol may help reduce the risk of heart disease."

8.13 Overview of Regulation

New foods intended for use in humans are subject to regulation by the FDA. U.S. law also recognizes two other categories of foods that are often thought to have a bearing on how the FDA can regulate health claims for functional foods. They are "foods for special dietary use" and "medical foods (Eberjard and Zajac 2010)."

8.14 Regulation as a Food Additive

A substance, when used as intended, is Generally Recognized as Safe (GRAS) by a panel of suitably qualified experts. The major drawback here is that the substance, say, a bioactive peptide, must exert a technical effect on the food to which it is added. In addition, the Food Additive Petition process is lengthy and expensive. It often requires an investment of millions to support safety testing and a lengthy evaluation period by the FDA (Eberjard and Zajac 2010).

8.15 Designation of a Substance as a Medical Food

GRAS determination is necessary, and the substance can only be administered at a hospital, clinic, or long-term-care facility and must be supervised by a medical doctor. Typical investments are on the order of \$20–30,000 U.S. The substance must also be a nutrient as defined by FDA (Eberjard and Zajac 2010).

8.16 Dietary Supplements

If the substance is not a recognized food ingredient, then GRAS designation is required. Adequate data supporting a function claim must also be provided. Developing scientifically sound and legally defensible structure–function claims is the most cost-effective route to market, requiring FDA notification and an investment usually of under $5,000.

8.17 Investigational New Drug–New Drug Application

The Investigational New Drug–New Drug Application approach is recommended by the FDA for substances for which therapeutic or diagnostic claims are intended. This may lead to a "drug" claim and "treatment" may be used in the claim. Drawbacks of this process include the fact that it is lengthy and expensive with a timeline of at least 10 years and it may cost hundreds of millions of dollars.

8.18 Nutrient Claim

Nutritional contents are regulated by the FDA and the "Nutrient Facts" box on all packaged foods comes under this claim. Caloric and fat content, vitamins, minerals, fiber, protein, and other specific parameters come under this claim.

8.18.1 Health Claims

Health claims and qualified health claims require premarket notification of and review by the FDA. Data requirements are rigorous, and functional food manufacturers rely on the FDA's list of previously approved claims for certain ingredients, unless potential market factors support undertaking an effort to seek new approval from the agency. These claims include calcium and vitamin D for osteoporosis, dietary lipids for cancer, saturated fats, cholesterol for heart risk, noncariogenic carbohydrate sweeteners for dental caries, and fiber for cancer treatment. Qualified claims include tomatoes for certain cancers, calcium for colorectal cancer, green tea for cancer, and selenium for cancer. The source of these and other health and qualified claims is the US FDA.

8.18.2 Food Additive Petition

In the event that a GRAS position cannot be reached, a Food Additive Petition is required, and the substance is subject to a premarket evaluation by the FDA.

8.19 For a Food Additive Petition the Following Are Required

(a) Identity of the substance
(b) Manufacturing process
(c) Specification for food grade material
(d) Stability of the added substance
(e) Intended technical effect and use
(f) Methodology for analysis of the added substance in food
(g) Consumer exposure

Companies seeking to market functional foods may petition the FDA for regulation for a food for special dietary use instead of for a health claim under the NLEA. However, the FDA has not recently made use of this authority and has instead withdrawn regulations for some foods for special dietary use (McNamara 1998). In the United States, there is a series of statutory provisions and regulatory policies that allow food and supplement companies to make structure–function claims. An example is, "Vitamin A is essential for normal vision." Many companies have marketed functional foods and their bioactive ingredients on the basis of structure–function claims, instead of seeking FDA approval for a health claim. Recently, the U.S. Food and Drug Administration initiated enforcement practices against companies that use functional food claims to overstate the benefits of their food products. In December 2008, the FDA warned Coca Cola that this labeling violated U.S. policy against marketing soda and other snack foods as nutritious. The makers of Lipton Green Tea were informed that the product's labeling is misleading because it suggests that Lipton tea is designed to treat or prevent disease.

8.20 Europe

European regulations on nutrition and health claims became effective on January 19th, 2007 and each of the E.U. member states now has a common regulation allowing health claims on foods. They may be considered the "gold standard" with regard to regulation and toughness regarding review. The use of bioactive components in food is regulated under the General Food Law (178/2002/EC), which assigns primary legal responsibility for the safety of the products on the market to the business operator. The European Food Safety Authority (EFSA) is in charge of the regulation. Article 2 of the General Food Law states that: "Food" means any substance or product, whether processed, partially processed or unprocessed, intended to be, or reasonably expected to be ingested by humans: excluding (a) feed; (b) live animals; (c) plants prior to harvesting; (d) medicinal products within Directive 2004/27/EC; (e) Cosmetics within Directive 76/768/EC; tobacco, narcotics or residues and contaminants. If a natural product is categorized as food or a food ingredient according to Article 2, it must be characterized and further defined as one of the following:

(a) Novel food (Directive 258/1997/EC)
(b) Foods for particular nutritional use, covering dietetic foods (Directive 89/398/EC)
(c) Food additives (Directive 89/107/EC)
(d) Food supplements (Directive 2002/46/EC)
(e) Flavorings (Directive 91/71/EC completing Directive 88/388/EC)

8.21 Regulation on Nutrition and Health Claims Made on Foods

There are a number of foods on the European market with labels bearing nutrition and health claims. These labels are regulated by Regulation 1924/2006/EC of December 2006 on nutrition and health claims made on foods. The main objectives of Regulation (EC) No. 1924/2006 of the European Parliament and of the Council of December 2006 on nutrition and health claims made on foods is to ensure high consumer protection against marketing and fair competition within the food industry. This regulation defines a health claim as "any claim that states, suggests or implies that a relationship exists between a food category, a food or one of its constituents and health." A reduction of disease risk claim is defined as "any health claim that states, suggests or implies that the consumption of a food category, a food or one of its components significantly reduces a risk factor in the development of a human disease". These claims along with claims dealing with children's development and health are dealt with specifically in Article 14, and other health claims are dealt with in Article 13 (Asp and Bryngelsson 2008).

The regulation applies to all nutrition and health claims made in commercial communications including:

(a) Labeling
(b) Advertising of food and promotional campaigns
(c) Trademarks
(d) Brand names
(e) Any information that may be interpreted as health or nutritional claims

In order to make a claim the following terms and conditions must be adhered to for labeling with nutrition and health claims.

(a) The physiological effects of the substance that is the subject of the claim must be scientifically substantiated.
(b) A significant amount of the functional ingredient must be available in an adequate daily portion of the food vehicle.
(c) The amount of the substance consumed regularly must be considered.
(d) The bioavailability of the substance must be known.

8.22 Health Claim

This claim relates to any claim that "states, suggests or implies that a relationship exists between a food category, a food or one of its constituents and health." Specific conditions and requirements for the use of health claims can be found in chap. IV of Regulation 1924/2006/EC.

8.23 Article 13

8.23.1 Article 13.1

This article deals with health claims describing or referring to:

(a) The role of a nutrient or other substance in growth, development, and the functions of the body
(b) Behavioral and psychological functions
(c) Weight control or reduction in the sense of hunger or increased satiety or a reduction in available energy from the diet

These claims should be supported by generally accepted scientific evidence and they should be well understood by the consumer.

8.23.2 Article 13.5

This claim relates to claims that are made "based on newly developed scientific evidence and/or which include a request for the protection of proprietary data." Applications under this claim require an extensive dossier. EFSA provides its judgment on a product or functional food component within 5 months following the request and a further 2 months is required by the commission following consultation with the member states.

8.24 Article 14

Article 14 relates to reduction of disease risk claims and claims referring to children's development and health. Claims submitted according to Article 14 undergo rigorous authorization procedures and an extensive dossier must be submitted. EFSA have published guidelines for dossier submission under Article 14. An extensive literature review and studies must be submitted to support any claim made under

Article 14. Submissions are made to the competent authority of the member states for further handling by EFSA and the commission.

European Union regulation provides the basis for allowance of health claims on foods. PASSCLAIM provides a scientifically robust tool for evaluating the quality of the data submitted in support of health claims on foods (Asp et al. 2008).

8.25 Scientific Substantiation

Scientific substantiation for any claim should comply with the following:

(a) The food or food component should be well characterized and comply with existing legislation.
(b) Claims should include human dietary intervention studies considering key target groups, and they should demonstrate the intended effect.
(c) Scientific data should be objective, balanced, and reproducible with appropriate statistical control.
(d) Scientifically validated biomarkers should be used where possible to evaluate reduction of disease state, for example.
(e) The assessment of significant scientific agreement should be based on critical interpretation of the data and on the application of scientific judgment.
(f) A nutrition or health claim should not be made if it is inconsistent with health principles or if it leads to a one-sided dietary practice.
(g) The use of health claims in beverages with alcoholic content of greater than 1.2% by volume is generally prohibited.

New health claims must first be authorized by the European Commission. Applications should be submitted to the national competent authority (in the case of Ireland, the Food Safety Authority of Ireland). EFSA decide if the claim can be permitted or not. The new claim must also appear on the E.U. community list.

Specific nutrient profiles for food are also important for the use of claims. The profiles include fat, saturated fatty acids, trans-fatty acids, sugars, and salt/sodium content. The nutrient profiles are also assessed by EFSA. Foods with inadequate nutrient profiles but that contain a functional component are prohibited.

8.26 Nutrition Claim in Europe

This relates to any claim that states, suggests, or implies that a food has a particular beneficial nutritional property due to:

(a) Energy
(b) The nutrients or other substances it contains or contains in reduced or increased portions or does not contain

8.27 Reduction of Disease Risk Claim

This relates to any claim that states, suggests or implies that the consumption of a food category, food or its constituents significantly reduces a risk factor in the development of disease.

8.28 Novel Foods and Novel Food Ingredients

Novel foods are defined and regulated by regulation 258/97/EC of 27 January 1997. This regulation applies to foods or food ingredients that were not used for human consumption in a considerable range before May 15th, 1997.

It includes the following.

(a) Foods and those ingredients isolated from microbes, algae, or fungi.
(b) Foods and food ingredients consisting of or isolated from plants and ingredients isolated from animals except those obtained by traditional propagating or breeding practices and having a long safe history of use.
(c) Foods and food ingredients having a new or intentionally modified primary molecular structure.
(d) Foods and food ingredients that are subject to new production processes not currently used that result in changes in the composition or structure of the foods or ingredients and may affect the nutritional profile, value, metabolism, or level of undesirable substances. This regulation does not apply to food flavorings for use in foodstuffs or extraction solvents used in the production of foodstuffs.

8.29 Recently Approved Health Claims in the European Union

8.29.1 Omega-3 fatty acids: Eicosapentaenoic acid (EPA) and Docosahexaenoic acid (DHA)

EFSA recently approved eicosapentaenoic acid (EPA) and docosahexaenoic acid (DHA) omega-3 fatty acids related to infant visual and brain health. From the 26th of May, 2011, the following claims may be made with regards to EPA and DHA in foods, including infant formula in Europe.

1. DHA intake contributes to the normal visual development of infants up to 12 months of age.
   A daily intake of 100 mg DHA is necessary to achieve this benefit, and the food must contain at least 0.3% of the total fatty acids as DHA.
2. DHA maternal intake contributes to the normal development of the eye of the fetus and breastfed infants.

The claim can be used only for food that provides a daily intake of at least 200 mg DHA, and information must be provided to pregnant and lactating women that 200 mg of DHA in addition to the recommended daily amount of omega-3 for adults (250 mg DHA and EPA) is necessary to achieve the benefit.

3. DHA maternal intake contributes to the normal brain development of the fetus and breastfed infants. The same conditions as for claim 2 also apply to claim 3. Other claims that have gained approval include walnuts and the improved function of blood vessels; the antioxidant effects of polyphenols found in olive oil on cholesterol; and the effect of caffeine on alertness and physical endurance.

8.30 Recently Approved Novel Food Claims in Europe

8.30.1 Chitin-Glucan from Aspergillus niger

Chitin-glucan from *Aspergillus niger*, which is a known fiber-boosting weight management ingredient gained EFSA approval after EFSA deemed it safe for use.

The Artinia-branded extract from the *Aspergillus niger* fungus is permitted for use in food supplements within the E.U.'s 27 member states at a maximum dose of 5 g/day.

References

Arai, S. 1996. Studies on functional foods in Japan-state of the art. *Bioscience, Biotechnology, and Biochemistry* 60: 9–15.

Arai, S., T. Osawa, H. Ohigashi, M. Yoshikawa, S. Kaminogawa, M. Watanabe, T. Ogawa, K. Okubo, S. Watanabe, H. Nishino, K. Shinohara, T. Esashi, T. Hirahara. 2001. A mainstay of functional food science in Japan – History, present status, and future outlook. *Bioscience, Biotechnology, and Biochemistry* 65(1): 1–13.

Asp, N.-G., and S. Bryngelsson. 2008. Health claims in Europe: new legislation and passclaim for substantiation. *The Journal of Nutrition* 138(6): 1210S–1215S.

Diplock, A.T., P.J. Aggett, M. Ashwell, F. Bornet, E.B. Fern, and M.B. Roberfroid. 1999. Scientific concepts of functional foods in Europe: consensus document. *British Journal & Nutrition*, 81 Supp. 1, S7–S27.

Diplock, A.T., P.J. Aggett, M. Ashwell, F. Bornet, E.B. Fern, and M.B. Roberfroid. 2000. Scientific concepts of functional foods in Europe: consensus document. In *Functional foods II*, ed. J. Buttriss and M. Saltmarsh, 8–59. Cambridge: The Royal Society of Chemistry.

Eberjard, J., and I. Zajac. 2010. Bringing functional foods to market: FDA's regulatory mandate drives the process. *Nerac White Paper*, pp. 2–11.

Fujita, H. 2001. Human study of sardine peptide on blood pressure. *Nutrition Research* 21: 1149.

Hata, Y., M. Yamamoto, M. Ohni, K. Nakajima, Y. Nakamura, and T. Takamo. 1996. A placebo-controlled study of sour milk on blood pressure in hypertensive subjects. *American Journal of Clinical Nutrition* 64: 767–771.

McNamara, S.H., and S.H. McNamara. 1998. So you want to market a food and to make health-related claims – how far can you go? What rules of law will govern the claims you want to make? *Food and Drug Law Journal* 53: 421–434.

Ministry of Health. 1996. *The regulation for health foods*. Beijing: China Standards.
Morinaga, Y., T. Shimizu, K. Sueki, and T. Hirahara. 2000. *Health claim on functional foods*. Japan: International Life Scientific of Japan.
Sekiya, S., Y. Kobayashi, and E. Kita. 1992. Effect of hydrolyzed casein on hypertension and its adverse effect. *Japanese Journal of Nutrition and Foods* 45: 513–517.
Shimizu, T. 2003. Health claims on functional foods: the Japanese regulations and an international comparison. *Nutrition Research Reviews* 16: 241–252.
Shimizu, T., Y. Morinaga, T. Tokunaga, S. Seki, K. Sueki, and T. Hirahara. 2001. *Functional food science in Japan*, 2–13. Japan: International Life Scientific of Japan.
Southon, S. 2000. Factors to consider when undertaking clinical trials for functional foods. In *Functional foods II, claims and evidence*, ed. J. Buttriss and M. Saltmarsh, 3–8. Cambridge: The Royal Society of Chemistry.
State Food and Drug Administration. 2005. The guideline of registration for functional food. Beijing: China Standards. http://www.sfda.gov.cn/syjz0461/syjz0461.htm.
Yamatoya, K. 1995. Effects of hydrolysed guar gum on frequency and characteristics of evacuation. *Journal of Japanese Society of Nutrition and Food Science* 46: 199–203.
Yang, Y. 2008. Scientific substantiation of functional food health claims in China. *The Journal of Nutrition* Supplement: 1199S–1205S.

Index

M. Hayes (ed.), *Marine Bioactive Compounds: Sources, Characterization and Applications*, DOI 10.1007/978-1-4614-1247-2,

MIX
Papier aus verantwortungsvollen Quellen
Paper from responsible sources
FSC® C105338

If you have any concerns about our products,
you can contact us on
ProductSafety@springernature.com

In case Publisher is established outside the EU,
the EU authorized representative is:
Springer Nature Customer Service Center GmbH
Europaplatz 3, 69115 Heidelberg, Germany

Printed by Libri Plureos GmbH
in Hamburg, Germany